Lecture Notes in Physics

Edited by J. Ehlers, München, K. Hepp, Zürich, and
H. A. Weidenmüller, Heidelberg
Managing Editor: W. Beiglböck, Heidelberg

23

Nuclear Structure Physics

Proceedings of the Minerva Symposium on Physics
held at the Weizmann Institute of Science
Rehovot, April 2–5, 1973

Edited by U. Smilansky, I. Talmi, and H. A. Weidenmüller
Weizmann Institute of Science, Rehovot/Israel

Springer-Verlag Berlin Heidelberg GmbH 1973

ISBN 978-3-540-06554-8   ISBN 978-3-540-37852-5 (eBook)
DOI 10.1007/978-3-540-37852-5

This volume is dedicated to
the memory of

<u>Prof. J. Hans D. Jensen</u>

a great scientist and a devoted friend

PREFACE

The Minerva Symposium on Physics took place at the Weizmann
Institute of Science, Rehovot, on April 2-5, 1973. It was held to
mark ten years of fruitful cooperation between German and Israeli
scientists, through the Minerva program under the leadership of
Prof. W. Gentner. The symposium was devoted to three branches of
physics - nuclear physics, elementary particle physics and geophysics.
The aim of the scientific program of the symposium was to review the
most recent developments in physics, with special emphasis on those
topics covered by the joint efforts of German-Israeli cooperation.
The choice of reports and speakers was guided by this aim.

The present volume contains the proceedings of the nuclear physics
section. Sessions were held on the following topics: (1) Individual
particle model, (2) Collective phenomena, (3) Interaction of the nucleus
with atomic electrons, (4) Nuclear Physics with the new heavy ion
accelerators.

A short time before the symposium all of us were shocked and deeply
grieved by the sad news of the untimely death of Prof. Hans Jensen.
The late Prof. Jensen was not only one of the most prominent figures
in modern nuclear physics, but also played a very important role in
creating and developing the German-Israeli scientific cooperation. As
a token of our gratitude and reverence, we dedicate these proceedings
to his memory.

On behalf of the organizing committee we would like to express
our gratitude to the Weizmann Institute staff for their invaluable
help in the preparation and during the course of the Symposium.
Special thanks are due to Mr. Yitzhak Berman and Miss Rachel Frankel
for their devoted help and efforts.

These proceedings were typed promptly and meticulously by
Mrs. Corinne Hasdai. We are greatly indebted to her.

The Editors

TABLE OF CONTENTS

VIII

LIST OF PARTICIPANTS

D. Agassi, The Weizmann Institute of Science, Rehovot, Israel

G. Alexander, Tel Aviv University, Tel Aviv, Israel

S. Amiel, Nuclear Research Center, Yavne, Israel

P. Armbruster, Gesellschaft für Schwerionenforschung MBH, Darmstadt, Germany

G. Assaf, The Weizmann Institute of Science, Rehovot, Israel

N. Auerbach, Tel Aviv University, Tel Aviv, Israel

R. Aviv, Tel Aviv University, Tel Aviv, Israel

Y. Avishai, The Negev University, Beer Sheva, Israel

Y. Bar-Tov, The Negev University, Beer Sheva, Israel

R. Bauminger, The Hebrew University, Jerusalem, Israel

J. Ben-David, Nuclear Research Center, Yavne, Israel

J. Benecke, Max-Planck-Institut für Physik und Astrophysik, München, Germany

K. Bethge, Zweites Physikalisches Institut der Universität Heidelberg,
        Heidelberg, Germany

M. Birk, The Weizmann Institute of Science, Rehovot, Israel

A. Blaugrund, The Weizmann Institute of Science, Rehovot, Israel

R. Bloch, Negev Research Institute, Beer Sheva, Israel

E. Bodenstedt, Institut für Strahlen-und Kernphysik der Universität Bonn,
        Bonn, Germany

H. Brafman, The Weizmann Institute of Science, Rehovot, Israel

F.W. Brasse, DESY, Hamburg, Germany

C. Broude, The Weizmann Institute of Science, Rehovot, Israel

J. Burde, The Hebrew University, Jerusalem, Israel

E. Cheifetz, The Weizmann Institute of Science, Rehovot, Israel

A. Citron, Institut für Experimentelle Kernphysik der Universität Karlsruhe,
        Karlsruhe, Germany

S.G. Cohen, The Hebrew University, Jerusalem, Israel

J. Daboul, The Negev University, Beer Sheva, Israel

S. Dado, Technion, Haifa, Israel

H. Dahmen, Institut für Theor. Physik der Universität Freiburg, Freiburg, Germany

A. Dar, Technion, Haifa, Israel

Y. Dar, Technion, Haifa, Israel

J. de Boer, Sektion Physik der Universität München, Garching, Germany

M. Deutschmann, III. Physikalisches Institut der Rhein.-Westf. Technischen
        Hochschule Aachen, Aachen, Germany

L. Diamant, The Weizmann Institute of Science, Rehovot, Israel

K. Dietrich, Physik-Department der Technischen Universität München,
        Garching, Germany

K. Dietz, Physikalisches Institut der Universität Bonn, Bonn, Germany

H.G. Dosch, Institut für Theor. Physik der Universität Heidelberg,
      Heidelberg, Germany

I. Dostrovsky, The Weizmann Institute of Science, Rehovot, Israel

Y. Dothan, Tel Aviv University, Tel Aviv, Israel

Y. Eisenberg, The Weizmann Institute of Science, Rehovot, Israel

A. El Goresy, Max-Planck-Institut für Kernphysik, Heidelberg, Germany

E. Esterman, Technion, Haifa, Israel

H. Fechtig, Max-Planck-Institut für Kernphysik, Heidelberg, Germany

Z. Fraenkel, The Weizmann Institute of Science, Rehovot, Israel

E. Friedman, The Hebrew University, Jerusalem, Israel

A. Gal, The Hebrew University, Jerusalem, Israel

Y. Gat, The Weizmann Institute of Science, Rehovot, Israel

G. von Gehlen, Physikalisches Institut der Universität Bonn, Bonn, Germany

W. Gentner, Max-Planck-Institut für Kernphysik, Heidelberg, Germany

J. Goldberg, Technion, Haifa, Israel

G. Goldring, The Weizmann Institute of Science, Rehovot, Israel

A. Gotsman, Tel Aviv University, Tel Aviv, Israel

E. Gradstein, Tel Aviv University, Tel Aviv, Israel

M. Gronau, Technion, Haifa, Israel

S. Gross, Geological Institute, Jerusalem, Israel

S. Gurvitz, The Weizmann Institute of Science, Rehovot, Israel

H. Harari, The Weizmann Institute of Science, Rehovot, Israel

U. Hauser, Erstes Physikalisches Institut-Universität zu Köln, Köln, Germany

M. Hoenig, Southeastern Mass. University, N. Dartmouth, Mass., USA*

L. Horowitz, Tel Aviv University, Tel Aviv, Israel

Y. Horowitz, The Negev University, Beer Sheva, Israel

G. Hortig, Max-Planck-Institut für Kernphysik, Heidelberg, Germany

J. Hufner, Fakultat für Physik der Universität Freiburg, Freiburg, Germany

A. Issar, The Hebrew University, Jerusalem, Israel

E. Jaeger, Universität Bern, Bern, Switzerland

A. Jaffe, The Hebrew University, Jerusalem, Israel

S. Kalbitzer, Max-Planck-Institut für Kernphysik, Heidelberg, Germany

R. Kalish, Technion, Haifa, Israel

U. Karshon, The Weizmann Institute of Science, Rehovot, Israel

H.A. Kastrup, Institut für Theoretische Physik, Aachen, Germany

A. Kaufman, The Weizmann Institute of Science, Rehovot, Israel

M.W. Kirson, The Weizmann Institute of Science, Rehovot, Israel

T. Kirsten, Max-Planck-Institut für Kernphysik, Heidelberg, Germany

F.S. Klein, The Weizmann Institute of Science, Rehovot, Israel

---

*Presently at the Dept. of Nuclear Physics, The Weizmann Institute

Z. Klein, Ministry of Agriculture, Jerusalem, Israel

Y. Kolodny, The Hebrew University, Jerusalem, Israel

S. Kruger, Tel Aviv University, Tel Aviv, Israel

H. Lehmann, II. Institut für Theoretische Physik, Hamburg, Germany

N. Levi, Technion, Haifa, Israel

H.J. Lipkin, The Weizmann Institute of Science, Rehovot, Israel

M. Luban, Bar Ilan University, Ramat Gan, Israel

R. Lüst, Max-Planck-Gesellschaft, München, Germany

G. Mack, Universität Bern, Bern, Switzerland

S. Mandel, The Hebrew University, Jerusalem, Israel

U. Maor, Tel Aviv University, Tel Aviv, Israel

A. Marinov, The Hebrew University, Jerusalem, Israel

E. Mazor, The Weizmann Institute of Science, Rehovot, Israel

O. Muller, Max-Planck-Institut für Kernphysik, Heidelberg, Germany

A. Nir, The Weizmann Institute of Science, Rehovot, Israel

A. Nissenbaum, The Weizmann Institute of Science, Rehovot, Israel

W. Noerenberg, Institut für Theoretische Physik der Universität Heidelberg,
    Heidelberg, Germany

S. Nussinov, Tel Aviv University, Tel Aviv, Israel

W. von Oertzen, Max-Planck-Institut für Kernphysik, Heidelberg, Germany

S. Ofer, The Hebrew University, Jerusalem, Israel

W. Paul, Physikalisches Institut der Universität Bonn, Bonn, Germany

H.C. Pauli, Institut für Theoretische Physik der Universität Basel,
    Basel, Switzerland

I. Pelah, Nuclear Research Center, Yavne, Israel

I. Pelly, The Negev University, Beer Sheva, Israel

D. Pelte, Erstes Physikalisches Institut der Universität Heidelberg,
    Heidelberg, Germany

L. Picard, The Hebrew University, Jerusalem, Israel

M. Popp, Institut für Strahlen und Kernphysik der Universität Bonn, Bonn, Germany

B. Povh, Erstes Physikalisches Institut der Universität Heidelberg,
    Heidelberg, Germany

J. Rafelski, Institut für Theoretische Physik der Universität Frankfurt/M,
    Frankfurt, Germany

A. Richter, Institut für Experimentalphysik der Ruhr-Universität Bochum,
    Bochum, Germany

A. Rinat, The Weizmann Institute of Science, Rehovot, Israel

H. Rollnik, Physikalisches Institut der Universität Bonn, Bonn, Germany

H. Romer, Physikalisches Institut der Universistät Bonn, Bonn, Germany*

B. Rosner, Technion, Haifa, Israel

H.R. Rubinstein, The Weizmann Institute of Science, Rehovot, Israel

F. Scheck, SIN, Villigen, Switzerland

---

*Presently at the Dept. of Nuclear Physics, The Weizmann Institute

D. Schildknecht, DESY, Hamburg, Germany

F. Schneider, Max-Planck-Gesellschaft, München, Germany

A. Schwimmer, The Weizmann Institute of Science, Rehovot, Israel

P. Seyboth, Max-Planck-Institut für Physik und Astrophysik, München, Germany

A. Shapira, The Weizmann Institute of Science, Rehovot, Israel

E. Skurnik, The Weizmann Institute of Science, Rehovot, Israel

U. Smilansky, The Weizmann Institute of Science, Rehovot, Israel

P. Soding, DESY, Hamburg, Germany

J. Sokolowski, The Weizmann Institute of Science, Rehovot, Israel

H.J. Specht, Sektion Physik der Universität München, Garching, Germany

K.H. Speidel, Institut für Strahlen und Kernphysik der Universität Bonn.
          Bonn, Germany

D. Start, University of Oxford, Oxford, England*

L. Stodolsky, Max-Planck-Institut für Physik und Astrophysik, München, Germany

L. Susskind, Tel Aviv University, Tel Aviv, Israel

I. Talmi, The Weizmann Institute of Science, Rehovot, Israel

R. Tieberger, The Negev University, Beer Sheva, Israel

I. Unna, The Hebrew University, Jerusalem, Israel

Z. Vager, The Weizmann Institute of Science, Rehovot, Israel

G. Veneziano, The Weizmann Institute of Science, Rehovot, Israel

G.A. Wagner, Max-Planck-Institut für Kernphysik, Heidelberg, Germany

G.J. Wagner, Max-Planck-Institut für Kernphysik, Heidelberg, Germany

H.A. Weidenmüller, Max-Planck-Institut für Kernphysik, Heidelberg, Germany

B. Wiik, DESY, Hamburg, Germany

G. Wolf, DESY, Hamburg, Germany

Y. Wolfson, The Weizmann Institute of Science, Rehovot, Israel

S.A. Wouthuysen, CERN, Geneva, Switzerland*

P. Wurm, Max-Planck-Institut für Kernphysik, Heidelberg, Germany

D. Yaalon, The Hebrew University, Jerusalem, Israel

D. Yaniv, Tel Aviv University, Tel Aviv, Israel

A. Yavin, Tel Aviv University, Tel Aviv, Israel

G. Yekutieli, The Weizmann Institute of Science, Rehovot, Israel

N. Zeldes, The Hebrew University, Jerusalem, Israel

Z. Zinnamon, The Weizmann Institute of Science, Rehovot, Israel

SINGLE PARTICLE PROPERTIES OF NUCLEI

Hans A. Weidenmüller[*]

Department of Nuclear Physics, Weizmann Institute of Science, Rehovot, Israel

Abstract. Landau's concept of quasiparticles is used to discuss certain single-particle properties of nuclei, and to relate experimental removal energies with calculated single-particle energies. The importance of single-particle motion in nuclei is stressed.

1. INTRODUCTION

In contrast to the other lectures presented in this volume, the present talk was addressed to a wide audience consisting of all participants in the Minerva Symposium, i.e. geophysicists, nuclear and particle physicists. It therefore was attempted to give a global review of single-particle properties of nuclei inasmuch as these give rise to quasiparticle behaviour. Landau's concept of the quasi-particle, basic for the entire review, is introduced in sect. 2. Evidence for quasiparticle behaviour in nuclei is presented in sects. 3 and 4. The theoretical concepts relating to quasiparticles are described in sect. 5, their interpretation and some recent numerical results are given in sect. 6. The importance of single-particle motion for other branches of nuclear physics is stressed in sect. 7.

2. REMINDER OF LANDAU's THEORY OF FERMI LIQUIDS [1-3]

Landau assumed that for sufficiently small excitation energies, the spectrum of an infinitely extended gas of interacting Fermi particles can be described in terms of excitations of quasiparticles, i.e. particles (or holes) with their associated polarization clouds. For sufficiently small excitation energy of the system, the density of quasiparticles is so small that they move independently. They can then be characterized by a momentum $\hbar\vec{k}$, and an effective mass m^*. Many thermodynamic properties of solids observed at low temperatures can be accounted

---

[*]Permanent address: Max-Planck Institut für Kernphysik, Heidelberg, Germany.

for by this simple picture of a Fermi gas of quasiparticles even if the two-particle interaction in the full Hamiltonian H is known to be strong.

Quasiparticles do not correspond exactly to eigenstates of H. Let $|0>$ be the ground state of the system of A particles, $|i>$ the eigenstates of the (A+1) particle system with energy ω_i, and $a^\dagger(\vec{k})$ the creation operator of a particle with momentum $\hbar\vec{k}$, with $|\vec{k}| > k_F$, the Fermi momentum. [For states with $|\vec{k}| \leq k_F$ we would instead consider "holes" or "quasiholes" and the associated operators $a(\vec{k})$]. Consider the quantity

$$S(\vec{k},\omega) = \sum_i |<i|a^\dagger(\vec{k})|0>|^2 \delta(\omega-\omega_i) \tag{2.1}$$

in the limit $A \to \infty$. If $a^\dagger(\vec{k})|0>$ were an eigenstate of H, i.e. if the Fermi gas consisted of free particles, $S(\vec{k},\omega)$ would be a delta function in ω. For a gas of interacting particles, $S(\vec{k},\omega)d\omega$ measures the probability density of finding the single-particle mode $a^\dagger(\vec{k})|0>$ in the energy interval $[\omega,\omega+d\omega]$ of the spectrum of H. The quasiparticle concept postulates that for fixed $\vec{k}$, $S(\vec{k},\omega)$ has the form shown in fig. 1. The position of the peak defines the quasiparticle energy $\text{Re}\omega(\vec{k})$, its full widths at half maximum (after background subtraction), the width $\Gamma(\vec{k})$ of the quasiparticle, related to its lifetime τ by $\tau = \hbar/\Gamma(\vec{k})$. The quasiparticle concept does not require the background in fig. 1 to vanish, it also allows that $S(\vec{k},\omega(\vec{k})) << 1$. It only requires the existence of a pronounced peak. Nuclear physicists would call $S(\vec{k},\omega)$ the "strength function", $a^\dagger(\vec{k})|0>$ a "doorway state",

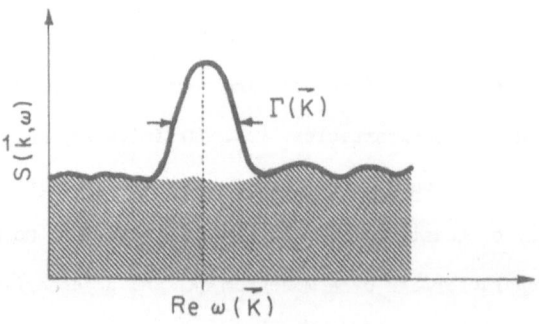

Fig. 1. The strength function $S(\vec{k},\omega)$ in the presence of a quasiparticle.

(Adapted from fig. 2, p. 71, of ref. 2)

$\Gamma(\vec{k})$ its "spreading width", and the area under the peak (without background) the "occupation probability" $P(\vec{k})$ of the doorway state.

The quasiparticle concept is only useful for experiments with time or energy resolutions that are neither too short nor too long. Addition of a particle to the ground state $|0\rangle$ creates the mode $a^{\dagger}(\vec{k})|0\rangle$ and leads to transients which correspond to the shaded area in fig. 1. The quasiparticle concept becomes operational only after the transients have died out, and is useful only for times $t \lesssim \tau$, i.e. until the quasiparticle decays. Times $t \gg \tau$ correspond to an energy resolution $\Delta E \ll \Gamma(\vec{k})$. The levels in the interval ΔE may have little to do with the quasiparticle since it is possible that $S \ll 1$. Only in experiments covering a sufficiently wide range of energies will it be possible to study the behaviour of $S(\vec{k},\omega)$ shown in fig. 1 and to detect quasiparticle features.

It is perhaps surprising that for sufficiently small excitation energy, the existence of quasiparticles is a rather general feature of infinite systems and does not depend upon the details of the interaction. Moreover, one can show that as ω approaches the Fermi energy ε_F, the width $\Gamma(\vec{k}) \propto (\omega(\vec{k}) - \varepsilon_F)^2$, so that the quasiparticle concept is valid to arbitrary accuracy near the Fermi surface. The rapid disappearance of $\Gamma(\vec{k})$ with $\omega(\vec{k}) \to \varepsilon_F$ is a consequence of the density of states near the Fermi surface.

3. QUASIPARTICLE BEHAVIOUR OF NUCLEI NEAR THE FERMI SURFACE

Since Landau theory applies in the limit $A \to \infty$, we expect it to work approximately for $A \gg 1$. The ideal nucleus for the study of independent particle motion with $A \gg 1$ is, of course, ^{208}Pb. It has a very small level density near the Fermi surface, so that the spreading width Γ is expected to be quite small. One finds, in fact, that the quasiparticle behaviour is concentrated on a single eigenstate of the system, so that $\Gamma=0$. [This does not imply $P = 1$, although one finds both experimentally and theoretically that P is close to one, $P \gtrsim 0.85$]. Clear-cut evidence for single-particle structure in ^{209}Pb is obtained by investigating the elastic and inelastic scattering of protons on ^{208}Pb, leading to states in the compound nucleus ^{209}Bi which are members of the same isospin

multiplet as the low-lying states in ^{209}Pb. Analysis of the angular distribution
and polarization of the elastically scattered protons gives the spins and parities
of the levels involved, as well as the partial proton widths from which the
coefficients P introduced above can be approximately determined. [This actually
involves a discussion of isospin violation in the reaction]. Fig. 2 shows the
excitation function for elastic proton scattering at a laboratory angle of 170°.

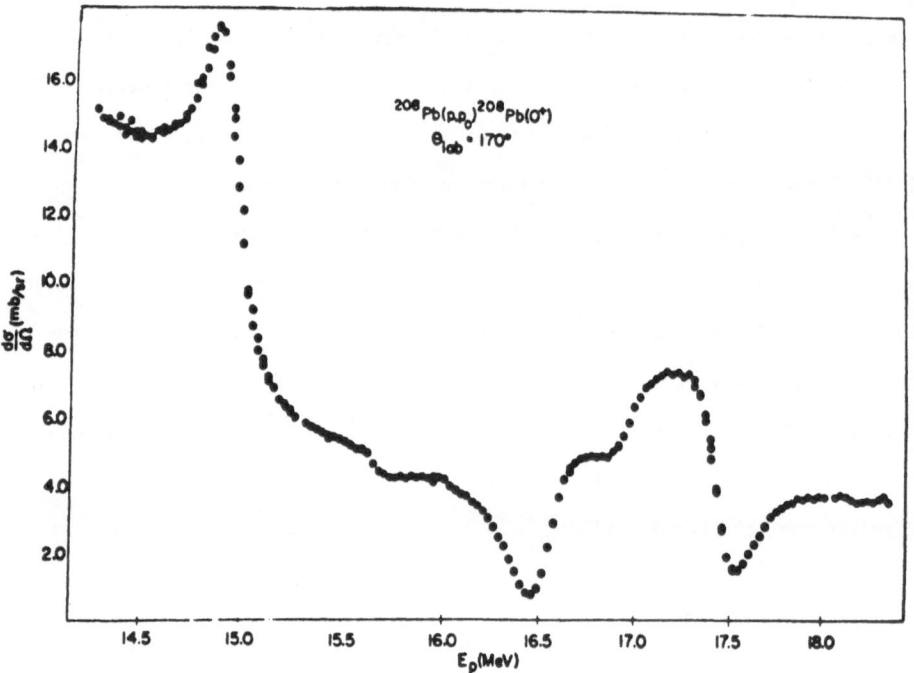

Fig. 2. Elastic proton scattering on ^{208}Pb (from C.F. Moore et al., Phys. Lett.
22, 616 (1966))

Resonances at 14.9, 15.7, 16.5, 16.9 and 17.4 MeV are clearly visible. These
correspond, respectively, to the single-particle states with quantum numbers
$g_{9/2}$, $i_{11/2}$, $d_{5/2}$, $s_{1/2}$ and, the last, to the unresolved doublet $g_{7/2}$ and $d_{3/2}$
in ^{209}Pb. Note that there is only one level for each value of spin and parity.

In contrast to infinite matter considered in Landau theory, nuclei have a
surface capable of vibrations and of rotations. How does the quasiparticle

interact with these modes of motion? This question was answered by de-Shalit[4] who assumed that the coupling was weak, so that the quasiparticle was entirely contained in the space spanned by a single-particle state coupled to a rotational state of given spin and parity. This model has turned out to be very successful. Another example for the interaction of quasiparticles with collective modes is given in fig. 3 below.

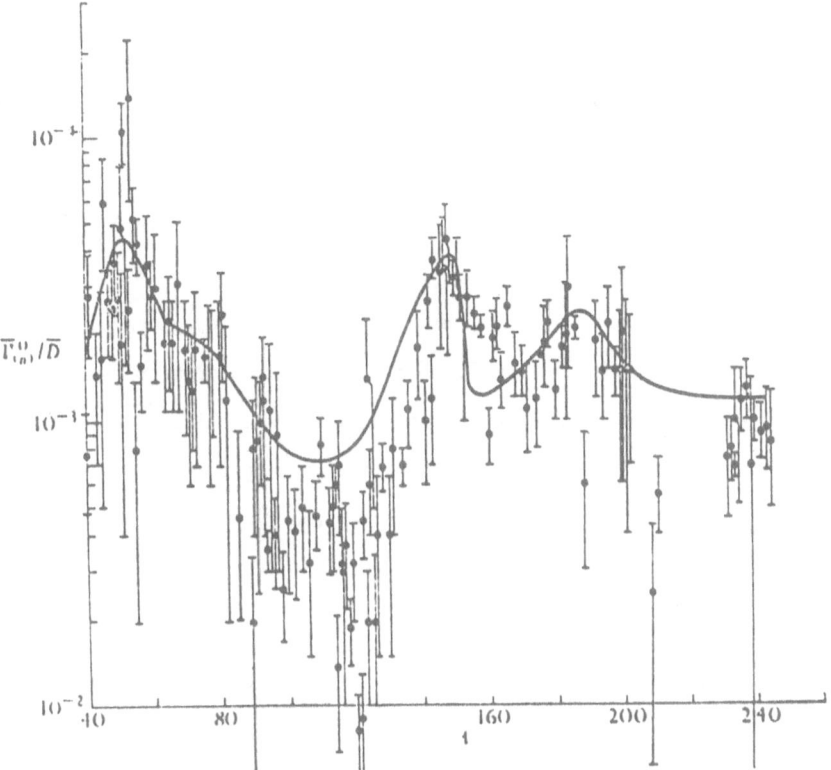

Fig. 3. The s-wave neutron strength function versus mass number. (from J.E. Lynn, The theory of neutron resonance reactions, Clarendon Press, Oxford 1968)

The Landau argument concerning the density of quasiparticles applies only to the infinite medium. For nuclei, quasiparticle interactions are often important. Migdal[5] adapted Landau theory to this new situation. The interaction between quasiparticles is to a large extend determined by symmetry considerations. A few parameters are fitted to the data, and a wealth of spectroscopic information can be understood. The investigation of nuclear spectra, moments, and transition rates in the frame of semi-phenomenological Landau-Migdal theory has remained an active area of research up to the present time.

The neutron strength function provides another beautiful example for usefulness of the quasiparticle concept in nuclei. In the elastic scattering of slow neutrons one finds a large number of resonances, owing to the fairly large level density near neutron threshold. A careful analysis of the data determines the parities and partial neutron widths Γ_n of these resonances and their average level spacing D. By taking out a suitable kinematical factor, one defines the reduced partial neutron width Γ_n^{red} and the strength function $D^{-1} < \Gamma_n^{red}>$ by an average over these. Since Γ_n^{red} is the square of a transition matrix element and D^{-1} is the level density, $D^{-1} <\Gamma_n^{red}>$ is proportional to S defined in eq. (2.1), except that we now refer to fixed values of the angular momentum ℓ of the neutron rather than to $\hbar\vec{k}$. Since it is very hard to measure the dependence of $D^{-1} <\Gamma_n^{red}>$ on energy, one plots it versus mass number. The nuclear density is roughly independent of A (saturation), so that the nuclear radius increases with A as $A^{1/3}$, and the number of bound single-particle states in the nuclear potential increases with it. We expect that for fixed ℓ, the function $D^{-1} <\Gamma_n^{red}>$ shows a peak for those values of A for which a quasiparticle state, being pulled into the well, passes neutron threshold. Fig. 3 shows the s-wave neutron strength function versus A. Two big bumps, at A $\approx$ 50 and at A $\approx$ 180, are clearly visible, corresponding to the 3s and the 4s level. The splitting of the peak at A $\approx$ 180 into two is a consequence of nuclear deformation [we are here in the region of strongly deformed nuclei] and provides another example for the coupling of the quasiparticle to surface modes. In contrast to the case of ^{208}Pb, we deal here with a true quasiparticle phenomenon, the strength of the single-particle states being spread over a large number of resonances.

4. DEEP-LYING QUASIHOLE STATES

This topic is discussed in detail at this Conference by G. Wagner[6]. Therefore, I confine myself to a few points relating to quasiparticle properties. Deep-lying hole states are created by reactions of the type (e,e'p), (p,2p), (d,^{3}He),

etc. For deep-lying states, the most extensive published data come from the

(p,2p) reaction. In the analysis, one assumes that such reactions, initiated with

bombarding energies of several 100 MeV, can be described as "direct" or "knock-on"

or "peripheral" processes involving only a single collision between incident

projectile and one of the target nucleons. Multiple scattering and compound

nucleus formation are only taken into account through a complex optical model

potential describing the motion of the projectiles before and after the collision.

Replacing the off-shell t-matrix element for the collision by its on-shell value,

one finds that the (p,2p) cross section is the product of the nucleon-nucleon cross

section and a quantity which is proportional to the strength function introduced

in eq. (2.1). Deep-lying quasihole states are therefore expected to lead to

peaks in the (p,2p) cross section as a function of transferred energy.

The strength function (2.1), taken at fixed momentum transfer $\hbar\vec{k}$, does

indeed display maxima at certain values of ℓ. In a given maximum, it has a

dependence on $\vec{k}$ typical of the square of the Fourier transform of a single-

particle function with fixed ℓ. We can thus identify the peaks in ω, some of

which are shown for fixed ℓ in fig. 4, with quasiparticle energies, and ascribe

to each peak an angular momentum. The high excitation energy in the residual

nucleus at which these peaks are observed suggests that we deal indeed with

quasiparticle (as opposed to single-particle) states. The energy resolution

has not yet been sufficiently good to prove this point, and to obtain values

for the spreading widths. The states found are the ones expected on the basis

of the shell-model. From the point of view of Landau theory, such findings were

not necessarily expected since this theory is known to generally apply only in

the vicinity of the Fermi surface.

The binding energy of the 1s quasiparticle state (this binding energy is

often called the nucleon removal energy) is 35±5 MeV for ^{12}C, rises slowly to

50±10 MeV for ^{40}Ca and thereupon seems to reach saturation at around 60±10 MeV

for A $\geq$ 60. A similar tendency, with smaller absolute values, is indicated for

higher-lying quasiparticle levels. The quasiparticle concept predicts a peak

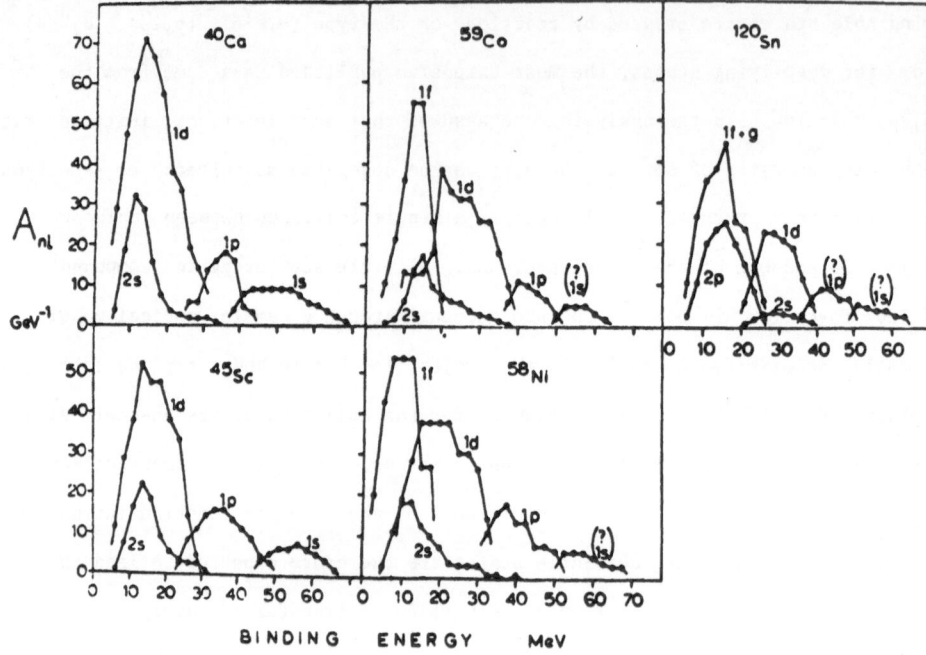

Fig. 4. The strength function for holes with fixed angular momentum in arbitrary

units (from A.N. James et al., Nucl. Phys. A138, 145 (1969)

in $S(\vec{k}, \omega)$. If the mixing of the 1 hole state with the more complicated modes

of motion is reasonably uniform, the peak will have a form as that shown in fig. 1.

If, on the other hand, other simple modes of motion are present, then the peak may

be split into two or more, as was shown in fig. 3. This is certainly a most

interesting question which can only be answered by future high-resolution

experiments.

5. THEORETICAL CONCEPTS

To compare the data shown with calculated values, it is necessary to adapt

the concepts of Landau theory to finite nuclei. We first review the situation

in infinite matter, and then indicate the necessary changes. These changes have

to do with the facts that spin, rather than linear momentum, is conserved for a

finite system, that the spectrum of an infinite system is smooth, while that of

a finite system may undergo fluctuations (caused by compound nuclear resonances), and that the form of the radial wave function for the quasiparticle need be determined, while in the infinite system, it is a plane wave.

For infinite systems, we consider the Green function for hole states,

$$G_{hole}(\vec{k},\omega) = <0|a^{\dagger}(\vec{k})[\omega^{+} + H]^{-1} a(\vec{k})|0> \quad , \quad |\vec{k}| \leq k_F \quad . \tag{5.1}$$

This function has a cut extending from minus the Fermi energy to minus infinity. The discontinuity across the cut is given by $S(\vec{k},\omega)$. One is thus led to identify peaks of $S(\vec{k},\omega)$ with poles of $G_{hole}(\vec{k},\omega)$ located on the second sheet. Such poles define quasiparticles. The real and imaginary parts of the location define $Re\omega(\vec{k})$ and $\frac{1}{2}\Gamma(\vec{k})$, respectively, the real part of the residue gives the occupation probability $P(\vec{k})$. Since G satisfies the Dyson equation, a quasiparticle pole is equivalently defined as the root of the equation

$$\omega = \varepsilon^{o}(\vec{k}) + \Sigma(\vec{k},\omega) \quad , \tag{5.2}$$

where $\varepsilon^{o}(\vec{k})$ is the unperturbed energy and $\Sigma(\vec{k},\omega)$ the self-energy. Eq. (5.2) serves as the starting point for perturbative methods to calculate ω.[21]

To adapt eq. (5.2) to the features of a finite system[7], we have (i) to replace the $a^{\dagger}(\vec{k})$, $a(\vec{k})$ appearing in eq. (5.1) by the creation and destruction operators $a^{\dagger}_{n\ell j}$, $a_{n\ell j}$ relating to a single-particle basis with angular momentum ℓ, spin $j = \ell \pm \frac{1}{2}$, and radial quantum number n [thereby, G_{hole} becomes a matrix in the indices n,n']; (ii) to average G_{hole} over a sufficiently wide energy interval so that all fluctuations are smeared out [the interval must be chosen large as compared to average spacing and width of the compound levels], so that Σ in eq. (5.2) is replaced by an appropriate average; (iii) to replace the scalar equation (5.2) by a matrix equation in the indices n,n'. The solutions of this equation give not only $Re\omega_{\ell j}, \frac{1}{2}\Gamma_{\ell j}$ and $P_{\ell j}$ [in a notation analogous to the case of infinite matter], but in addition yield the quasiparticle wave function because of the matrix nature of eq. (5.2).

The definition of single-particle energies through eq. (5.2) is not the only one available. Koltun[8], modifying a proposal by Baranger[9], suggested [we give his definition here for an infinite system, with obvious modifications necessary for the finite system] that the (real) single-particle energy $E(\vec{k})$ be defined by

$$E(\vec{k}) \; = \; \frac{- \int S(\vec{k},\omega)\omega d\omega}{\int S(\vec{k},\omega)d\omega} \; = \; \frac{- <0|a^{\dagger}(\vec{k})[H,\;a(k)]|0>}{<0|a^{\dagger}(\vec{k})\;a(k)|0>} \; = \; \frac{\varepsilon(\vec{k})}{n(\vec{k})} \qquad (5.3)$$

The numerator in eq. (5.3) is a weighted sum over the strength function, the denominator accounts for the fact that the state $\vec{k}$ is not completely occupied in the ground state $|0>$. The definitions (5.2) and (5.3) differ, as shown below. The definition (5.3) has the advantage that it leads to a sum rule relating the $\varepsilon(\vec{k})$ with the total binding energy of the system as in Hartree-Fock theory. It requires, however, a precise knowledge of the function $S(\vec{k},\omega)$ not only at, but also outside of the peaks. This knowledge is available only if it is possible to separate $S(\vec{k},\omega)$ from the background due to other reactions like, f.i.,multiple scattering. The definition (5.2), on the other hand, is applicable directly to the data, provided only that $S(\vec{k},\omega)$ displays quasiparticle maxima. It is for this reason that the definition (5.2) seems preferable to me, although the calcula-tion of $\omega(\vec{k})$ is more difficult than that of $E(\vec{k})$. These two quantities are related,

$$\omega(\vec{k}) \; = \; E(\vec{k}) \; + \; \Delta(k) \; - \; \frac{i}{2} \; \Gamma(\vec{k}) \; . \qquad (5.4)$$

The real shift $\Delta(\vec{k})$ has roughly the same magnitude as $\Gamma(\vec{k})$, see below.

6.   INTERPRETATION.   CALCULATED VALUES

Both the expressions (5.2) and (5.3) can be evaluated with the help of perturbation theory, and of nucleon-nucleon potentials fitted to the nucleon-nucleon scattering data. Because of the strong repulsion at short distances, the perturbation series is reordered by summing certain ladder diagrams in the way introduced by Brueckner. The numerous technical problems of this procedure are not mentioned here.

The energies $E(\vec{k})$ defined by eq. (5.3) are identical with the self-consistent single-particle energies introduced in renormalized Brueckner-Hartree-Fock theory[10]. They are also identical with the single-particle energies defined in the frame of the renormalized cluster expansion of Jastrow theory[11,12]. There, the cluster

expansion is subjected to the Ritz variational problem. Lagrange multipliers

are introduced to take account of the normalization of the wave function, expressed

as a condition of constraint on the renormalized cluster expansion. The sum of

these Lagrange multipliers and $\frac{\hbar^2 k^2}{2m}$ defines the $E(\vec{k})$.

The physical meaning of the $E(\vec{k})$ is displayed in fig. 5. They are given by

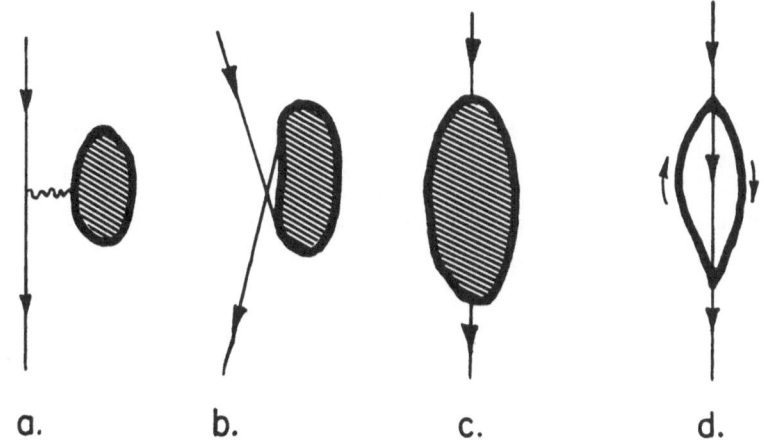

a.   b.   c.   d.

Fig. 5. Diagrammatic contributions to $\omega(\vec{k})$ [a and b] and to $\omega(k)$ [also c]:

A special case of (c) is displayed in (d)

the sum of all linked diagrams with an incoming and an outgoing hole line which

have the form of fig. 5a or fig. 5b. These diagrams take account of all inter-

actions of the hole which lead to an average field. Interactions not included

in $E(\vec{k})$, but included in $\omega(\vec{k})$ are the dispersion-type diagrams which have the

form of fig. 5c, a special case of which is shown in fig. 5d. They correspond

to the real or virtual decay of the hole state into $(n+1)$-hole n-particle con-

figurations, a process by means of which the hole state defined in eq. (5.3)

becomes a quasihole. The energies $\varepsilon(\vec{k})$ appearing in eq. (5.3) differ from both

$E(\vec{k})$ and $\text{Re}\omega(\vec{k})$. The difference between $\varepsilon(\vec{k})$ and $\text{Re}\omega(\vec{k})$ comes from the back-

ground, i.e. the shaded area in fig. 1. Its distribution over energy largely

determines the values of the $\varepsilon(\vec{k})$. Since these quantities fulfill a sum rule[8] relating them with the total binding energy, they are known[13] to be on the average 3-5 MeV larger in magnitude than the corresponding $|Re\omega(\vec{k})|$. [This fact alone shows that Hartree-Fock calculations cannot be used to understand simultaneously binding energies and removal energies in nuclei]. The difference $\Delta(\vec{k})$ between $Re\omega(\vec{k})$ and $E(\vec{k})$ [see eq. (4.4)] has been calculated for infinite nuclear matter[14,15]. Taking into account recent corrections due to occupation probabilities, one finds[7] $\Delta(\vec{k})$ = +18 MeV at the bottom and $\Delta(\vec{k})$ = +3 MeV at the top of the Fermi sea. Qualitatively similar results are expected for finite nuclei, so that $|E(\vec{k})| > |Re\omega(k)|$ by a significant amount. Estimates for the widths[16] when similarly corrected give[7] $\Gamma \approx 8$ MeV and $\Gamma \approx 15$ MeV for the 1s states in ^{12}C and ^{40}Ca, respectively. A recent reexamination[17] of these values has shown, however, that they depend sensitively upon the details of the calculation (the gap in the spectrum).

In table 1 we compare the experimental removal energies (ERE) and the $E(\vec{k})$ found[10] from a renormalized Brueckner-Hartree-Fock calculation (RBHF) for the low-lying states in ^{40}Ca. The close coincidence of the values contradicts the estimates for $\Delta(\vec{k})$ quoted above and suggests that the calculated $|E(\vec{k})|$ are too small. This is also indicated by the facts that RBHF calculations give too little total binding energy and too large nuclear radii. The last column shows the values of a semiphenomenological density-dependent Hartree-Fock (DDHF) calculation[18]. The agreement between DDHF and ERE is expected and turns out to be good. However, the connection between DDHF and nucleon-nucleon forces is only indirect. Application of DDHF to nuclei involves the adjustment of at least two parameters. In this sense, the $E(\vec{k})$ (and, hence, the $\omega(\vec{k})$), i.e. the single-particle energies in nuclei, are not yet quantitatively understood in terms of the nucleon-nucleon force.

TABLE 1.

Single-particle energies in ^{40}Ca

Level	ERE	RBHF	DDHF
1s	-50 ± 11	-55	-43
1p	-34 ± 6	-35	-27
1d	-12.6	-17	-11.8

7. OUTLOOK

We have seen how Landau's quasiparticle concept is modified when it is applied to finite systems, how important it is for the description of independent-particle features in nuclei, and how far beyond the immediate vicinity of the Fermi surface its validity is established by the data. This last point is understood quantitatively in the frame of Brueckner theory where it is shown that the strong two-body correlations caused by the nucleon-nucleon force affect the single-particle motion only over short distances (healing). On the average, particles move quite independently. The short-range correlations lead to the background in fig. 1 and to the fact that the area under the peak is $P(\vec{k}) \simeq 0.85$. These short-range forces are expected to be seen only in experiments with bombarding energies of $\gtrsim 100$ MeV. Hence, single-particle motion dominates the low-energy features of nuclei.

This fact is clearly seen not only when nuclear spectra are interpreted in terms of interacting quasiparticles. It is also borne out by the fact that nuclear level densities can, with minor modifications, be well described in terms of the level density of a gas of free particles moving independently under the influence of a central potential. Even collective properties can be understood in the frame of a single-particle model. This is demonstrated most impressively by Strutinsky's[19] shell-correction method. Here, it is assumed that the fission process can be semiquantitatively described by the liquid drop model, with parameters taken from the semiempirical mass formula, and that the most important corrections to this model, which determines the potential energy of the drop at each stage of the fission process, comes from the shell correction. This accounts for the fact that the level density near the Fermi surface is not uniform but shows shell structure which depends on the deformation, a fact not taken into account in the (classical) liquid drop model. It has been shown that this simple quantal correction to the liquid drop model, a correction based upon the assumption of independent-particle motion, yields a very successful theory capable of describing many features of the fission process.

The importance of single-particle motion in nuclei was established 24 years ago by Mayer and Haxel, Jensen and Suess[20]. I have tried to show how successful this concept is for the interpretation of data even at high excitation energies, far from the Fermi surface. It underlies most of the work done in Nuclear Physics to-day, and it will be interesting to see how it can be combined with the new properties of nuclei we expect to find with beams of higher energy and fine energy resolution, and with beams of heavy ions. I have perhaps not emphasized sufficiently that the unified understanding of nuclear phenomena with the help of the single-particle model is based on models, not on an exact theory: Different classes of phenomena often require the independent fitting of a few parameters. This raises the question of whether we can find a theoretical formulation which encompasses all these approaches.

This work was done while the author was a visitor at the Weizmann Institute of Science. He would like to thank Professor Igal Talmi and the other members of the Department of Nuclear Physics for their warm and generous hospitality.

REFERENCES

1. Landau, L.D.: JETP (Sov. Phys.) 3, 920 (1956); 5, 101 (1957); 8, 70 (1959).

2. Nozieres, P.: Theory of Interacting Fermi Systems, Benjamin, New York 1964.

3. Abrikosov, A.A., Gorkov, L.P., Dzyaloskinski, I.E.: Methods of quantum Field Theory in Statistical Physics, Prentice Hall, New York 1963.

4. de-Shalit, A.: Phys. Rev. 122, 1530 (1961).

5. Migdal, A.B.: Theory of Finite Fermy Systems and Applications to Atomic Nuclei, Interscience, New York 1967.

6. Wagner, G.J.: Contribution to this Conference.

7. Engelbrecht, C.A., Weidenmüller, H.A.: Nucl. Phys. A184, 385 (1972).

8. Koltun, D.S.: Phys. Rev. Lett. 28, 182 (1972).

9. Baranger, M.: Nucl. Phys. A149, 225 (1970).

10. Davies, K.T.R., McCarthy, R.J.: Phys. Rev. C4, 81 (1971).

11. da Providencia, J., Shakin, C.M.: Phys. Rev. C4, 1560 (1971).

12. Schäfer, L.: Private communication.

13. Becker, R.L.: Phys. Lett. 32B, 263 (1970).

14. Brueckner, K.A., et al.: Phys. Rev. 118, 1438 (1960).

15. Köhler, H.J., Phys. Rev. 137B, 1145 (1965).

16. Köhler, H.J.: Nucl. Phys. 88, 529 (1966).

17. Monga, K.J., et al.: Private communication.

18. Negele, J.W.: Phys. Rev. C1, 1260 (1970).

19. Strutinski, V.M.: Arkiv Fysik 36, 629 (1967); Nucl. Phys. A95, 420 (1967); Nucl. Phys. A122, 1 (1968).

20. Goeppert-Mayer, M.: Phys. Rev. 75, 1969 (1949); Haxel, O, Jensen, J.H.D., Suess, H.E.: Phys. Rev. 75, 1766 (1949).

21. Another approach, based directly on Green function theory, has early been taken by R. Puff, A.S. Reiner and L. Wilets, Phys. Rev. 149, 778 (1966).

SINGLE-NUCLEON REMOVAL ENERGIES

Gerhard J. Wagner

Max-Planck-Institut für Kernphysik, Heidelberg

and

C.E.N. Saclay, BP 2, 91190-Gif-sur-Yvette, France[†]

Abstract. A survey is given of measurements of nucleon separation energies from occupied shells. Particular emphasis rests on the results of pick-up experiments and recent quasi-free scattering experiments which have considerably improved our knowledge of nuclear hole states. So, in ^{40}Ca the average separation energy of 1s protons has been found to be 50 MeV (rather than 80 MeV). Similarly, the average separation energy of a $1p_{1/2}$ proton from ^{28}Si is 16 MeV (rather than 25 MeV). Recently, one has obtained evidence for an increase of the 1p spin-orbit splitting from 6.5 MeV in ^{16}O to about 15 MeV in ^{28}Si. Results from pick-up and knock-out experiments are now in full agreement.

The relation between the experimental removal energies and single particle energies from many-body theories is briefly discussed. It is demonstrated that the inclusion of density dependent interactions improves the results of Hartree-Fock calculations considerably. Eventually, the use of model independent sum rules is recommended for future applications.

1. INTRODUCTION

The ground state properties of a stable nucleus may be summarized by listing the mass, the r.m.s. radius (and possibly some higher moments of the charge distribution), the spin, and the electric and magnetic moments. All these quantities reveal little of the microscopic structure of the nucleus. Transfer reactions, in particular pick-up and knock-out of single nucleons are more intimately connected to many-body wave functions which describe the degrees of freedom of individual

---

[†] Present address.

nucleons in the target ground state. A familiar example is the close connection of spectroscopic factors from such reactions to the coefficients of fractional parentage of the shell model.

Extrapolating from the findings on a few active nucleons outside an inert core the shell model postulated the existence of occupied shells within this core. The shell model was more than ten years old when this ingenious extrapolation obtained a brilliant confirmation by the pioneering (p,2p)-experiments of Tyrén et al.(ref. 77 and refs. given there). The lower part of fig. 1 which is taken from this work contains a spectrum from the ^{16}O(p,2p) reaction which exhibits three prominent groups which are attributed to the $1p_{1/2}$, $1p_{3/2}$ and $1s_{1/2}$ shells.

On the theoretical side the a posteriori justification of the shell model is attempted through Hartree-Fock (HF) and Brueckner-Hartree-Fock (BHF) calculations. The results of these complicated calculations which include single particle energies for various shells clearly have to be subject to an experimental verification. In section 2 we shall briefly summarize the methods for the determination of the relevant experimental quantities, namely (i) the removal energies, (ii) the orbital characteristics and (iii) the occupation numbers of the various shells. The intricate problem of the relation between the experimental removal energies and the theoretical single-particle energies will be discussed in section 4 after we have presented a summary of the proton removal energies from light and medium weight nuclei (section 3).

The last review of proton separation energies[69] appeared four years ago. It showed (see fig. 2) 1s and 1p separation energies increasing rapidly with the target mass, while the 2s and 1d shells exhibited a completely different "saturation" behaviour, which led to a 20 MeV gap between the outer and inner shells in ^{40}Ca. We shall see that this difficulty (among others) has been resolved during the past years which have seen a drastic improvement and revision of our knowledge. This was achieved partly by (p,2p)[36,51] - and (e,e'p)[15] - experiments with good momentum resolution and partly by (d,^{3}He) - reactions with superior energy resolution. The latter have become possible due to the availability of deuteron beams from isochronous cyclotrons with energies between 30 and 80 MeV. While there are excellent review articles on (p,2p)[36] - and (e,e'p) reactions[1] the important contribution of pick-up reactions to the problem of removal energies is not generally known[*]. This may justify the emphasis which shall be laid on the pick-up data in the following.

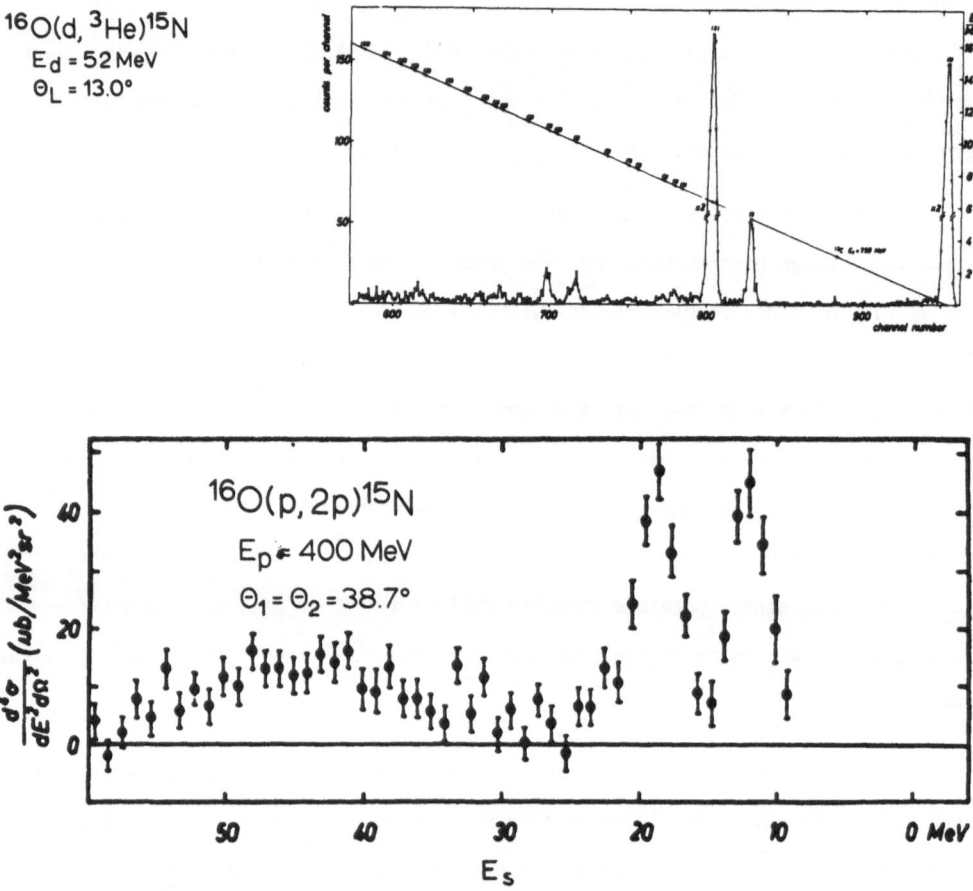

$^{16}O(d, ^3He)^{15}N$
$E_d = 52\,MeV$
$\Theta_L = 13.0°$

$^{16}O(p, 2p)^{15}N$
$E_p \approx 400\,MeV$
$\Theta_1 = \Theta_2 = 38.7°$

Fig. 1. Comparison of spectra obtained in a (p,2p)-experiment[77] and a (d,^{3}He)-experiment[28] on ^{16}O. The abscissa scales are identical

[*] Also it has been incorporated only partly into a recent review[37] which appeared after completion of this manuscript.

2. EXPERIMENTAL METHODS AND ANALYSIS

2.1. Determination of Removal Energies

2.1.1. By quasi-free Scattering Reactions

In quasi-free scattering (QFS) reactions, i.e. in (e,e'p) - and (p,2p) - reactions in regions where the impulse approximation is valid, one determines the kinetic energies T_1 and T_2 of the two outgoing particles and their scattering angles, hence their momenta k_1 and k_2. Since most of the relevant experiments have been performed at incident lab energies T_0 above 150 MeV, a more typical value being 400 MeV, the prevailing detector systems are magnetic spectrometers. Other techniques used in some older experiments are summarized in ref. 69.

The removal or separation energy of the knocked-out proton, defined as the energy needed to promote the proton into the continuum with zero kinetic energy, is then obtained as

$$E = T_o - (T_1 + T_2) - T_{A-1} \tag{1}$$

where T_{A-1} is the recoil energy of the final nucleus which is generally negligible. Note that E > 0 by this definition.

The typical energy resolution today is still 2-5 MeV, with a best value of 1.2 MeV obtained in the (e,e'p) - experiments at Saclay[15]. One hopes to achieve a resolution of 600 keV through the use of two new broad range spectrographs at the 600 MeV electron linac at Saclay in the course of 1973. Structures in the missing energy spectra have been observed up to separation energies $E \lesssim 60$ MeV, in exceptional cases up to 80 MeV[1].

2.1.2. By Pick-up Reactions

Pick-up experiments are experimentally much simpler as they require only the determination of the energy and scattering angle of one particle for a kinematically complete experiment. The measurements are generally performed with telescopes (for particle identification) consisting of cooled Si surface barrier detectors. The typical energy resolution is 100 keV. A summary of the methods used by various groups may be found in table I which for brevity is restricted[*] to the proton

---

[*] A more complete compilation may be found in ref. 81.

pick-up reactions (t,α) and (d,³He) as far as they have contributed to the results
presented in this article.

The upper part of fig. 1 demonstrates the advantage of the improved energy
resolution and statistical accuracy obtainable in pick-up reactions. One notices
a significant excitation of the $5/2^+$, $1/2^+$ doublet at 5.3 MeV in ^{15}N by pick-up
of $1d_{5/2}$ and $2s_{1/2}$ protons from ^{16}O which is a consequence of ground state correlations in ^{16}O. Another deviation from the simple shell model which becomes visible
due to the improved energy resolution is the splitting of the $1p_{3/2}$ hole strength
into at least three final states at 6.32, 9.94 and 10.70 MeV in ^{15}N [58].

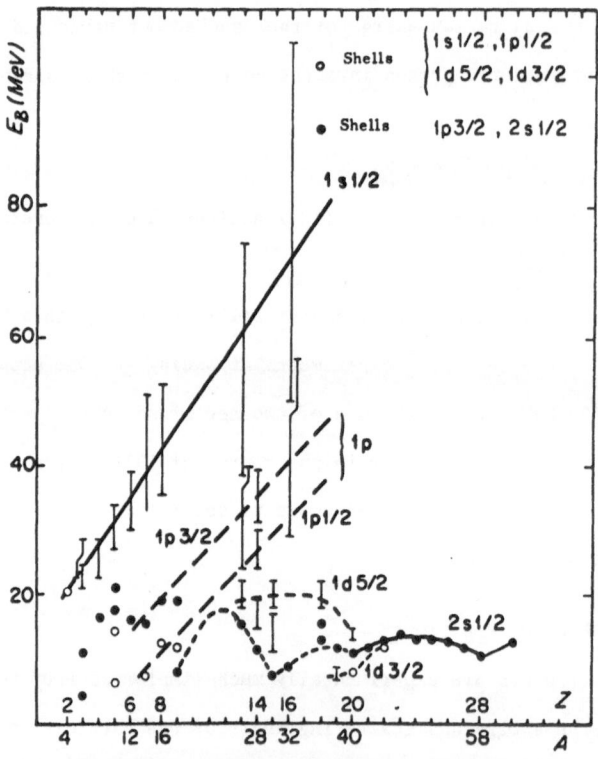

Fig. 2. Summary of proton separation energies as available in 1968 (from ref. 69)

The ratio of the areas below the dominant $1p_{1/2}$ (g.s.) and $1p_{3/2}$ (6.32 MeV)
peaks is close to unity (with a slight angular dependence of the ratio) and thus
differs from the ratio of shell model occupation numbers by a factor of about two.

TABLE I

Proton pick-up from occupied shells. Only references which
provided results utilized in the present articles are given.

Incident Energy (MeV)	Accelerator	Detector System	Energy Resolution (keV)	Target Nuclei	Ref.
		$(d,{}^{3}He)$ - Reactions			
29	Univ. of Michigan cyclotron	Magnetic spectrograph	40	${}^{45}Sc$	26)
34	Oak Ridge Nat'l. Lab. Cyclotron	Semiconductor Telescope	80 - 120	${}^{16}O$, ${}^{40}Ca$ ${}^{23}Na$, ${}^{27}Al$ ${}^{28}Si$ ${}^{48,50}Ti$, ${}^{51}V$, ${}^{54}Fe$	30) 85) 84) 63)
40	Berkeley cyclotron	Semiconductor Telescope	100	${}^{22}Ne$	14)
52	Karlsruhe cyclotron	Semiconductor Telescope	300 - 400[*]	${}^{12}C$, ${}^{14}N$ ${}^{19}F$, ${}^{20,22}Ne$ ${}^{27}Al$ ${}^{48,50}Ti$, ${}^{51}V$ ${}^{63,65}Cu$	32) 39) 80) 31) 55)
			80 - 100	${}^{12,13,14}C$ ${}^{15}N$ ${}^{16}O$ ${}^{17}O$ ${}^{18}O$ ${}^{24,25,26}Mg$ ${}^{28,29,30}Si$ ${}^{31}P$, ${}^{32}S$, ${}^{36}Ar$ ${}^{38}Ar$	60) 40) 28) 59) 29) 49) 57) 20)
80	Orsay Synchro-cyclotron	Magnetic Spectrometer	100-350	${}^{23}Na$, ${}^{24}Mg$ ${}^{27}Al$, ${}^{28}Si$	3)
82	Jülich cyclotron	Semiconductor Telescope	300[*]	${}^{12}C$	42)
		(t,α) - Reactions			
13	Aldermaston Tandem	Multiangle Spectrograph	20	${}^{40,42,44,46,48}Ca$	71)
15	Los Alamos Tandem	Single Semiconductor	40 - 55	${}^{58,60,62,64}Ni$	11)

[*] Without momentum analysis of the beam

This is true for the (d,^{3}He) - as well as the (p,2p) - spectrum which demonstrates the importance of absorptive processes even for high energy (p,2p) measurements. (See also ref. 52). It is not surprising that for large separation energies these absorptive processes affect the pick-up reactions, which involve complex particles in one or both channels, to a larger extent than the QFS reactions. They limit the usefulness of pick-up reactions to separation energies below 25 MeV, or 35 MeV in an exceptional case[42].

2.2. Determination of Orbital Characteristics

2.2.1. By Pick-up Reactions

As is well known the shape of the angular distributions obtained in pick-up reactions is characteristic for the angular momentum of the transferred particle. The angular distributions from the ^{24}Mg(d,^{3}He)^{23}Na reaction (fig. 3) may serve as an example. One can easily distinguish ℓ = 0, 1 and 2 transfer. Moreover, one finds in the entire mass region of 12 $\lesssim$ A $\lesssim$ 60 that the second minimum is slightly more pronounced for j = ℓ + 1/2 transfer than for j = ℓ - 1/2. But this "j - effect" is not sufficiently strong to serve as a safe basis for j - value determinations. The solid and dashed lines of fig. 3 are angular distributions calculated with the distorted wave Born approximation (DWBA) using different optical parameters[49].

The pronounced structure of the angular distributions, the success of the DWBA calculations and their insensitivity to the optical model parameters are typical for reactions with a marked surface localization of the transition amplitude which is the result of good angular momentum matching $|k_i R - k_f R| \sim \ell$ (ref. 74). In that respect the (d,^{3}He) - reaction turns out to be superior to the (p,d) - (refs. 6, 44, 75), (^{3}He,α) - (ref. 74)[*] and (t,α) - (ref. 11) reactions for the interesting region of Q-values.

---

[*] Within a small "window" of slightly negative Q-values one finds also very pronounced diffraction patterns in the (^{3}He,α) - reaction [10].

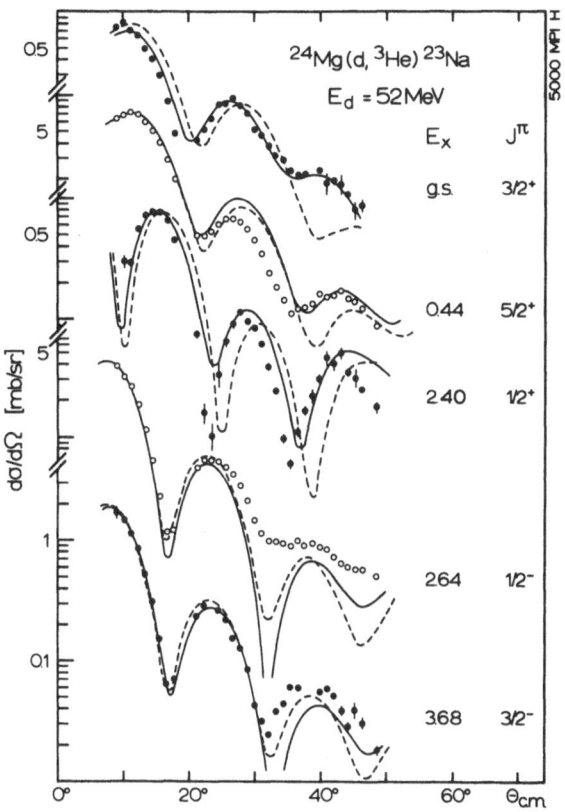

Fig. 3. Angular distributions from the ^{24}Mg(d,^{3}He)^{23}Na reaction for different ℓ
and j transfer (from ref. 49). The solid and dashed curves are the
results of DWBA calculations with different optical potentials

In general one has no difficulties in determining the ℓ-values of the transferred nucleon, in particular with the (d,^{3}He) reaction. Yet we should mention
at this point a longstanding DWBA problem[38] which still needs to be solved if one
wants to make full use of the virtues of the pick-up reactions: the DWBA gives only
a reasonable fit to the data, as far as the positions of the maxima and minima of
the angular distributions are concerned, if the nucleon is transferred from or
into states near the Fermi surface. This is demonstrated in fig. 4 for the case of
pick-up of $1p_{1/2}$ protons. The fit is satisfactory only for ^{16}O and deteriorates
systematically with increasing number of (2s,1d) nucleons above the 1p-shell.
Analog effects exist also for other shells[38,72]. Clearly, one can have little

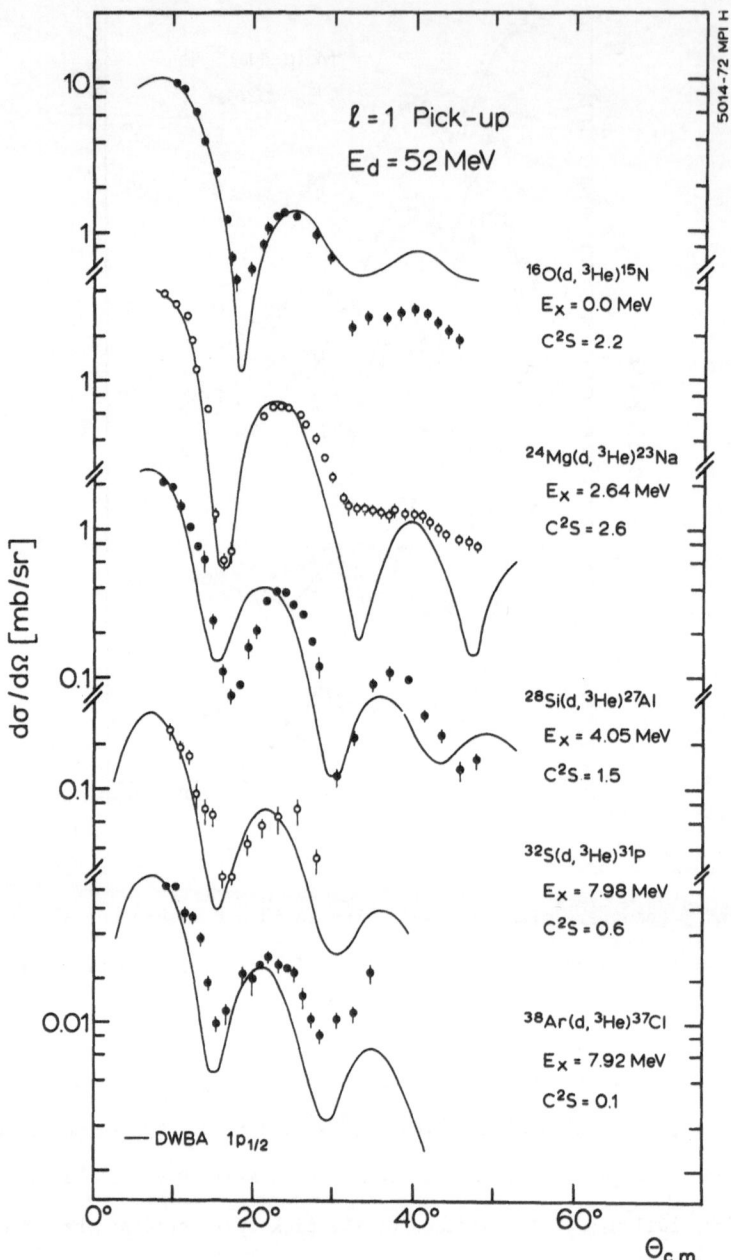

Fig. 4. Angular distributions of the strongest ℓ=1 transitions observed in (d, ³He)
reactions from (2s, 1d)-shell nuclei (from ref. 20). The DWBA curves
(solid lines) exhibit a stronger shift towards smaller angles with
increasing target mass than the data. Note that the peak cross sections
drop much faster than the spectroscopic factors C²S

confidence in spectroscopic factors extracted with such bad fits. Due to the extreme surface character of the reaction, modifications of the bound state wave function fail to change the shape (not the height) of the angular distributions.

2.2.2. By QFS Reactions

QFS reactions owe their reputation as a powerful spectroscopic tool to the fact that the cross-section measured for a given separation energy E and recoil momentum $q_R = k_o - (k_1 + k_2)$ may be factored into the free p-p or e-p cross section and a factor which basically depends on the target wave function (for details see ref. 37). Certainly one has to consider off-shell corrections to the free cross sections and distortions of the protons waves. But one has frequently found that the plane wave impulse approximation (PWIA), where the momentum q of the struck proton before the collision is $q = -q_R$, gives a reasonable fit to the ("distorted") momentum distributions which follow directly from the data after division by the free cross sections. An inclusion of distortion effects in the WKB-approximation[50] tends to reduce the PW-cross sections with only minor changes of the shape of the momentum distributions, i.e. the distortion is mainly absorptive and not refractive. To the extent that this is true or that one will be able to account for the distortions exactly one can extract from the data the nuclear structure factor, namely the "diagonal" spectral function[27,23] of the target ground state $|A>$

$$S(q,E) = <A|a^+(q) \cdot \delta(H - E - W_A) \cdot a(q)|A> \qquad (2)$$

which after division by the number of protons with momentum q in the target ground state

$$n(q) = <A|a^+(q) \cdot a(q)|A> \qquad (3)$$

has the simple interpretation of being the joint probability $P(q,E)$ of finding a proton with momentum q in the target nucleus and the total energy $W_{A-1} = W_A + E_s$ of the final nucleus after removal of this proton. (The operators $a(q)$ and $a^+(q)$ annihilate and create, respectively, a proton with momentum q). In an individual particle model without residual interactions this spectral function for a given eigenenergy $E = \varepsilon_{n\ell j}$ will apparently be the square of the single particle wave function $|\psi_{n\ell j}(q)|^2$ multiplied by the number of protons $N_{n\ell j}$ in the corresponding orbit. It is this property of the QFS cross section which has been utilized since the experiments by Tyrén et al.[77]

One notices that the cross section depends only indirectly on the angular momentum. In fact, of all possible angular correlations only $\ell=0$ is qualitatively distinct from all others, namely by a minimum, at $q=0$. In practice one has encountered difficulties in distinguishing $\ell=1$ and $\ell=2$ knock-out[2] which has caused some misinterpretations of (p,2p) - data with considerable consequences on the 1p separation energies. These misinterpretations were later corrected by studies of (d,^{3}He) reactions[79,49,3] (see section 3.2).

On the other hand, the QFS reactions offer the unique chance to study fairly directly single particle wave functions of the knocked out protons. It is here, and not so much with the energy resolution, that the younger generation of QFS experiments[36,51,15] has achieved major progress: the momentum resolution has been improved to typically 10 MeV/c.

This leads to very impressive demonstrations of the individual particle features of the nuclei investigated. Fig. 5 which has been prepared using the ^{40}Ca(e,e'p)-data of the Saclay group[*] shows the cross sections as a function of recoil momentum in various ranges of separation energies. With increasing separation energy one finds successively distributions which are compatible with the assumptions of $1d_{3/2}$, $2s_{1/2}$, $1d_{(5/2)}$, 1p and $1s_{1/2}$ knock-out. One may expect that the radiation corrections which still have to be made will reduce the background at high separation energies. A comparison with a similar figure by James et al.[36] demonstrates the progress achieved by the improved energy resolution in the (e,e'p) experiment.

[*] I am grateful to the authors of ref. 15 for the permission to use their data prior to publication.

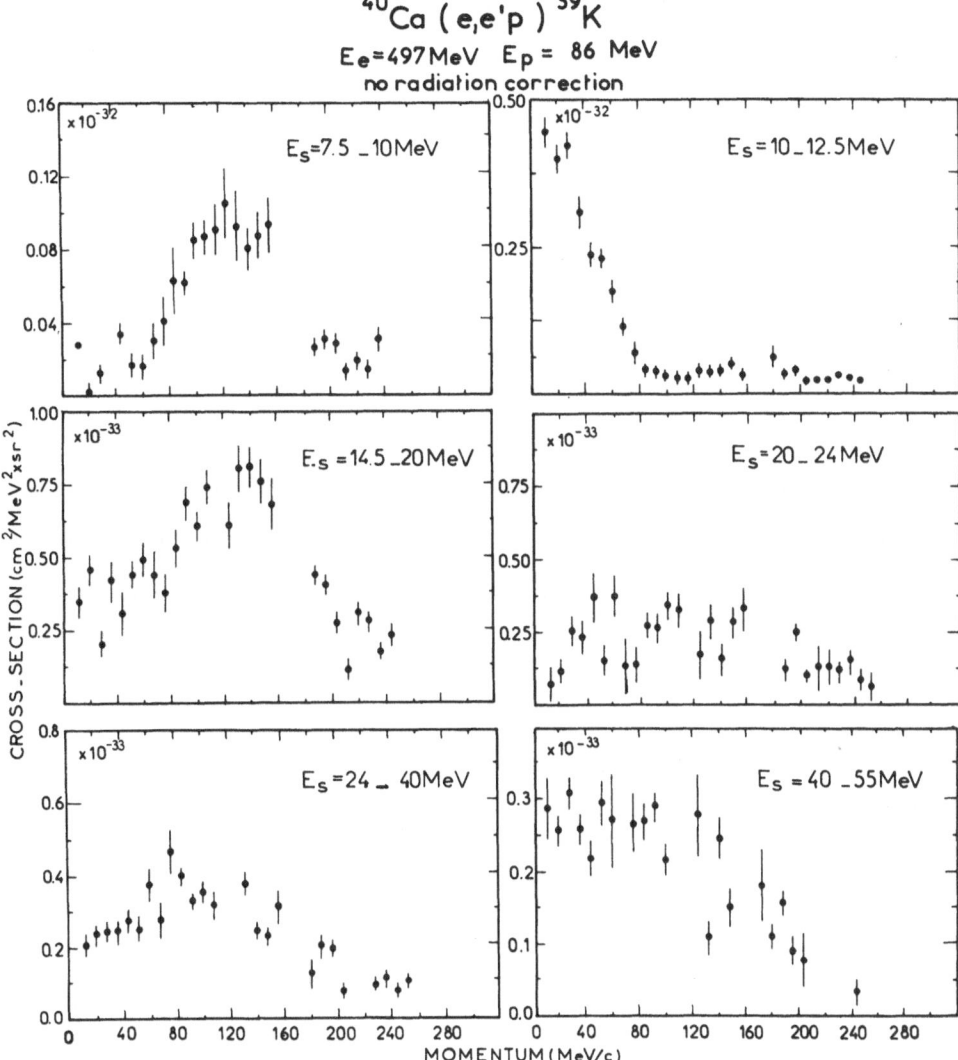

Fig. 5. Recoil momentum distributions as obtained in ^{40}Ca(e,e'p)-experiment

(ref. 15) for various regions of separation energies

2.3. Occupation Numbers

2.3.1. Quasi-free Scattering

Provided one has determined the spectral function S(q,E) the number of

knocked-out protons results from an integration over E (see eq. 3) and over q.

Ideally, the result should equal the number Z of target protons. Deviations from this number indicate the extent to which one has failed to take into account the absorptive processes. In the PWIA-analysis of the (p,2p)-data at 285 MeV[36] one has obtained "reduction factors" of 0.2 for ^{12}C to 0.02 for ^{208}Pb. A preliminary analysis of the Saclay (e,e'p)-results[15] yields reduction factors of 0.3 for ^{12}C and 0.2 for Ni. The DWIA-analysis of the CERN-data[51] brings the occupation numbers within typically a factor two of the shell model values with occasionally much larger deviations for the inner shells.

Even though the plane wave basis implied in eq. (2) appears to be the most proper one for the QFS-experiments, it is only recently that one has discussed the virtues of this basis[45] (see sect. 4.4). So far, the prevailing basis for discussions was the shell model. In fact, we have already considered the ^{40}Ca(e,e'p)-results in these terms (see sect. 2.2.2). The transformation from the plane wave into the shell model basis is being made by decomposing the measured momentum distributions S(q,E=E') for a fixed energy E=E' into momentum distributions calculated in a shell model basis

$$S(q,E=E') = \sum_{n\ell j} A_{n\ell j}(E') |\psi_{n\ell j}(q)|^2 \tag{4}$$

To keep the number of basis states small, i.e. identical to the normally occupied orbits, one adjusts the single-particle potential for the various orbits separately to reproduce the measured momentum distributions and disregards the resulting non-orthogonality. An example of such a decomposition of the Liverpool (p,2p)-data[36] will be given in fig. 13.

2.3.2. Pick-up Experiments

As mentioned above, pick-up reactions with good angular momentum matching are strongly surface localized reactions. Thus one can only measure the amplitude of the single particle wave functions at the nuclear surface. Experimental spectroscopic factors are determined by comparing the measured angular distributions with DWBA-cross sections.

In complete analogy to eq. (2) one defines[54] spectroscopic factors in a shell model basis for a pick-up reaction leading to discrete final states |A-1,f>

$$c^2 S_{n\ell j} \equiv |<A-1,f \, |a_{n\ell j}|A>|^2 \tag{5}$$

c^2 is an isospin Clebsch-Gordan coefficient. Obviously, one can only expect an agreement of the experimentally determined spectroscopic factors with this defini- tion if among others the following usual approximations for the bound state wave function in the DWBA amplitude are not too bad:

(i) The single particle wave function is calculated in a Woods-Saxon well whose depth is adjusted to reproduce the experimental separation energy. This method,which only guarantees the exact asymptotic shape of the wave function,is at best valid for spectroscopic factors close to the shell model limit (2j+1) in which case the single particle concept is justified a posteriori. More basic calcula- tions of the single particle wave functions have been proposed but rarely applied due to their complexity[21,67].

(ii) For unbound hole states the coupling to the decay channels is neglected. Since in pick-up reactions one has not observed a significant broadening of the final states even when excited several MeV above the particle thresholds (see e.g. fig. 12) this neglect appears to be justified.

Summing the spectroscopic factors eq. (5) for a given $n\ell j$-transfer over all final states $|f\rangle$ yields immediately the familiar sum rule[54]

$$\sum_f c^2 S_{n\ell j} = N_{n\ell j} \tag{6}$$

where $N_{n\ell j}$ is the number of target protons within the given shell model orbit. In spite of the many DWBA problems the application of this sum rule frequently yields

TABLE II

Quantities and typical precision presently obtainable from quasi-free scattering and pick-up experiments. (for details see section 2).

Method	Separation Energy (MeV)	Orbital Quantities	Occupation Numbers
Q.F.S.	$\Delta E_s \sim 2 - 5$	$\ell = 0, \ell \neq 0$	$\Delta N_{n\ell j}/(2j+1) \lesssim 5$
	$E_s \lesssim 60$	$P(q, E_s)$	
Pick-up	$\Delta E_s \sim 0.1$	$\ell, (j)$	$\Delta N_{n\ell j}/(2j+1) \lesssim 0.3$
	$E_s \lesssim 25$		

very reasonable results. E.g. in the $^{16}O(d,^3He)^{15}N$ reaction[28,58] one finds

$N_{1p_{1/2}}$ = 2.2 and $N_{1p_{3/2}}$ = 3.8 as compared to 2 and 4 from the simple shell model.
Also for active shells where detailed comparisons between experimental and theoretical spectroscopic factors have been possible one has frequently observed satisfactory agreement (see e.g. refs. 32 and 39).

A brief summary of sections 2.1 - 2.3 is contained in table II.

2.4. Mean Removal Energies

From the reasonable spectroscopic factors we infer that pick-up reactions as well as knock-out reactions at sufficient incident energies proceed preferentially as direct reactions. This means (as eq. (5)shows) that the population of a state $|A-1,f>$ by the reaction is proportional to the percentage of the "λ-hole state"

$$|A - 1,\lambda> = a_\lambda |A>/<A|a_\lambda^+ a_\lambda |A> \tag{7}$$

This hole state is generally not realized in nature but distributed over many final states. The basis λ may be a plane wave basis (see eq. (2)) or a shell model basis (see eq. (5)). The energy $W_{A-1,\lambda}$ of this hole state is by definition given by the expectation value of the many-body Hamiltonian H. The corresponding (positive) "mean separation energy" is then

$$<E(\lambda)> = <A|a_\lambda^+[H,a_\lambda]|A>/<A|a_\lambda^+ a_\lambda |A> \tag{8}$$

We want to determine this mean separation energy, which is often called "centroid separation energy" since it is related to the experiment by

$$<E_\lambda> = \int^f P(\lambda,E) \cdot E \cdot dE \tag{9}$$

or in a notation more familiar for the discussion of pick-up results

$$<E_\lambda> = \sum_f C^2 S_\lambda \cdot E / \sum_f C^2 S_\lambda . \tag{9'}$$

To the extent that both reactions measure the hole state components $a_\lambda|A>$ of the final states they should yield the same results. After the many discussions in this context which have been prompted by some misinterpretations of (p,2p)-data (see sections 2.2.2 and 3.2) it is worth mentioning that in all cases where comparable data exist the results from pick-up and QFS reactions agree perfectly.

It has also been discussed [ref. 13 and refs. given therein] whether reactions which proceed slowly compared to the lifetimes of the hole states with respect

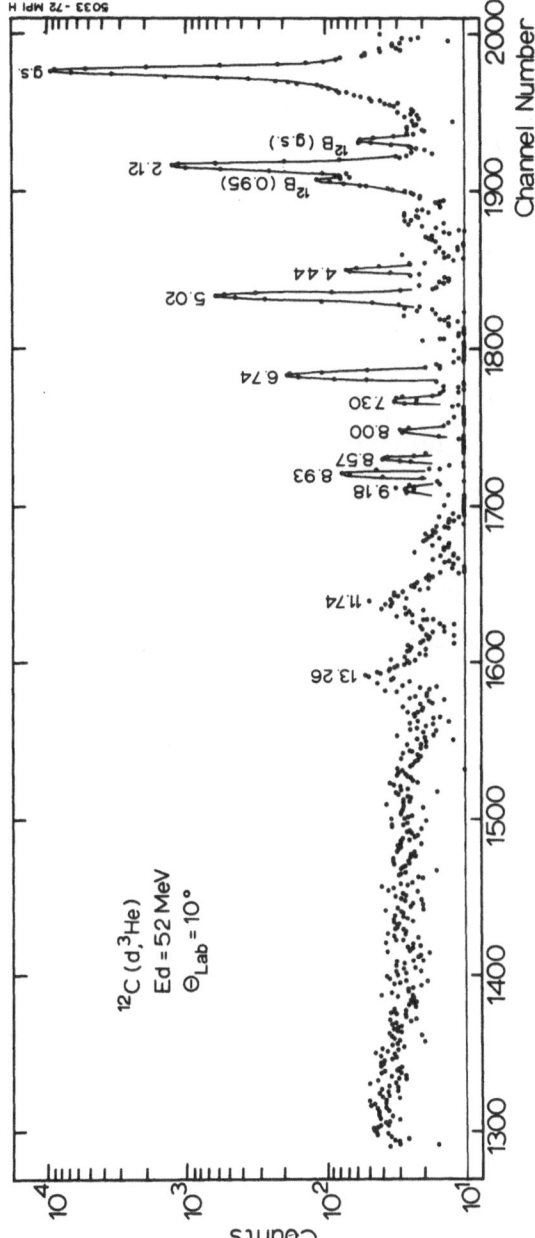

Fig. 6. Spectrum from the $^{12}C(d, ^3He)^{11}B$ reaction showing the excitation of "pick-up forbidden" $5/2^-$ and $7/2^-$ states at 4.44 and 6.74 MeV respectively

to re-arrangement processes might yield smaller mean separation energies. In the extreme case of adiabatic removal one would only excite the hole state of given spin and parity with the smallest separation energy. It seems difficult to define such an adiabatic hole state in the continuum region. At the moment there is no evidence[*] for a shift of the centroid separation energies as a function of reaction velocity. To really observe such shifts would require a consistent analysis of the same reaction performed over a large range of incident energies. At the moment one rather has to be aware of two-step processes which through inelastic excitations simulate an increase of the separation energies. So, the $^{12}C(d,^{3}He)^{11}B$ reaction (see fig. 6) shows e.g. excitation of $5/2^-$ (4.44 MeV) and $7/2^-$ (6.74 MeV) states at the level of 1% to 10% of the ground state strength[**] which are "pick-up forbidden" unless the ground state correlations of ^{12}C provide sufficient f^2-components.

Before regarding the results one should be aware of several fundamental difficulties which one encounters:

(i) The analysis in terms of the shell model implies the knowledge of principal quantum numbers. These quantities are generally "assigned" on the basis of plausibility arguments.

(ii) One can never be certain that one has not missed some weak strengths at high separation energies. The neglect of such strengths affects the centroid energies seriously.

(iii) Baranger's definition of single particle energies[7] contains in addition to the pick-up term (eq. (8)) an analog stripping term. Aside from divergence problems arising from this term in the case of hard core interactions[73,19] one faces the experimental difficulty of having to treat stripping into unbound states. Therefore we prefer the definition eq. (8) which for occupied shells agrees with Baranger's definition.

---

[*] The experimental errors are too large to take the examples suggested in ref. 13 seriously.

[**] The relative excitation depends strongly on the scattering angle (see ref. 32).

3. PROTON HOLE STATES IN LIGHT AND MEDIUM WEIGHT NUCLEI

The presentation of data will be confined to proton pick-up from light and medium weight nuclei where more data and many-body calculations exist than for heavy nuclei.

3.1. On $1d_{3/2}$ and $2s_{1/2}$ Hole States in f-p Nuclei

The study of proton pick-up from occupied $2s_{1/2}$ and $1d_{3/2}$ shells is simplified by a small fractionation of the hole strengths[31,55,71]. In even target nuclei one finds essentially one strong $\ell=0$ transition which is also seen in (p,2p)-experiments[70] and (e,e'p)-experiments (see fig. 5). In addition at low separation energies one has generally one strong $\ell=2$ transition. The two peaks are commonly identified with the $2s_{1/2}$ and $1d_{3/2}$ shells. On the 5% to 10% level of this strengths one finds one or two additional $\ell=0$ or $\ell=2$ transitions up to two MeV higher. These do not shift the centroid energies by more than 200 keV. Separated by several MeV one finds several groups with $\ell=2$ transfer which are attributed to the $1d_{5/2}$ shell. The spectroscopic factors for $d_{5/2}$ transfer do not add up to the shell model sum rules in contrast to the $2s_{1/2}$ and $1d_{3/2}$ shells. Therefore in fig. 7 we have plotted the mean separation energies of $2s_{1/2}$ and $1d_{3/2}$ protons only for target masses from 40 to 65. One observes a general increase of the binding energies with the mass, with a "modulation" superimposed which leads to a maximum near A=50. Also one notices a crossing over of the two shells, with the $2s_{1/2}$ protons more strongly bound than the $1d_{3/2}$ shell in ^{40}Ca and vice versa in Ni and Cu target nuclei.

A Woods-Saxon potential which is to describe these separation energies as eigenenergies obviously has to be supplemented by additional charge and isospin independent terms[33]. We prefer instead a description offerred by Bansal and French[4] for that purpose, since it is more transparent and has proven very useful as "poor man's Hartree-Fock".

The separation energy of a $1d_{3/2}$ or $2s_{1/2}$ proton from the ^{40}Ca-core of a $f_{7/2}$ nucleus is according to that method given by

$$E(Z,A) = E(^{40}Ca) - (A-40) \cdot \overline{E} + \frac{T_o}{2} \overline{\Delta E} - (Z-20) \cdot \varepsilon_c \qquad (10)$$

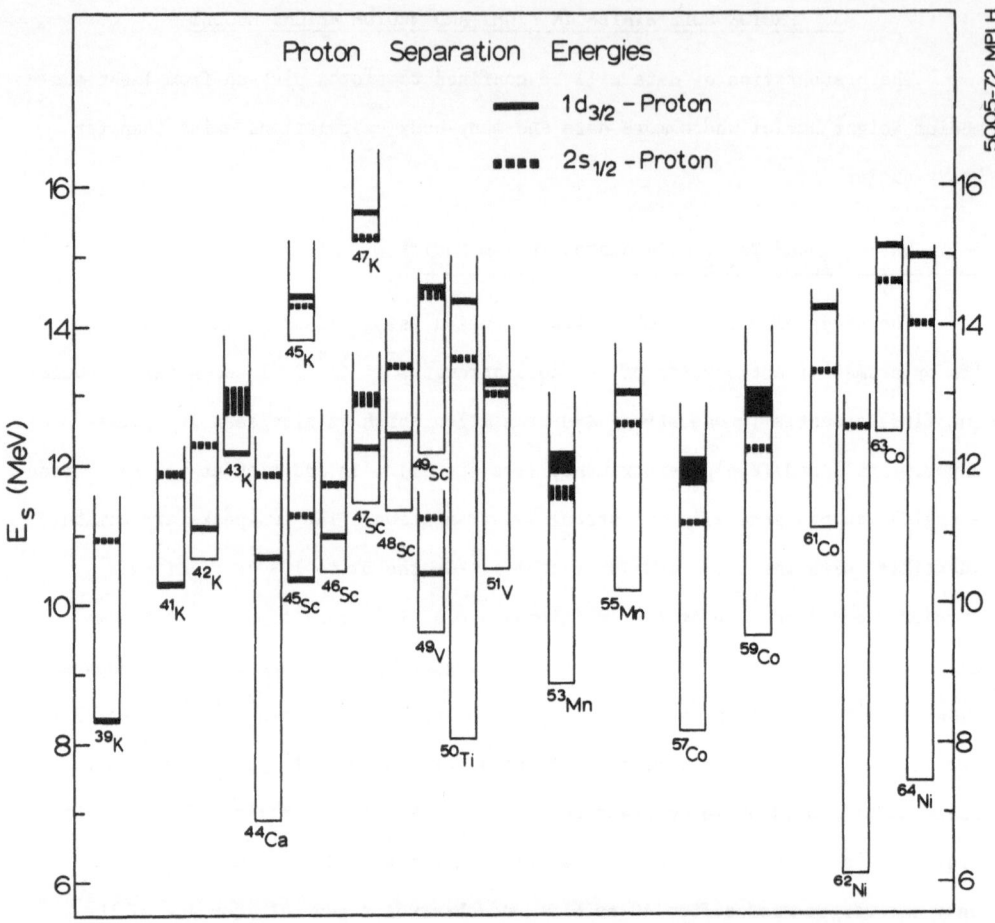

Fig. 7. Mean separation energies of $1d_{3/2}$ (solid bars) and $2s_{1/2}$ (dashed bars)
         protons from f-p shell nuclei. The "level schemes" beginning at the
         ground state separation energies are denoted by the final nuclei of the
         pick-up reaction. The widths of the bars represent estimated uncertainties
         (from ref. 82)

$\bar{E}$ and $\overline{\Delta E}$ are the isoscalar and isovector parts of the space-spin monopole interaction between a $f_{7/2}$ particle and a $1d_{3/2}$ or $2s_{1/2}$ particle. We determined the
two parameters $\bar{E}$ and $\overline{\Delta E}$ from a simultaneous least square fit to the proton and
neutron separation energies with the average Coulomb interaction fixed to $\epsilon_c = 350$
keV. The result of this two-parameter fit is given in fig. 8 for the 2s-proton
hole states. It is good enough to prove that one has treated the essential phenom-

ena by this simple calculation. Moreover, the values of $\bar{E}$ and $\bar{\Delta E}$ determined in this way agree reasonably well[82] with the proper spin and isospin averages of diagonal particle hole matrix elements as obtained from shell model calculations with Talmi fits or from realistic forces.

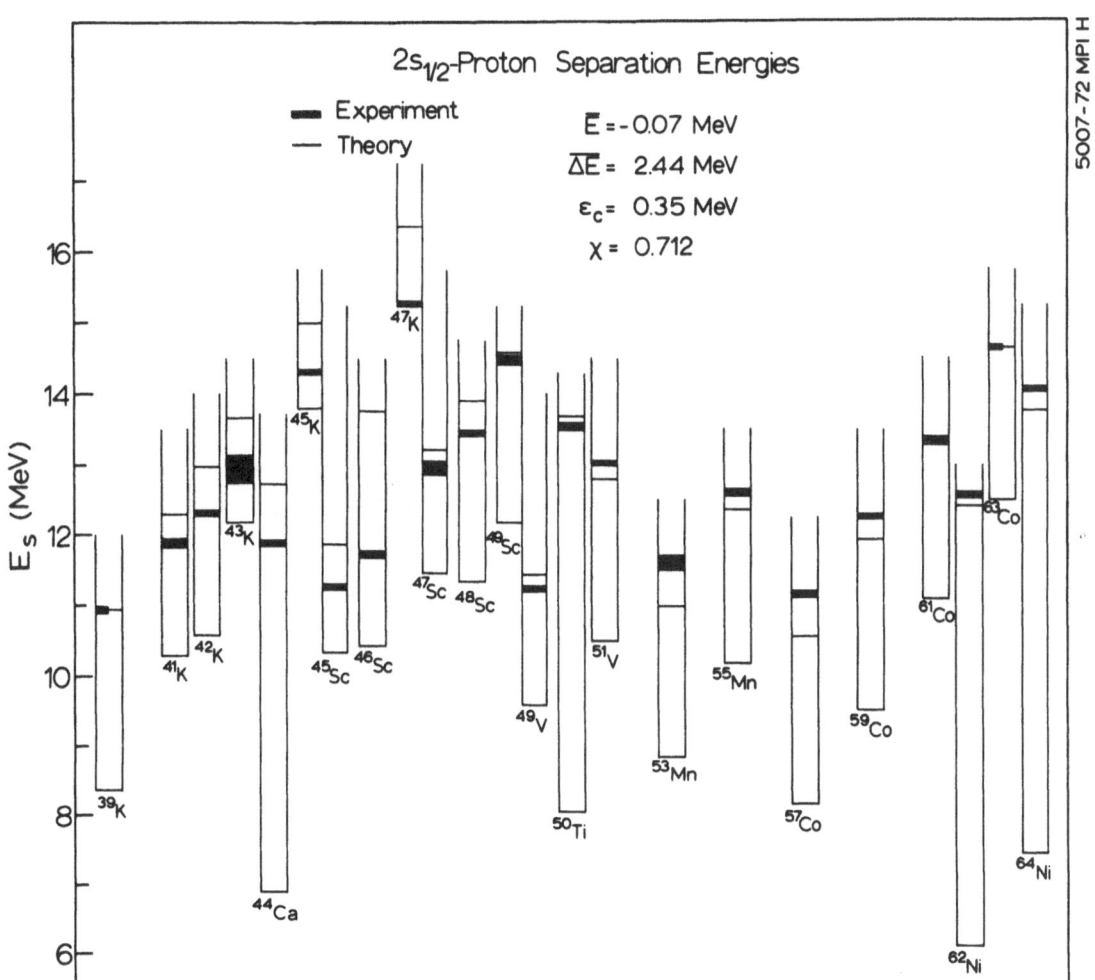

Fig. 8.   Results of a two-parameter fit to the mean removal energies for $2s_{1/2}$ protons obtained with the method of Bansal and French[4] (from ref. 82)

3.2.   On 1p Hole States in (2s,1d)-Nuclei

The deformation which characterizes most (2s,1d)-nuclei leads to a mixing of
the $1p_{3/2}$ and $1p_{1/2}$ shells as is well known from the Nilsson model.   While in pick-
up from ^{16}O (see fig. 1) we observed essentially one strong $1p_{1/2}$ and $1p_{3/2}$ peak
each, the situation in ^{18}O is already more complex[29].   And in the ^{22}Ne(d,^{3}He)^{21}F-
reaction (see fig. 9) one finds in addition to the $2s_{1/2}$ and $1d_{5/2}$ transitions from
the active shells, a total of at least five ℓ=1 transitions.

Fig. 10, which contains ^{24}Mg(p,2p) and ^{24}Mg(d,^{3}He) spectra with identical
abscissa scales, explains why the dominant ℓ=1, j=1/2 strength at 6.24 MeV could
be missed in the (p,2p)-reaction due to the presence of a close $1/2^+$ state.   As a
result, the $1p_{1/2}$ centroid separation energy was deduced to be nearly 10 MeV higher
than it is believed now.   Unfortunately, a similar situation in ^{28}Si seemed to

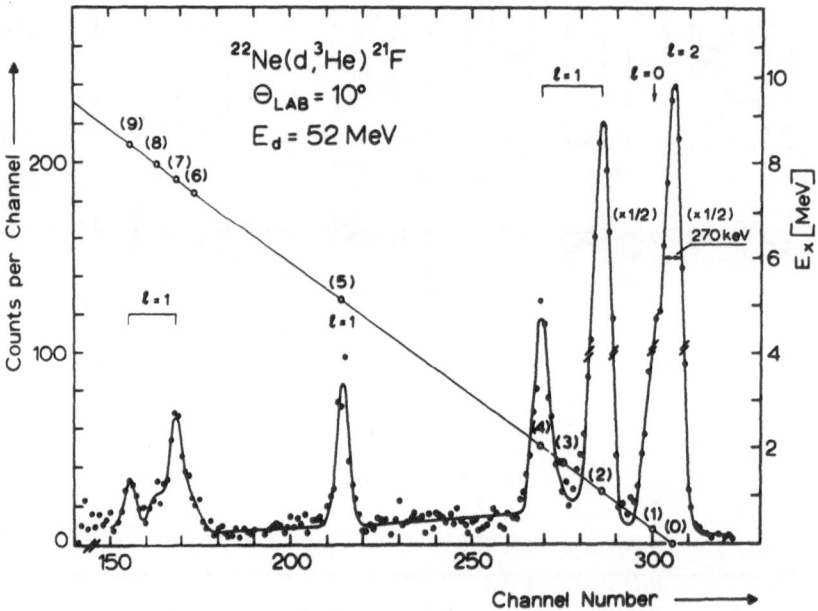

Fig. 9.   Spectrum of the ^{22}Ne(d,^{3}He)^{21}F reaction showing the excitation of at
least 5 ℓ=1 transitions corresponding to pick-up from the 1p shell (from
ref. 39).   The bad energy resolution is the result of the energy spread
of the deuteron beam before installation of an analysing magnet at the
Karlsruhe cyclotron

confirm this misinterpretation which determined the behaviour of the $1p_{1/2}$ shell in fig. 2.

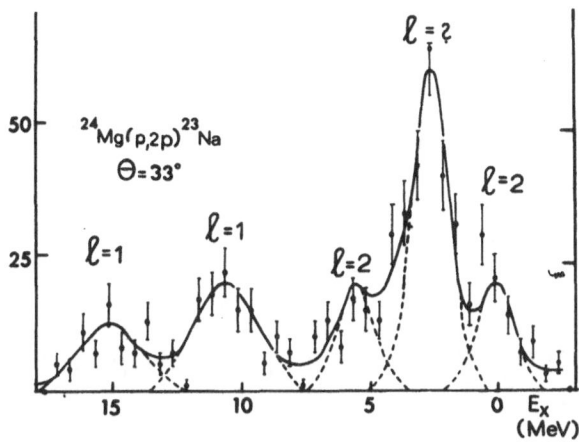

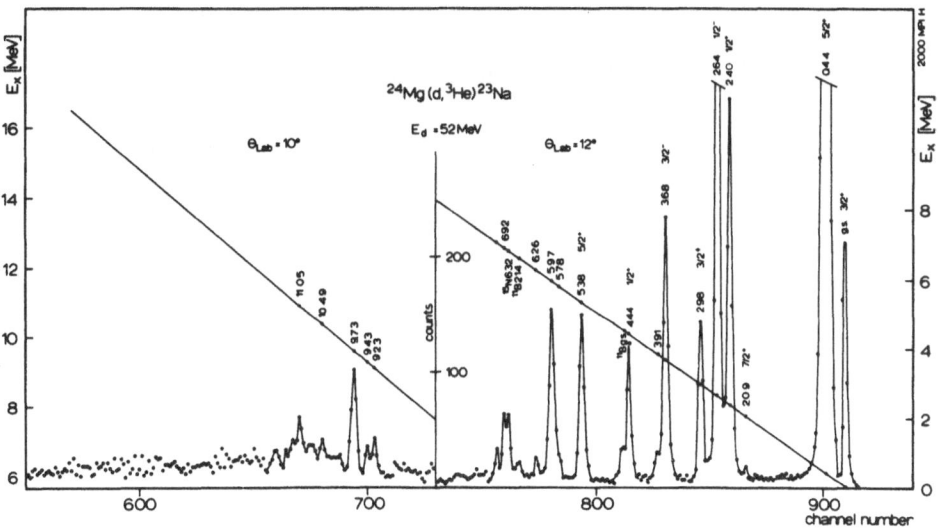

Fig. 10. Comparison of a ^{24}Mg(p,2p) spectrum at 156 MeV [2] and a ^{24}Mg(d,^{3}He) spectrum at 52 MeV [49] with the same energy scales. The (d,^{3}He)-spectrum shows that the group at E_x = 2.5 MeV, which could not be identified in the (p,2p) experiment and which was attributed to the (2s,1d)-shell, carries the major part of the $1p_{1/2}$ strength

Fig. 11 contains a summary of (nearly) all $\ell=1$ spectroscopic factors and the corresponding separation energies which have been measured over the past years. If one focuses attention on the strong $1p_{1/2}$ transition with the smallest separation energy, the systematic relation between the ground state of ^{15}N and the corresponding hole states in ^{19}F, ^{23}Na, ^{27}Al and ^{31}P becomes obvious. In fact, as has been noticed simultaneously for neutron pick-up[44] the corresponding separation energies for $\ell=1$ pick-up from T=0 targets increase linearly with the mass number A. The

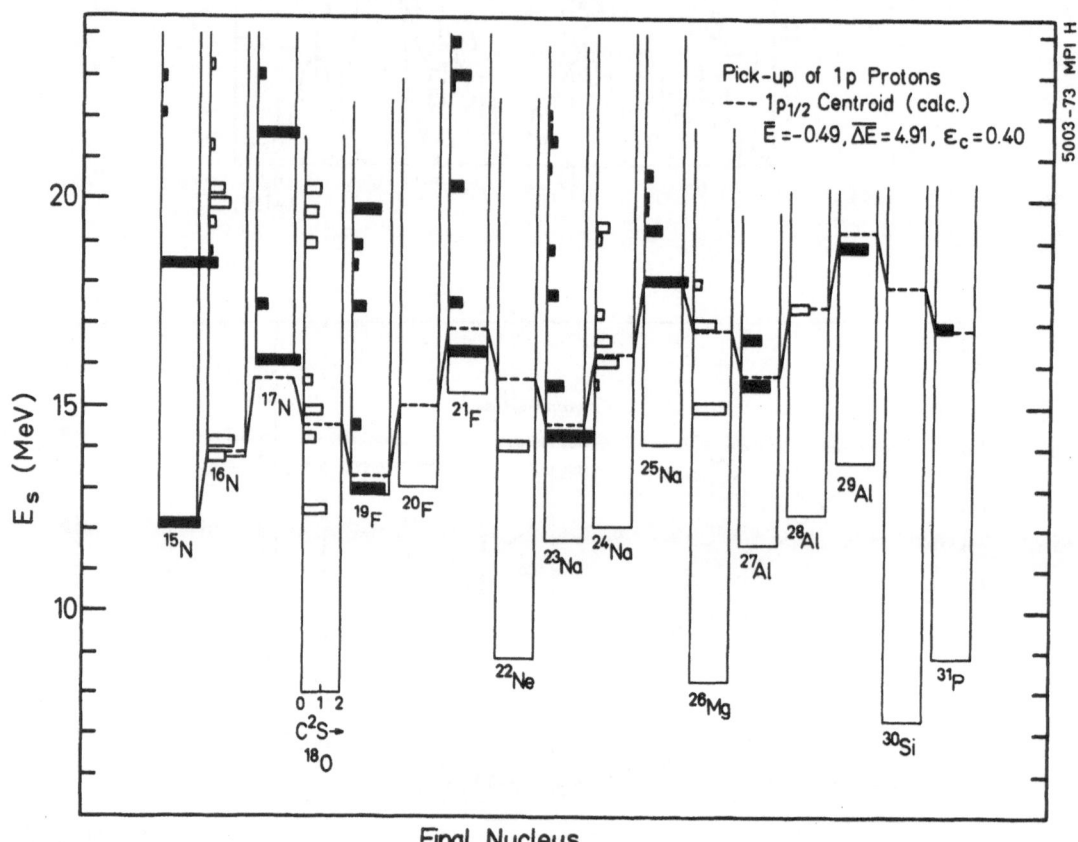

Fig. 11. Summary of $\ell=1$ spectroscopic factors and the corresponding separation energies observed in (d,^{3}He) reactions leading to the final nuclei indicated. Spectroscopic factors obtained with J=0 targets are denoted by black bars. The connected dashed lines which reproduce fairly well the separation energies for the lowest $\ell=1$, (j=1/2) transition from even target nuclei are the result of a parameter-free calculation

isovector part of the particle-hole interaction expresses itself in the additional binding energy for $1p_{1/2}$ protons which one notices for T=1/2 and T=1 targets. (The situation for T=1/2 targets is complicated by the spin-spin interaction which produces additional fractionation of the strength). The T-dependence is clearly demonstrated by the connected dashed bars in fig. 11 which are the result of a Bansal-French type calculation[79]. It is essential to point out that the parameters of this calculation have not been adjusted but were taken from an independent shell model calculation for the ^{16}O-region[87].

We consider these data as evidence for a $1p_{1/2}$ separation energy of about 16 MeV in ^{28}Si[79]. In 1968 this was in contrast (i) to the old (p,2p) data which had resulted in a value of 25 MeV (see fig. 2) and (ii) to the Hartree-Fock calculations existing at that time. Recently, we have succeeded in observing ℓ=1 pick-up from ^{38}Ar at a separation energy of only 18 MeV [20]. These measurements in the upper (2s,1d)-shell are complicated since the cross section drops considerably with increasing mass and separation energy (see fig. 4). At first sight these 18 MeV (which do not necessarily represent the centroid $1p_{1/2}$ separation energy) seem to disagree with the average 1p separation energy of more than 30 MeV deduced from (p,2p)-experiments in this mass region. A possible explanation comes from a 1p-spin orbit splitting strongly increasing with A. In this respect it is interesting that the $1p_{3/2}$ strength in fig. 11 vanishes somewhat more rapidly than expected for a constant spin-orbit splitting.

Evidence for such an increasing spin-orbit splitting has been obtained quite unexpectedly in a recent study of the ^{17}O(d,t)^{16}O reaction[59] where we have been able to locate the full $1p_{3/2}$ hole strength for T=0 and T=1 states in ^{16}O. The isospin was determined by a parallel ^{17}O(d,^{3}He)^{16}N measurement (see fig. 12). Since the states excited via ℓ=1 pick-up from ^{17}O are dominant $1d_{5/2}$ particle- $1p_{1/2}$ (or $1p_{3/2}$) hole states we could obtain the isoscalar ($\bar{E}$) and isovector ($\overline{\Delta E}$ of eq. (3.1)) parts of the $d_{5/2}$-$p_{1/2}$ and $d_{5/2}$-$p_{3/2}$ interactions. Surprisingly, and in contrast to the effective interactions for the ^{16}O-region[25,66,34], we have found that the average interaction of a $1d_{5/2}$ neutron with a $1p_{3/2}$ neutron is twice as attractive ($\bar{E} \simeq -1.2$ MeV) than with a $1p_{1/2}$ neutron (E $\simeq -0.5$ MeV, see fig. 11). As a result

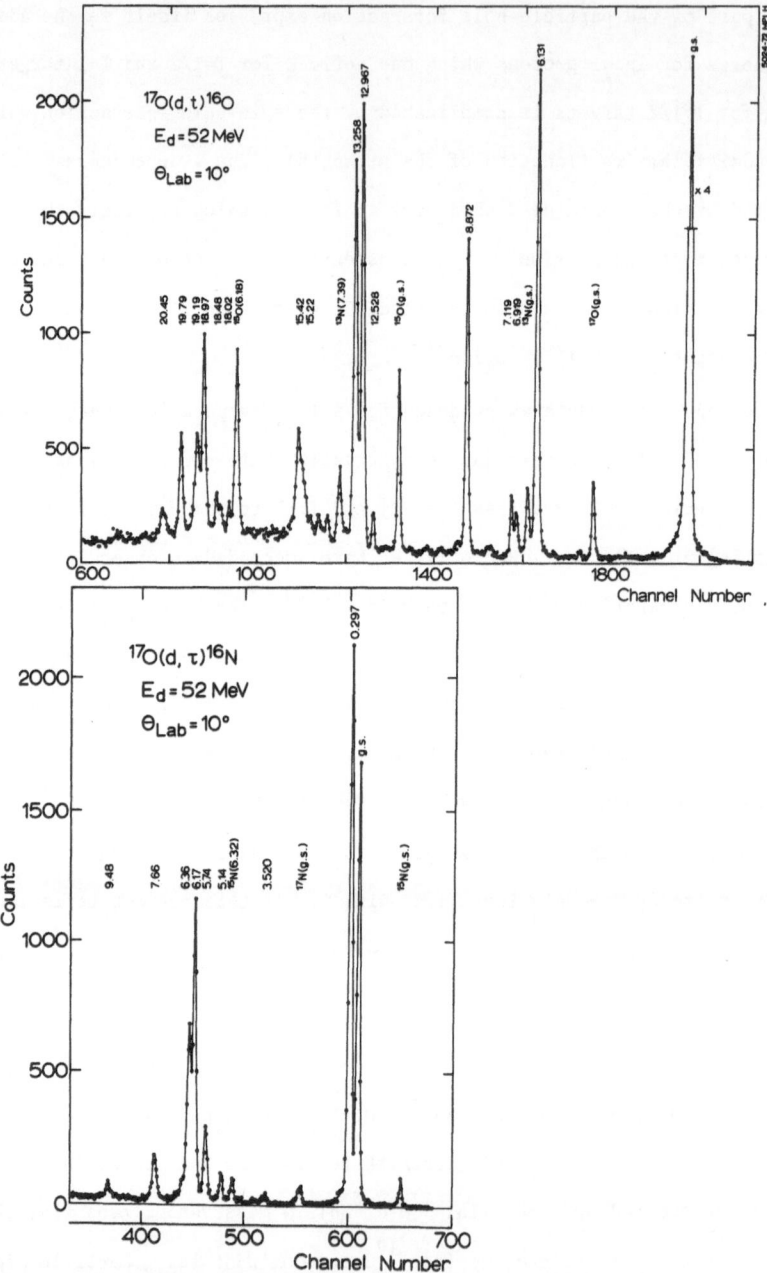

Fig. 12. Spectra from the $^{17}O(d,t)^{16}O$ and $^{17}O(d,^{3}He)^{16}N$ reactions at 10°. Nearly all groups result from pick-up from the $1p_{1/2}$ or $1p_{3/2}$ shells of ^{17}O. The energy scales were chosen such that analog states of the A = 16 nuclei came to lie on top of each other, which allows an easy identification of the T=1 states in ^{16}O

the 1p spin-orbit splitting should increase from 6.5 MeV in ^{16}O to about 14 MeV in ^{28}Si. By taking the weighted average of $1p_{1/2}$ and $1p_{3/2}$ separation energies we would predict a 1p-separation energy from ^{28}Si of 25 MeV in good agreement with the recent (p,2p)-data of Landaud et al.[51], which give 27 ± 2 MeV (see also fig. 14).

3.3. On 1s-hole States

So far, one has not been able to locate 1s centroid separation energies[*] by pick-up reactions. The 1s centroid separation energies have been obtained from QFS reactions in a semi-quantitative manner. In the case of light nuclei where the shells do not overlap they have been taken directly from the spectra (see e.g. fig. 1) and have been identified with the positions of the peaks[**] in the cross section. In the case of heavier nuclei one has performed a decomposition of the measured momentum distributions into shell model momentum distributions (see eq. (4)). The result for the Liverpool data is given in fig. 13. This represents the clearest demonstration of the shell model structure, e.g. of ^{40}Ca, which is available. A 1s separation energy from ^{40}Ca of about 50 MeV is deduced from this figure in agreement with the CERN (p,2p)-data and in contrast to the 80 MeV from the Frascati (e,e'p)-experiments[1], where no angular correlations had been performed to justify the ℓ-assignment. Again one has rather taken the peak 1s amplitude than the centroid energy which is not identical for the non-symmetrical peak shape which is seen in fig. 13. One has to await a distorted wave analysis of the data to see whether the skewed peak shape is only the result of an increased absorption at larger missing energies or of more fundamental origin [see e.g. ref. 68].

3.4. Summary of Mean Separation Energies

Fig. 14 represents an attempt to summarize our present knowledge on proton separation energies from light and medium weight nuclei. Only those data have been

---

[*] Observation of some 1s strength in the ^{12}C(d,^{3}He ^{11}B) reaction has been reported recently[42].

[**] Strictly speaking, the energy of the "quasi-hole states" thus determined differs from the mean separation energy (see ref. 83 for details).

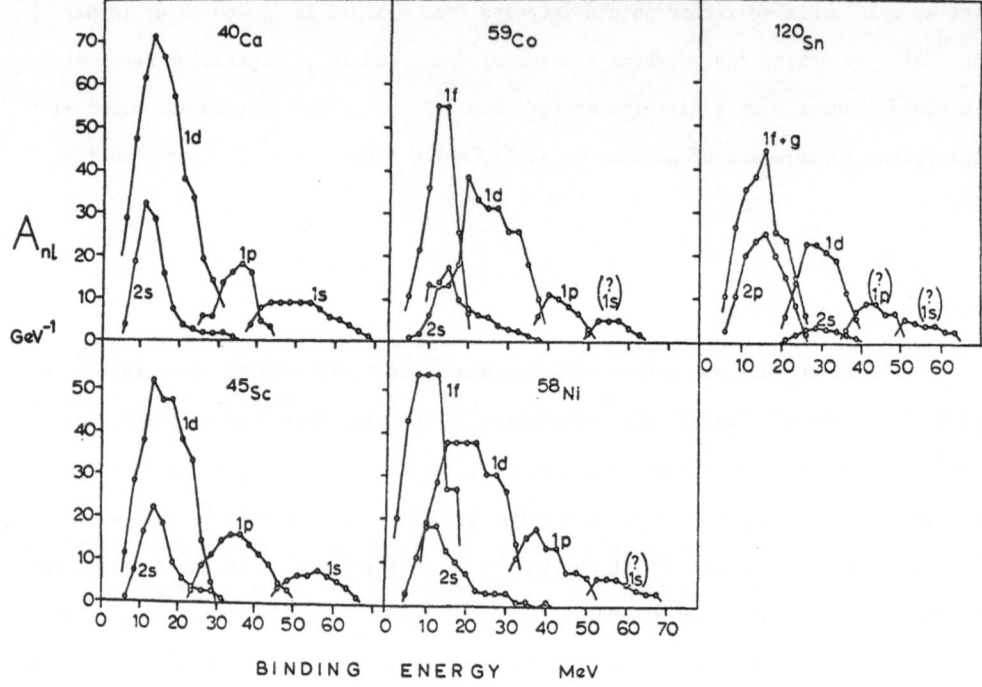

Fig. 13. Decomposition of the recoil momentum distributions measured in (p,2p)-experiments at 385 MeV into pure harmonic oscillator distributions. The amplitudes A_{nl} (see eq. (4)) have been evaluated in 5 MeV bins (from ref. 36)

admitted where (1) the angular momentum of the transferred proton has been measured, and (2) there was a high probability that the major part of the total hole strength had been observed. The latter was assumed to hold if (i) the energy covered was very large as,e.g., in the (p,2p)- and (e,e'p)- experiments; or (ii) the DWBA spectroscopic factors agree within 20% with the shell model sum rule as e.g. in the case of the $^{16}O(d,^{3}He)^{15}N$ reaction; or (iii) the single particle strength was not split within a range of several MeV as,e.g.,the 2s-strength in the $^{40}Ca(d,^{3}He)^{39}K$ spectrum. It is encouraging that experiments which fulfilled criteria (1) and (2) so far have agreed within the errors. While the thick bars of fig. 14 represent the peak width at half maximum the thin error bars show estimated uncertainties resulting,e.g.,from difficulties in the 1p-1d discrimina-

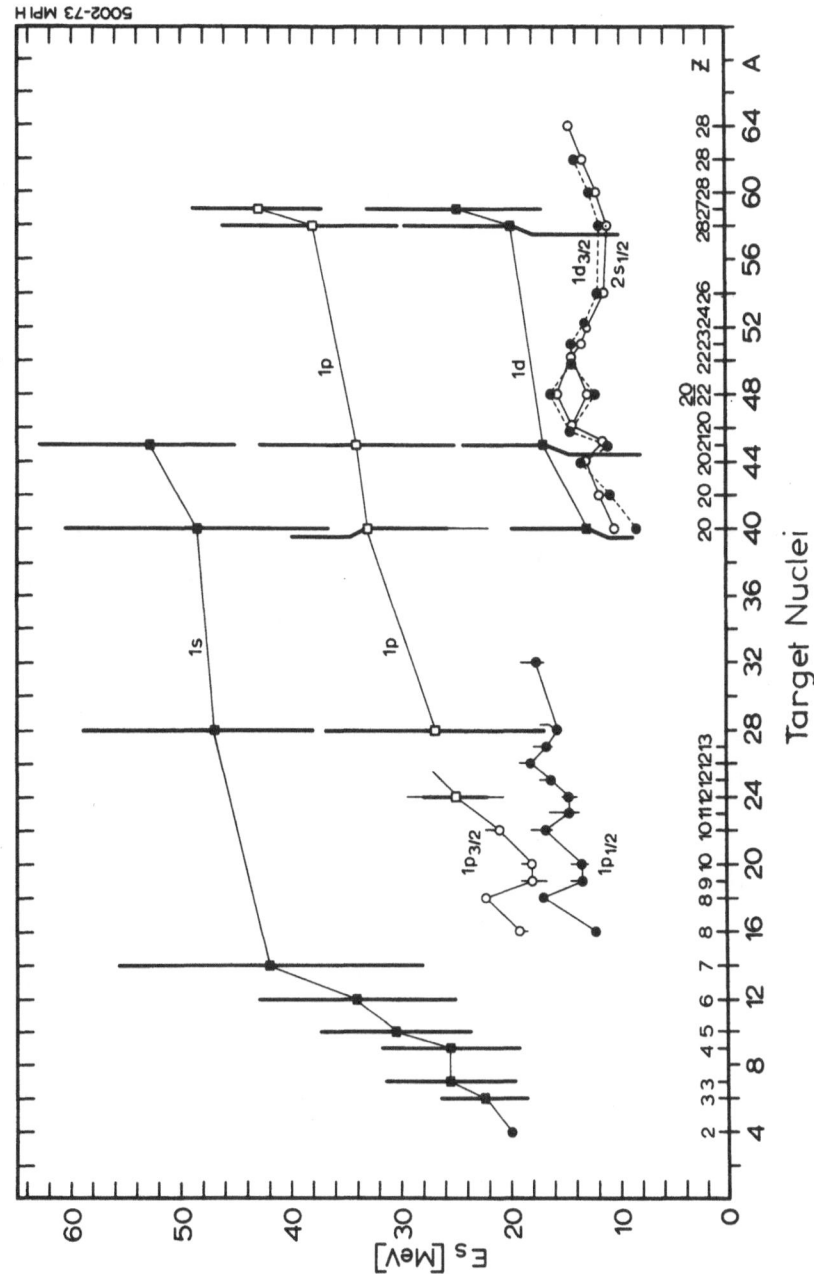

Fig. 14. Summary of mean proton separation energies from light and medium weight nuclei as available presently. The points and circles are the results from pick-up experiments, the full and open squares those from QFS experiments. The thick bars characterize the FWHM of the broad hole strength distributions. The thin bars represent estimated systematical uncertainties

tion in the ^{40}Ca(p,2p) experiment[36] or in the $1p_{1/2}$ - $1p_{3/2}$ discrimination in the (d,^{3}He) experiments (see fig. 11).

The results may be summarized as follows:

(i) The 1s separation energy increases by about 2 MeV per additional target nucleon for A=4 to A=16. Above A=16 one notices saturation phenomena; the average increase is only about 0.3 MeV per additional nucleon.

(ii) The $1p_{1/2}$ separation energy shows an increase of 0.25 MeV per nucleon. For $1p_{3/2}$ protons we have some evidence for a much stronger increase of the separation energy, namely about 0.8 MeV per nucleon in the lower (2s,1d) shell. The resulting increase of the 1p spin orbit splitting from 6.5 MeV in ^{16}O to about 14 MeV in ^{28}Si is in good agreement with the (p,2p)-data for the average 1p separation energy.

(iii) The average increase of the $2s_{1/2}$ and $1d_{3/2}$ separation energies is about 0.15 MeV per nucleon. A comparison between the pick-up results on the $1d_{3/2}$ shell and the QFS results on the average 1d separation energy suggests also a 1d spin-orbit splitting increasing with increasing target masses.

A detailed comparison between figs. 2 and 14 does not seem very fruitful with respect to the drastic changes. However, it is gratifying that the different general behaviour of the 1s and 1p shells on the one hand and the 2s and 1s shells on the other which was visible in fig. 2 has now disappeared.

The saturation behaviour of the 1s shell lets us already suppose that the level spacings for heavier target nuclei are much smaller. This has been verified for several Sn isotopes and some nuclei of the lead region. Fig. 15 shows five peaks, representing the five shells which have been observed in the ^{208}Pb(d,^{3}He)^{207}Tl reaction[56]. The spectrum from the ^{209}Bi(d,^{3}He)^{208}Pb reaction shows that a higher energy resolution is needed for most studies with heavy nuclei. So this project has to be left to the hopefully not too distant future.

Immediate progress may be expected from the investigation of the (d,t) - reactions which we have started (see fig. 12) at Karlsruhe. We are convinced that a systematic study of Coulomb energy differences of hole states will tell us about the radial dependence of the hole wave functions. We can hope to get

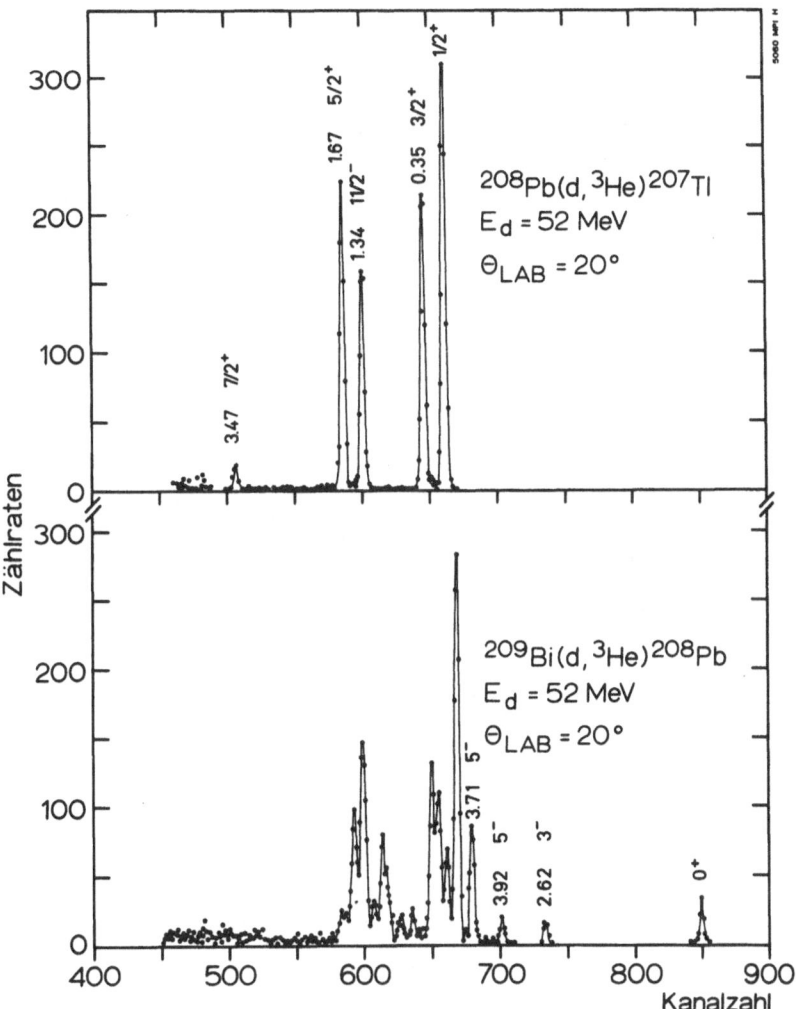

Fig. 15. Spectra of the (d, ³He) reactions on ²⁰⁸Pb and ²⁰⁹Bi (from ref. 56)

(e,e'p)-spectra with a decent resolution and very good statistics in the near
future from Saclay.

4.   REMOVAL ENERGIES AND RELATED THEORETICAL QUANTITIES

The intricate relation between experimental removal energies and related
quantities in various many-body theories has recently been illuminated from various
points of view in several detailed papers (refs. 48, 8, 19, 23 and 46). This
permits a brief discussion here.

4.1. The Dilemma of Conventional Hartree-Fock Calculations

In a given external potential well U(r) the separation energy E_i of a bound nucleon is identical to the negative of its eigenvalue ε_i. In phenomenological shell model calculations the choice of the single particle potential is a matter of convenience. The arbitrariness can be accounted for by the choice of the residual interactions. So the eigenvalues of a potential U(r) can be meaningful only to the extent that U is the result of a self-consistent calculation. But even then the identity of removal energies and eigenvalues which is known as Koopmans' theorem[43] does hold only if the potential well does not change due to the removal of the nucleon. Other than in the case of atomic electrons[53] the neglect of re-arrangement processes is by no means justified for the atomic nucleus.

In Hartree-Fock (HF) and Brueckner-Hartree-Fock (BHF) approximation the total binding energy W^{HF} can be written as

$$W^{HF} = \frac{1}{2} \sum_i^{occ} (<i|t|i> + \varepsilon_i^{HF}) - T_{cm} \tag{11}$$

where the sum runs over all normally occupied eigenstates $|i>$ with expectation values $<i|t|i>$ of the kinetic energy and eigenvalues ε_i^{HF} of the HF single particle equations

$$t|i> + U|i> = \varepsilon_i^{HF}|i> \tag{12}$$

$$<i|U|i> = \sum_{j=1}^{A} <ij|\bar{v}|ij> . \tag{13}$$

Here, $\bar{v}$ is the antisymmetrized two-body interaction and T_{cm} is a correction for the centre of mass motion of the nucleus. Köhler[47] was the first one to notice a severe violation of this equation when inserting the negative of mean separation energies $- < E_i >$ in the place of the eigenvalues ε_i. With kinetic energies taken from a harmonic oscillator potential the resulting total binding energies W/A were too low by several MeV per nucleon. The difference between the observable $< E_i >$ and the single particle energy needed to satisfy eq. (11) is referred to as "re-arrangement energy"

$$\varepsilon_i^{RA} = -(\varepsilon_i^{HF} + < E_i >) \tag{14}$$

which hence is also a model dependent quantity. An improved treatment of the kinetic energies*) in a Woods-Saxon potential[22] did not alter the situation appreciably.

This indicates a principal dilemma of the conventional first order Hartree-Fock calculations. These theories do not provide an appreciable re-arrangement energy themselves. Calculations of the HF (or "orbital") re-arrangement energies[62] yield values between 0.2 and 1 MeV per nucleon. So within this accuracy one has to apply Koopmans' theorem, i.e. to identify Hartree-Fock single particle energies with the mean separation energies if one stays strictly within the model. But since we have seen that inserting the experimental separation energies results in a severe violation of eq. (11) we conclude that a conventional Hartree-Fock calculation can never simultaneously reproduce the total binding energy and the experimental separation energies, no matter what forces one uses. A few examples of such failures are given in fig. 16.

4.2. Possible Cures of the Dilemma

To solve the dilemma outlined above many conceptually different ways to include higher order corrections have been discussed in the literature. Here we shall mention only two methods which were particularly successful as far as the comparison with the experiments goes, namely (i) renormalized Brueckner-Hartree-Fock (RBHF) calculations and (ii) Hartree-Fock calculations with density dependent interactions (DDHF).

4.2.1. Renormalized Brueckner-Hartree-Fock Calculations

In RBHF calculations (ref. 18 and refs. given therein) one takes into account the partial depletion of normally occupied orbits. This firstly affects the HF-potential in first order and secondly provides the required change in the relation between single-particle energies and the total energy

$$W^{RBHF} = \frac{1}{2} \sum_i \left\{ (1 + d_i)\epsilon_i + (1 - d_i) <i|t|i> \right\} - T_{cm} \qquad (15)$$

---

*) A substantial improvement has recently been reported by Miller[61] who used an OBEP interaction in a relativistic Hartree calculation.

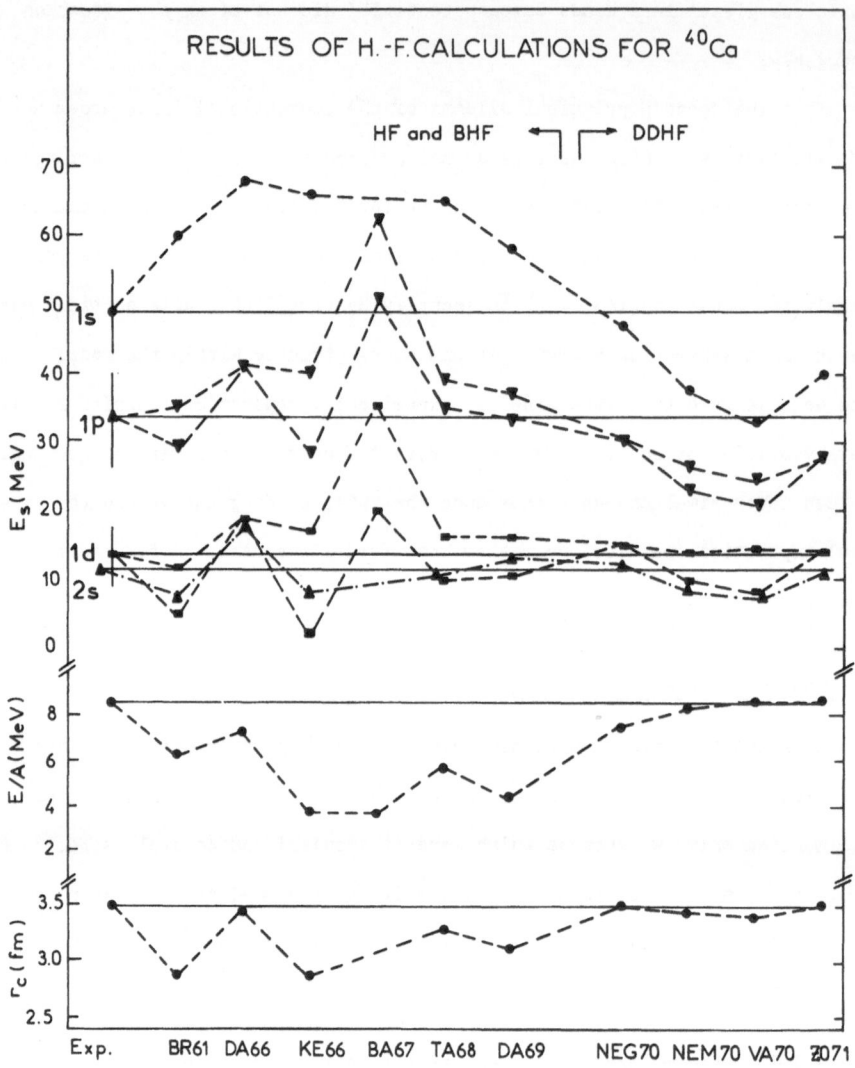

Fig. 16. Results of various Hartree-Fock calculations for ^{40}Ca. The r.m.s. charge radii r_c, the total binding energy per nucleon E/A and the separation energies E_s are compared to the experimental results which are denoted by horizontal lines

Identifying the single-particle energies with experimental separation energies[*)]

Becker[9] has treated the depletion factors d_i as parameters such as to make the re-arrangement energy zero. In this way he determined average depletion factors $<d>$ for several light nuclei and arrived at reasonable values between 15% and 30% for spherical and deformed nuclei, respectively.

4.2.2. Density Dependent Hartree-Fock Calculations

If one introduces into the Hartree-Fock theory forces which explicitly depend on the density of the surrounding nuclear matter one obtains single particle equations[48]

$$t|i> + U|i> + U^R|i> = \epsilon_i^{HF}|i> \tag{16}$$

which contain in addition to the conventional terms (13) a so-called re-arrangement potential U^R. This results from the minimization of the total energy $\delta<\psi|H(\rho)|\psi>=0$ with respect to the wave functions $|i>$ which determine the density ρ. Since the total energy is still given by

$$W^{DDHF} = \sum_i^{occ} <i|t|i> + \frac{1}{2} \sum_{ij}^{occ} <ij|\bar{v}|ij> \tag{17}$$

the introduction of the ϵ_i from eqs. (13) and (16) leads to

$$W^{DDHF} = \frac{1}{2} \sum_i^{occ} (<i|t|i> + \epsilon_i^{HF} - <i|U_i^R|i>) \tag{18}$$

which now contains an explicit expression for the re-arrangement energies. Now, we can apply Koopmans' theorem to good accuracy[48] and identify the E_i with the $-\epsilon_i$ without violating eq. (18).

Fig. 16 shows the charge radii, the total binding energies per nucleon and separation energies for ^{40}Ca as obtained from experiment and various HF calculations. This demonstrates clearly (i) the failure of the conventional HF calculations and (ii) the success of the concept of density dependent forces, in reproducing simultaneously the large total binding energies and the small separation energies required by the experiments. It has been pointed out[73] that the single

---

[*)] The 1p$_{3/2}$ separation energies for ^{28}Si and ^{32}S ought to be larger by several MeV in the light of our recent results (see section 3.2) with a resulting reduction of the depletion factor $<d>$.

particle energies from BHF calculations should be multiplied with the occupation probabilities $(1-d_i)$ before comparison with the mean removal energies (as defined by eq. (18)) is made. This has not been done in quoting ref. 17 in fig. 16. But for the inner shells where the disagreement is largest these occupation probabilities are probably closest to unity.

4.3. Calculations of the Spectral Function

With respect to the many difficulties in relating the experimental removal energies to theoretical single particle energies it would be very desirable to find predictions for mean removal energies instead of single particle energies in the theoretical papers. But even then the width of the averaging interval used to determine the mean separation energy could remain a model dependent quantity. These difficulties can be overcome if the ambitious attempts[86] to calculate directly the special functions (eq. (2)) are successful. These applications of the Shapiro graph technique would allow a comparison with experimental results in any energy interval desired since they are not restricted to yield only the first moment of the spectral distributions. Presently one meets serious divergence problems[24] for separation energies above 20 MeV, so a comparison with experiment would be premature.

Probably some of the difficulties may be avoided in less ambitious attempts to calculate single particle energies as defined by Engelbrecht and Weidenmüller (refs. 23 and 83). There, the single particle energies are identified with the positions of the maxima which survive an energy averaging procedure of the spectral function. Since these maxima practically coincide with the peak cross sections one apparently also avoids some of the difficulties encountered in the determination of the mean removal energies (eq. (8)).

4.4. Model Independent Analysis of Nucleon Removal Energies

Provided one will be able to solve all problems arising from the optical potentials for the protons and extract the spectral function $S(q,E)$ as defined by eq. (2) from the QFS data one will be able to utilize the data in a completely model independent way. The best existing approximation to a spectral function is given in fig. 17 which shows the corss sections from the ^{12}C(e,e'p) data of Saclay[15]. It differs from the spectral functions in two aspects. Firstly, the free proton-electron cross section, which is a slowly varying function in the q-E plane, has not been divided out. Secondly, the distortion of the outgoing proton waves has not yet been accounted for. But the kinematical conditions of the experiment were chosen such as to fix the energy of the protons to about 86 MeV. In that way one achieves a distortion more or less independent of the separation energy. "Spoiled" by our knowledge of the shell model we are immediately inclined to detect the $1p_{3/2}$ and $1s_{1/2}$ shells. And we find roughly the same reduction factors of 0.3 for both shells.

But if we forget about the shell model and take the spectral function as what it is, namely the joint probability of finding a proton of given momentum and separation energy, we can calculate the average kinetic energy $T_m = <q^2/2m>$, and the average removal energy $E_m = <E>$. The average is taken of all protons irrespective of their shell model properties. Starting from a very general Hamiltonian which includes only two-body forces Koltun[45] has shown that these quantities are now strictly[*)] related to the total energy by

$$W/A = \frac{1}{2} (T_m - E_m). \tag{19}$$

This equation reminds us of the HF-equation (11) or rather of the RBHF equation (15) since no assumptions on fully occupied shells have to be made and depletions are automatically taken into account. The most interesting point now of this sum rule is that its right hand side needs to be supplemented by the additional term $-\{ \frac{1}{2} <A|H_3|A> / A \}$ i.e. half of the average contribution of the three-body forces per nucleon, if these do not vanish. So an inequality found in the application of eq. (19) may be indicative of the existence of three body forces.

Koltun's plane wave analysis[45] of the data of James et al.[36] seems to be very optimistic with regard to the large reduction factors (see section 2.3.1)

[*)] One has to take neutron-proton differences into account properly (see ref. 45 for details).

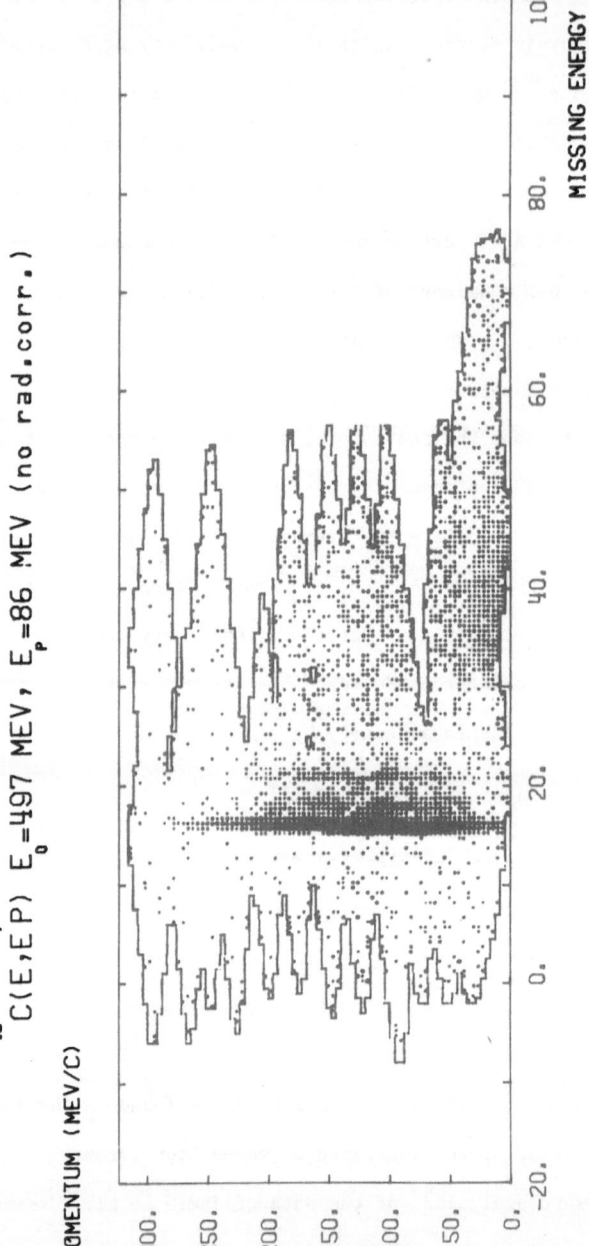

Fig. 17. Cross sections (denoted by the intensity of the data points) of the $^{12}C(e,e'p)$ experiment (ref. 15) plotted in the recoil momentum vs. missing energy plane. Some regions have not been scanned by the spectrometer settings. Background subtraction but no radiative corrections have been made

which are likely to be energy dependent. A preliminary analysis of the recent (e,e'p) data on ^{12}C and ^{40}Ca from Saclay suggests that the right hand side of eq. (19) is too small, which doulc indicate repulsive (!) three body forces. But this is little more than speculation at the moment. More probably, parts of the spectral function beyond the observed region (which certainly covers the normal shell model frame) will account for the missing binding energy.

5. CONCLUSION

We have seen that in spite of the many years which have been devoted to one-nucleon transfer reactions, still many problems of experiments and analysis remain to be solved. We are approaching a quantitative level which will enable us to gain a lot of nuclear structure information which cannot be obtained from other experiments.

We have seen a rapid change in our knowledge on nucleon separation energies during the last years. Today, we can claim a reasonable agreement between theory and experiment. But the same statement may be found in Riou and Ruhla's paper from 1969 when the experimental situation was completely different. This should be a warning and a good reason to proceed in this field and expect further surprise in the near future.

ACKNOWLEDGEMENT

The author wants to express his sincere thanks to his coworkers at the Karlsruhe cyclotron, in particular to Dr. G. Mairle, for help on many occasions. The continuous interest of Prof. U. Schmidt-Rohr in this work is acknowledged. I wish to thank him and Profs. P. Brix and W. Gentner for partly supporting my stay at the C.E.N. Saclay. I am grateful to Profs. W.A. Friedman and H.A. Weidenmüller for enlightening discussions. Finally, I wish to thank Drs. J. Mougey and C. Tzara for their warm hospitality during my stay at Saclay.

REFERENCES

1. Amaldi, U., Jr.: Suppl. Nuovo Cim. I 5, 1225 (1967).

2. Arditi, M., Doubre, H., Riou, M., Royer, D., Ruhla, C.: Nucl. Phys. A103, 319 (1967).

3. Arditi, M., Bimbot, L., Doubre, H., Frascaria, N., Garron, J.P., Riou, M., Royer, D.: Nucl. Phys. A165, 129 (1971).

4. Bansal, R.K., French, J.B.: Phys. Lett. 11, 145 (1964).

5. Bassichis, W.H., Kerman, A.K., Svenne, J.P.: Phys. Rev. 160, 746 (1967).

6. Bachelier, D. Bernas, M., Brissaud, I., Détraz, C., Radvanyi, P.: Nucl. Phys. A126, 60 (1969).

7. Baranger, M.: Nucl. Phys. A149, 225 (1970).

8. Bassichis, W.H., Strayer, M.R.: Annals of Phys. 66, 457 (1971).

9. Becker, R.L.: Phys. Lett. 32B, 263 (1970).

10. Bhatia, T.S., Daehnick, W., Wagner, G.J.: Phys. Rev. C5, 111 (1972).

11. Blair, A.G., Armstrong, D.D.: Phys. Rev. 151, 930 (1966).

12. Brueckner, K.A., Lockett, A.M., Rotenberg, M.: Phys. Rev. 121, 255 (1961).

13. Brueckner, K.A., Meldner, H.W., Perez, J.D.: Phys. Rev. C6, 773 (1972).

14. Butler, G.W., Cerny, J., Cosper, S.W., McGrath, R.L.: Phys. Rev. 166, 1096 (1968).

15. Bussière, A., Mougey, J., Ho, Phan Xuan, Priou, M., Sick, I.: Lett. Nuovo Cim. 2, 1149 (1971) and private communication

16. Davies, K.T.R., Krieger, S.J., Baranger, M.: Nucl. Phys. 84, 545 (1966).

17. Davies, K.T.R., Baranger, M., Tarbutton, R.M., Kuo, T.T.S.: Phys. Rev. 177, 1519 (1969).

18. Davies, K.T.R., McCarthy, R.J.: Phys. Rev. C4, 81 (1971).

19. Dieperink, A.E.L., Brussard, P.J., Cusson, R.Y.: Nucl. Phys. A180, 110 (1972).

20. Doll, P.: Diplomarbeit, MPI für Kernphysik, Heidelberg (1972).

21. Elton, L.R.B., Swift, A.: Nucl. Phys. A94, 52 (1967).

22. Elton, L.R.B.: Phys. Lett. 25B, 60 (1967).

23. Engelbrecht, C.A., Weidenmüller, H.A.: Nucl. Phys. A184, 385 (1972).

24. Faessler, A., Kusuno, S.: Verhandlungen der DPG (1973) p. 178.

25. Gillet, V., Vinh Mau, N.: Nucl. Phys. 54, 321 (1964).

26. Gray, W.S., Wei, T., Polichar, R.M.: Bull. Am. Phys. Soc. 13, 697 (1968) and private communication.

27. Gross, D.H.E., Lipperheide, R.: Nucl. Phys. A150, 449 (1970).

28. Hartwig, D.: Dissertation, Karlsruhe (1971).

29. Hartwig, D., Kaschl, G.Th., Mairle, G., Wagner, G.J.: Z. f. Phys. 246, 418 (1971).

30. Hiebert, J.C., Newman, E., Bassel, R.H.: Phys. Rev. 154, 898 (1967).

31. Hinterberger, F., Mairle, G., Schmidt-Rohr, U., Turek, P., Wagner, G.J.: Z. f. Phys. 202, 236 (1967).

32. Hinterberger, F., Mairle, G., Schmidt-Rohr, U., Turek, P., Wagner, G.J.: Nucl. Phys. A106, 161 (1968).

33. Hodgson, P.E.: Predeal Lectures (1972); Hodgson, P.E., Millener, D.J., Private communication.

34. Hsieh, S.T., Lee, T.Y., Chen-Tsai, C.T.: Phys. Rev. C4, 105 (1971).

35. Jacob, G., Maris, Th.A.J.: Revs. Mod. Phys. 38, 121 (1966).

36. James, A.N., Andrews, P.T., Kirkby, P., Lowe, B.G.: Nucl. Phys. A138, 145 (1969).

37. Jacob, G., Maris, Th. A.J.: Revs. Mod. Phys. 45, 6 (1973).

38. Kaschl, G.Th., Wagner, G.J., Mairle, G., Schmidt-Rohr, U., Turek, P.: Phys. Lett. 29B, 167 (1969).

39. Kaschl, G.Th., Wagner, G.J., Mairle, G., Schmidt-Rohr, U., Turek, P.: Nucl. Phys. A155, 417 (1970).

40. Kaschl, G., Mairle, G., Mackh, H., Hartwig, D., Schwinn, U.: Nucl. Phys. A178, 275 (1971).

41. Kerman, A.K., Svenne, J.P., Villars, F.M.H.: Phys. Rev. 147, 710 (1966).

42. Knöpfle, K.T., Rogge, M., Schwinn, U., Turek, P.; Mayer-Böricke, C.: Verhandlungen der DPG (1973) p. 178.

43. Koopmans, T.: Physica 1, 104 (1934).

44. Kozub, R.L.: Phys. Rev. 172, 1078 (1968).

45. Koltun, D.S.: Phys. Rev. Lett. 28, 182 (1972).

46. Koltun, D.S.: Proc. Symposium "Present Status and Novel Developments in the Many-Body Problem", Rome (1972), to be published.

47. Köhler, H.S.: Nucl. Phys. **88**, 529 (1966).

48. Köhler, H.S., Lin, Y.C.: Phys. **A167**, 307 (1971).

49. Krämer, E., Mairle, G., Kaschl, G.: Nucl. Phys. **A165**, 353 (1971).

50. Kullander, S., et al.: Nucl. Phys. **A173**, 357 (1971).

51. Landaud, G., et al.: Nucl. Phys. **A173**, 337 (1971).

52. Lim, K.L., McCarthy, I.E.: Phys. Rev. **133B**, 1006 (1964).

53. Lindgren, I.: Phys. Lett. **19**, 382 (1965); Rosen, A., Lindgren, I.: Phys. Rev. **176**, 114 (1968).

54. Macfarlane, M.H., French, J.B.: Revs. Mod. Phys. **32**, 567 (1960).

55. Mairle, G., Kaschl, G.Th., Link, H., Mackh, H., Schmidt-Rohr, U., Wagner, G.J., Turek, P.: Nucl. Phys. **A134**, 180 (1969).

56. Mairle, G., Heusler, A.: private communication.

57. Mackh, H.: Dissertation, Heidelberg (1971).

58. Mairle, G., Wagner, G.J.: Z. f. Phys. **258**, 321 (1973).

59. Mairle, G., Wagner, G.J.: Contribution to the International Conference on Nuclear Physics, Munich (1973).

60. Mairle, G., Schwinn, U., Wagner, G.J.: to be published.

61. Miller, L.D.: Phys. Rev. Lett. **28**, 1281 (1972).

62. Müther, H., Goeke, K., Faessler, A.: Z. Phys. **253**, 61 (1972).

63. Newman, E., Hiebert, J.C.: Nucl. Phys. **110**, 366 (1968).

64. Negele, J.W.: Phys. Rev. **C1**, 1260 (1970).

65. Nemeth, J.E., Vautherin, D.: Phys. Lett. **32B**, 561 (1970).

66. Perez, S.M.: Nucl. Phys. **A136**, 599 (1969).

67. Philpott, R.J., Pinkston, W.T., Satchler, G.R.: Nucl. Phys. **A119**, 241 (1968).

68. Pittel, S., Austern, N.: Phys. Rev. Lett. **29**, 1403 (1972).

69. Riou, M., Ruhla, C.: Progress in Nuclear Physics **II**, 195 (1969).

70. Ruhla, C., Arditi, M., Doubre, H., Jacmart, J.C., Liu, M., Ricci, R.A., Riou, M., Roynette, J.C.: Nucl. Phys. **A95**, 526 (1967).

71. Santo, R., Stock, R., Bjerregard, J.H., Hansen, O., Nathan, O., Chapman, R., Hinds, S.: Nucl. Phys. **A118**, 409 (1968).

72. Sherr, R., Bayman, B.F., Rost, E., Rickey, M.E., Hoot, C.G.: Phys. Rev. 139, B1272 (1965).

73. Shakin, C.M., Da Providencia, J.: Phys. Rev. Lett. 27, 1069 (1971).

74. Stock, R., Bock, R., David, P., Duhm, H.H., Tamura, T.: Nucl. Phys. A104, 136 (1967).

75. Sundberg, O., Källne, J.: Ark. Fys. 39, 323 (1969).

76. Tarbutton, R.M., Davies, K.T.R.: Nucl. Phys. A120, 1 (1968).

77. Tyrén, H., Kullander, S., Sundberg, O., Ramachandran, R., Isacsson, P., Berggren T.: Nucl. Phys. 79, 321 (1966).

78. Vautherin, D., Brink: D.M., Phys. Lett. 32B, 149 (1970).

79. Wagner, G.J.: Phys. Lett. 26B, 429 (1968).

80. Wagner, G.J., Mairle, G., Schmidt-Rohr, U., Turek, P.: Nucl. Phys. A125, 80 (1969).

81. Wagner, G.J.: Habilitationsschrift, M.P.I. Heidelberg (1970)-V8.

82. Wenneis, S.: Dissertation, Heidelberg (1971).

83. Weidenmüller, H.A.: Proc. of the 1. Minerva Symposium, preceding article.

84. Wildenthal, B.H., Newman, E.: Phys. Rev. 167, 1027 (1968).

85. Wildenthal, B.H., Newman, E.: Phys. Rev. 175, 1431 (1968).

86. Wille, U., Lipperheide, R.: Nucl. Phys. A189, 113 (1972).

87. Zamick, L.: Phys. Lett. 19, 580 (1965).

88. Zofka, J., Ripka, G.: Nucl. Phys. A168, 65 (1971).

EFFECTIVE INTERACTIONS IN NUCLEI

Michael W. Kirson

Department of Nuclear Physics, Weizmann Institute of Science, Rehovot, Israel

In recent years, a number of talks and articles have been devoted to a
thorough review of the microscopic theory of nuclear effective interactions[1]. It
would thus be wasteful to summarize again here what is already freely available
in the literature. This survey will concentrate instead on a few specific areas
in which the theory might encounter problems and where further study would be
useful. However, in an effort to keep this work more or less self-contained,
some brief descriptions of the background and framework of the calculations will
be interpolated where needed.

The essential problem is to define and calculate, from the bare nucleon-
nucleon interaction, an effective interaction which can partially account for
observed nuclear spectra despite drastic limitation of the number of degrees
of freedom of the nucleus. The framework is the nuclear shell model; the tool,
many-body perturbation theory[2]. It is assumed that the nuclear single-particle
model, with a small number of active nucleons confined to a few single-particle
orbits, can give a reasonable description of low-energy nuclear spectra, provided
the interaction between nucleons is renormalized to take into account the effects
of the many neglected configurations. This is the conclusion that can be drawn
from the great success of the shell model. The neglected configurations are
those in which some or all of the active nucleons are outside the set of orbitals
considered, and those in which the supposedly inert core is excited by transferring
passive nucleons to higher single-particle orbitals (fig. 1). From the point of
view of perturbation theory, the neglected configurations define a set of intermediate states which can be virtually excited by the interaction between nucleons.

The intermediate states can be roughly divided into two classes - those of
high unperturbed excitation energy, and those of low energy. This distinction is
useful because the nucleon-nucleon interaction can be regarded as consisting of

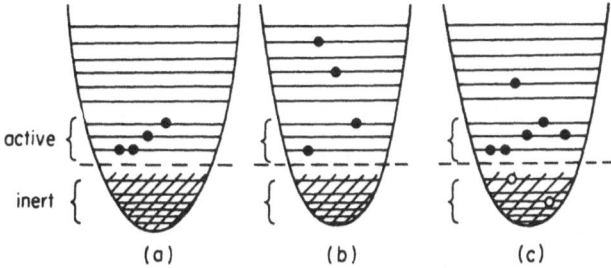

active {

inert {

(a)   (b)   (c)

Fig. 1. Model-space states and intermediate states. The cross-hatched orbitals
represent the (occupied) inert core. Open circles denote holes in the core.
(a) A typical model-space state. (b) Some valence nucleons are excited
out of the set of active orbitals. (c) Some nucleons are excited out of
the "inert" core

two parts - the short-range part and the long-range part - which affect the high-
and low-energy intermediate states respectively. The strong short-range part of
the force includes the repulsive core and, in the case of a hard-core force, the
strong compensating attraction immediately beyond the core (fig. 2). Short range

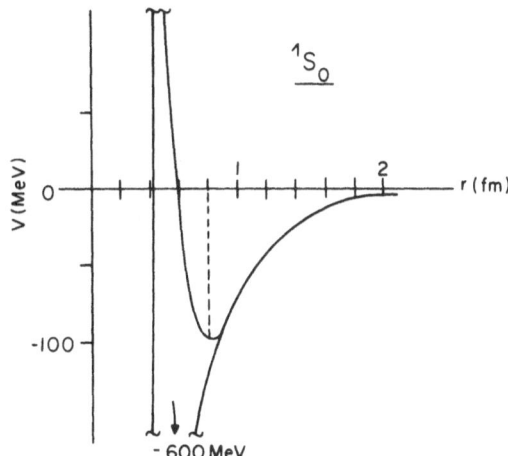

Fig. 2. Typical nucleon-nucleon $^1S_0$ interactions. The soft core interaction
becomes attractive around r=0.6 fm and has a maximum attraction of about
100 MeV. The hard core interaction is infinitely repulsive inside about
0.4 fm, and has a maximum attraction at that point of about 600 MeV

corresponds to large momentum components and hence, roughly, to high excitation energies. This part of the force is characterized by distances of the order of the core radius, around 1/2 fermi. Since the average separation of nucleons in nuclei is about 1 fermi, the ratio of the "interaction volume" to the "occupation volume" is about 1/8 for the short-range force, so that the nucleus is a low-density system relative to this part of the force.

On the other hand, the longer-range part of the nuclear force, characterized by distances of the order of the pion Compton wavelength, has much smaller momentum components and hence affects mainly the intermediate states of low excitation energy. This part of the force is associated with collective nuclear phenomena - it "sees" the nucleus as a high-density system.

For the treatment of the short-range force, cluster techniques are appropriate. The first improvement over an independent-particle description would be an independent-pair picture, in which the interaction between any pair of particles is treated as accurately as possible, the dynamic effects of all other particles being neglected. This picture would be corrected by the introduction of three-particle clusters, and so forth. In a low-density system, the simultaneous interaction of three particles would be much less probable than that of a pair of particles, and so on, so that the cluster approach should converge very rapidly. Calculations in infinite nuclear matter have indicated that the three-body terms are small compared to the two-body terms[3], so that the independent-pair approximation should be a good one. A check of this result in finite nuclei would be very welcome.

The effect of high-energy neglected configurations in the renormalization of the nucleon-nucleon interaction is therefore described by the introduction of a Brueckner reaction matrix G in place of the bare nucleon-nucleon interaction (fig. 3). The G-matrix results from considering the total interaction between a pair of nucleons, limiting the virtual intermediate states of the pair to those of high excitation energy, excluding those single-particle orbitals already occupied by other nucleons, and taking into account the average field produced by the other nucleons and the total excitation energy of the nucleus.

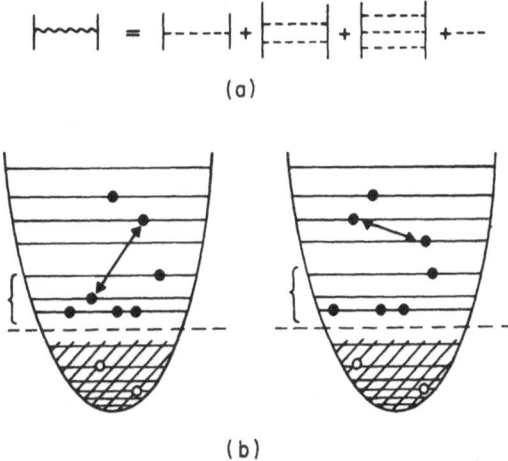

(a)

(b)

Fig. 3.   Brueckner G-matrix (independent-pair approach). (a) The G-matrix sums

two-body nucleon-nucleon interactions to all orders. (b) Illustration of

intermediate states included in the G-matrix. All nucleons are "frozen"

except the two which interact (connected by the arrows)

Once the high-energy intermediate states have been handled by the introduction

of the G-matrix, the low-energy intermediate states, excited predominantly by the

relatively weak long-range part of the force, may be investigated by means of

perturbation theory.[*] A very simple heuristic derivation of the basic equation

is the following[**].

The set of configurations included in the shell-model calculation span a

subspace of the Hilbert space. This subspace is called the model space, The

operator P is defined as the projection operator onto the model space. Its com-

plementary projector is denoted Q, so P+Q=1. Suppose the model-space is d-

dimensional. A set of d eigenstates of the total nuclear Hamiltonian, Ψ_i,

---

[*]   A recent preprint by J.P. Vary, P.U. Sauer and C.W. Wong indicates that the
long-range part of the effective tensor force significantly excites quite high-
energy intermediate states. This makes at least some perturbation theory calculations converge slowly with intermediate-state energy and casts doubt on all
previous perturbative calculations, which have concentrated only on the lower
intermediate states.

[**] This simple derivation is due to J. Rajewski.

together with their eivenvalues E_i, is selected. The projections of these states on the model space are $P\Psi_i$. An operator Y is defined by the formal requirement $YP\Psi_i = Q\Psi_i$. The effective Hamiltonian W in the model space is now required to satisfy $WP\Psi_i = E_i P\Psi_i$ for all states Ψ_i. Operating on the Schrödinger equation $H\Psi_i = E_i \Psi_i$ with P and using 1 = P+Q, one obtains

PHP. $P\Psi_i$ + PHQ. $Q\Psi_i = E_i P\Psi_i = W.P\Psi_i$. But $Q\Psi_i = YP\Psi_i$, so that <u>W = PHP + PHQ. Y.</u> Operating on the Schrödinger equation with Q, one finds

QHP.$P\Psi_i$ + QHQ.$Q\Psi_i = E_i Q\Psi_i = E_i Y.P\Psi_i = Y.W.P\Psi_i$, so that <u>YW = QHP + QHQ.Y.</u> These are the Schucan-Weidenmüller[4] equations for W.

If H is separated into a one-body operator H_o, whose eigenstates are the shell-model configurations, and a perturbation V, then, since P commutes with H_o, W may be written as $H_o + \mathcal{V}$, where $\mathcal{V}$ satisfies the equations $\mathcal{V}$ = PVP+PVQ.Y and $Y\mathcal{V}$ = QVP + QVQ.Y + QH_oQ.Y - Y.PH_oP. If these are expanded in powers of V, the resulting expansion for $\mathcal{V}$ is the Bloch-Horowitz-Brandow folded-diagram expansion[2].

A tremendous amount of effort has gone into calculating some of the terms in this perturbation expansion. The number of diagrams increases rapidly with the order of the diagram, so relatively complete calculations have been done

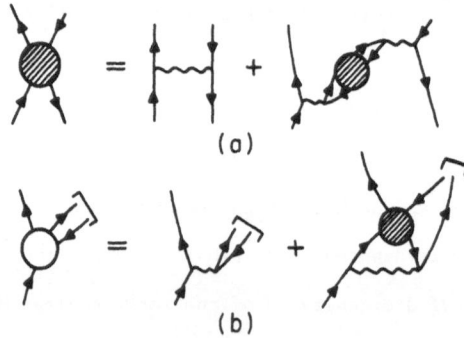

Fig. 4. Consistent treatment of particle-hole interactions. (a) The interaction between particles and holes includes a part arising from the exchange of interacting particle-hole pairs (self screening). (b) The coupling of valence particles to particle-hole pairs in the core also includes a part involving the particle-hole interaction

only up to third order[5]. The results obtained depend somewhat on the way in which the nucleon-nucleon interaction is converted into a G-matrix, in particular on which intermediate two-body states are included in the calculation of G and which are included in the perturbation expansion. However, there is no indication that second or even third order contributions in perturbation theory are small. The large terms in higher order are easily identified, and turn out to involve the excitation of particle-hole states in the supposedly-inert core. Selective summations of terms involving the particle-hole interaction to all orders lead to strongly collective effects, due to the appearance of collective core vibrations, but these effects disappear when the particle-hole interaction is treated consistently wherever it appears[6]. This damping is caused by a self-screening of the particle-hole interaction (allowing the particle and hole to interact by the exchange of particle-hole vibrations) and by a change in coupling of active nucleons to core vibrations due to the exchange of core vibrations (fig. 4). It appears reasonable to assert that the large terms in low orders of perturbation theory do not build up even further in higher orders, but remain reasonably small.

Unfortunately, any such assertion based on partial summations presumes that rearrangement of the perturbation expansion is legitimate, and this presupposes that the convergence of the expansion is such that the results of partial summations are physically meaningful. This supposition is strongly challenged by the empirical observation of so-called intruder states in nuclear spectra. The prototype example is the deformed 4p2h states in A=18 nuclei (fig. 5).

In the shell-model approach, the mass-18 nuclei are commonly described as two active nucleons in the sd-shell, outside an inert ^{16}O core. These are the model-space states. The unperturbed 4p2h states lie at an energy of at least 2$\hbar\omega$, or about 30 MeV, while the 2p states lie in the range of 0 to 10 MeV (the $d_{5/2}$ single-particle energy being taken as zero). The mutual interaction of the 4 sd-shell particles in the 4p2h state produces deformed states quite low in energy, so that the spectra of both ^{18}O and ^{18}F contain low-lying deformed 4p2h states among the perturbed 2p states[7]. In simple terms, the 4p2h states intrude into the model space when the interaction between particles is turned on. This may be

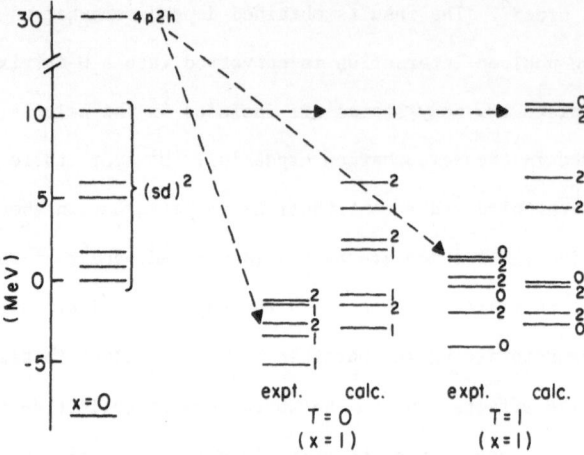

Fig. 5.   Intrusion of 4p2h states into the $(sd)^2$ model space in A=18 nuclei.   The
unperturbed spectrum (x=0) shows the 4p2h states well above the $(sd)^2$
states.   With the full perturbation (x=1), the experimental spectra contain low-lying deformed 4p2h states which are not reproduced by the
calculations

expressed in terms of the analytic behaviour of the effective interaction in the
complex x-plane, where x is the strength factor in the interaction, i.e. H(x) =
H_0 + xV.   At x=0, the 4p2h states lie above the model-space states; at x=1, some
4p2h states lie below some model-space states.   Although the eigenvalues of H(x)
cannot cross on the real x-axis, a pair of branch-points appears in the complex
x-plane when intruder states appear below model-space states[4].   These branch-points
define the circle of convergence of the perturbation expansion, and it is clear
that the physical situation (x=1) lies outside the circle of convergence.   Further,
since the eigenvalues of H(x) are continuous for real x, and do not cross, a
consistent calculation starting with the lowest eigenstates of H(0) must produce
the lowest eigenstates of H(1), a criterion satisfied by none of the perturbation-
theory calculations to date.

     These results do not invalidate all perturbation theory calculations.   If the
intruder states are not included as intermediate states in perturbation theory,
then the calculation is performed essentially in a world where the intruders do not

exist, hence do not intrude and hence allow the perturbation expansion to converge and to be consistent. But the effect of the intruders on the model-space states is not included, so any detailed comparison with experiment is meaningless. More encouragingly, it can be shown that suitable partial summations allow the analytic continuation of the perturbation expansion beyond the intruder branch-points. This can be illustrated in a simple two-dimensional model.[*]

Consider a two-dimensional Hilbert space spanned by the eigenvectors $|1> \equiv \binom{1}{0}$ and $|2> \equiv \binom{0}{1}$ of the unperturbed Hamiltonian $H_o = \begin{pmatrix} 0 & 0 \\ 0 & 1 \end{pmatrix}$. Let the perturbation be $V = \begin{pmatrix} u_1 & w \\ w & u_2 \end{pmatrix}$, with $u_1 > u_2$. The eigenvalues of $H(x) \equiv H_o + xV$ are $\lambda_{\pm}(x) = \frac{1}{2}[1 + (u_1+u_2)x] \pm \frac{1}{2}\sqrt{[1 - (u_1-u_2)x]^2 + 4w^2x^2}$. In the absence of coupling ($w=0$), these intersect at $x = \frac{1}{u_1-u_2}$. For w small but non-zero, they do not intersect for any real x, but the resulting pair of continuous non-interacting eigenvalue curves, $\lambda_\uparrow(x)$ and $\lambda_\downarrow(x)$, "flip" from λ_+ to λ_-, or vice versa, at $x = \frac{1}{u_1-u_2}$ (fig. 6). There are thus two ways of characterizing the eigenvalues of $H(x)$ for real x - in terms of the continuous curves $\lambda_\uparrow(x)$ and $\lambda_\downarrow(x)$,

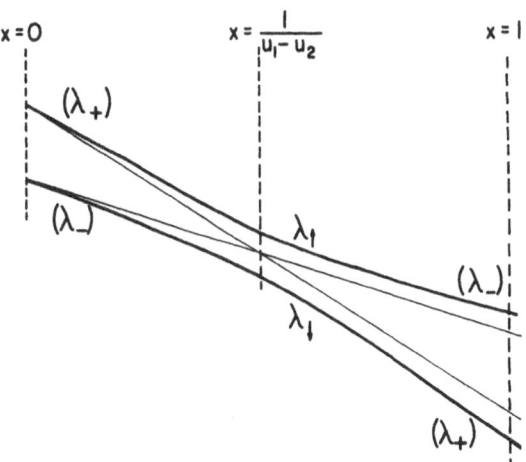

Fig. 6. Eigenvalues of the 2x2 model as a function of the (real) strength parameter x. For $x < \frac{1}{u_1-u_2}$ the upper curve $\lambda_\uparrow$ is equal to λ_+, the lower curve $\lambda_\downarrow$ to λ_-. For $x > \frac{1}{u_1-u_2}$, $\lambda_\downarrow$ is λ_+ and $\lambda_\uparrow$ is λ_-

---

[*] This simple 2x2 model, a 3x3 model and a semirealistic SU_3-like model have been studied in detail by P. Schaefer. His results will shortly be submitted for publication.

or in terms of the discontinuous curves $\lambda_{\pm}(x)$. These are related to different choices of cuts in the complex x-plane (fig. 7). At the complex branch-points

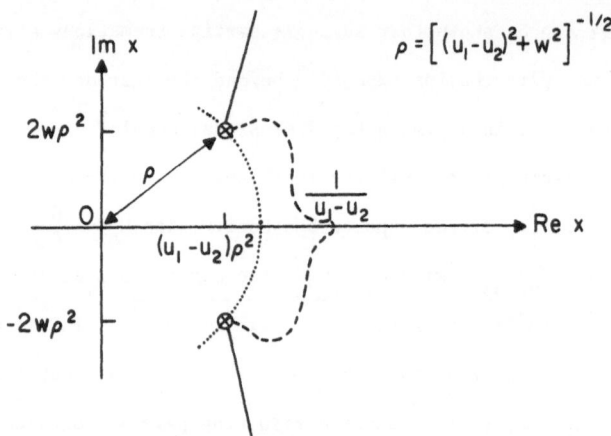

Fig. 7. Analytic structure of the 2x2 model in the complex x-plane. The branch points are connected by a cut, for which two possible choices are shown. The radius of convergence of the perturbation expansion is ρ

$x = \dfrac{1}{u_1 - u_2 \pm 2iw}$, $\lambda_{\pm}$ coincide. If the cut joining the branch-points crosses the real x-axis, the appropriate choice of eigenvalues is $\lambda_{\pm}$ (discontinuous for real x). If the cut avoids the real x-axis, passing through the point at infinity, the appropriate choice of eigenvalues is $\lambda_{\uparrow\downarrow}$ (continuous for real x). Whichever choice is made, the perturbation expansion in powers of x converges only inside the circle of radius $\dfrac{1}{\sqrt{(u_1 - u_2)^2 + 4w^2}}$. If the perturbation series serves as a starting point, some rearrangement must be sought allowing one to cross the circle of convergence, and the set of eigenvalues produced will then depend on the type of rearrangement chosen.

Some insight is obtained by studying the eigenvectors of H(x) (fig. 8). The ratio of the amplitude of state $|1\rangle$ to that of state $|2\rangle$ in the eigenvectors corresponding to $\lambda_{\pm}(x)$ is $\dfrac{2wx}{1 - (u_1 - u_2)x \pm \sqrt{[1 - (u_1 - u_2)x]^2 + 4w^2x^2}}$. For λ_{++}, this starts at zero for x=0, increases to unity (equal mixing) at $x = \dfrac{1}{u_1 - u_2}$, and then decreases again for increasing x, though it deviates from zero even at infinite x by an amount proportional to the relative magnitude of w. Thus the choice $\lambda_{\pm}$

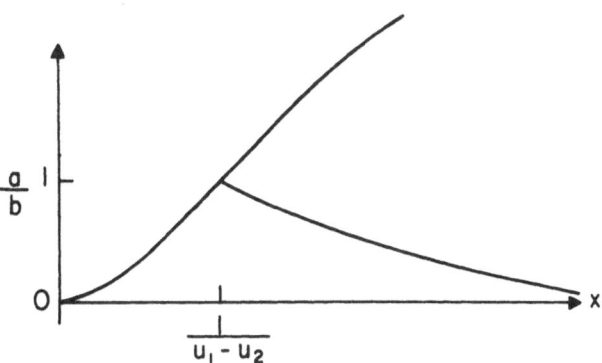

Fig. 8. Components of the eigenvectors in the 2x2 model. The upper curve shows

the ratio of amplitudes of lower and upper unperturbed states in the eigen-

vector belonging to the continuous eigenvalue $\lambda_\uparrow(x)$. The lower curve

shows the same quantity for the eigenvector belonging to the eigenvalue

$\lambda_+(x)$. The two curves coincide for $x \leq \dfrac{1}{u_1 - u_2}$

corresponds to selecting those eigenvalues whose eigenvectors retain their "un-

perturbed character" (predominantly $|1>$ or predominantly $|2>$) for large x. For

$\lambda_\uparrow$, the proportion of state $|1>$ in the eigenvector starts from zero at x=0,

increases to unity at $x = \dfrac{1}{u_1 - u_2}$, and continues to increase with increasing x.

Thus the choice $\lambda_{\uparrow\downarrow}$ corresponds to selecting those eigenvalues whose eigenvectors

change their "unperturbed character" on passing through $x = \dfrac{1}{u_1 - u_2}$.

In terms of actual nuclear calculations, it is clear that calculations in

a 2p model space which miss completely the 4p2h intruder states in the experiment-

al spectrum are consistent if they correspond to the $\lambda_\pm$ choice, i.e. to a cut

x-plane where the cut crosses the real x-axis. They must thus correspond to

analytic continuation of the perturbation series around the outside of the branch

points. The simple two-dimensional model helps to elucidate the nature of the

rearrangements required to achieve this result. It is not possible to reproduce

all the details of the analysis here, but the salient points will be briefly

described.

For a perturbation treatment, the model space is chosen as the one-dimensional

space spanned by the state $|1>$. The terms in the perturbation expansion are of two types - "normal" terms, in which $|1\rangle$ does not occur as an intermediate state; and "folded" terms, in which $|1\rangle$ does occur as an intermediate state. In this simple case, the set of normal terms constitute a geometric series, whose sum is $u_1 x + \frac{w^2 x}{1+u_2 x}$ (fig. 9(a)). This sums to all orders the effect of the "intruder"

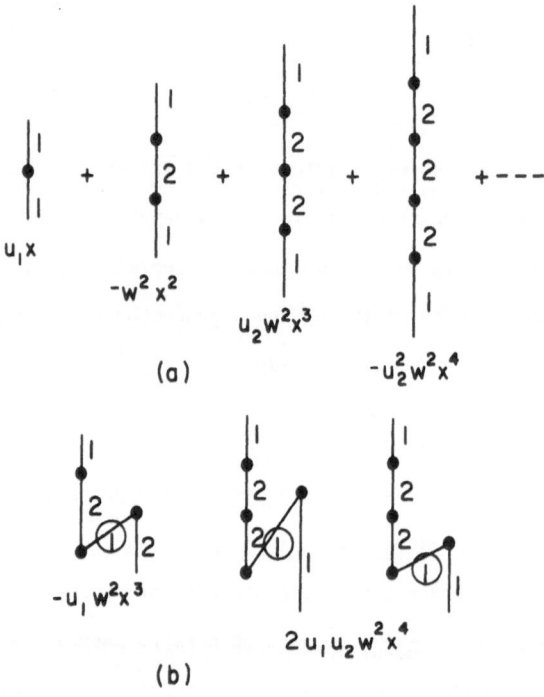

$$u_1 x \qquad -w^2 x^2 \qquad u_2 w^2 x^3 \qquad -u_2^2 w^2 x^4$$

(a)

$$-u_1 w^2 x^3 \qquad 2 u_1 u_2 w^2 x^4$$

(b)

Fig. 9. Diagram representation of the perturbation expansion of the 2x2 model.

(a) Normal terms. (b) Some "folded" terms

state $|2\rangle$, except for folded terms, and is well-defined everywhere except for a pole at $x = -\frac{1}{u_2}$. If there are included with these "normal" terms all folded terms in which the state $|1\rangle$ occurs any number of times as an intermediate state, but before any intermediate state $|2\rangle$, then the sum is altered to $u_1 x + \frac{wx}{1+(u_2-u_1)x}$, which has a pole at $x = \frac{1}{u_1-u_2}$ (fig. 9(b)). The fact that this pole coincides with the point at which $\lambda_\downarrow$ flips from λ_- to λ_+ is not accidental - it indicates that it is the state involved in the branch-point singularity, with its associated cut, that has been partially summed to all orders. The remaining folded terms in the perturbation expansion, with intermediate states $|1\rangle$ occurring between

intermediate states $|2\rangle$, can also be regrouped into a series in terms of the

renormalized interaction $\frac{wx}{1+(u_2-u_1)x}$. This series converges everywhere outside

a region containing the pole and the branch points, and hence constitutes an

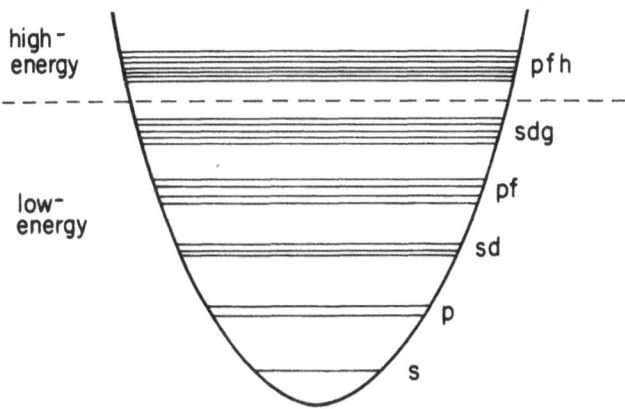

Fig. 10.   A possible truncation of single-particle harmonic oscillator states into

high-energy and low-energy states, appropriate to A = 18

analytic continuation of the original series beyond and outside the branch-points,

which is exactly the desired result.   It is unfortunately not clear that the

"physical" point x=1 lies outside the region where the rearranged series diverges.

It is generally true that the "linear" approximation to the perturbation

series (the term of lowest order in the interaction, or in the renormalized inter-

action) serves as a fair approximation to the eigenvalue belonging to the state

of "character" $|1\rangle$ even beyond the radius of convergence of the series.   The

prescription suggested by this analysis is thus to perform infinite partial summa-

tions taking the intruder states into account to all orders, and to stop at that

point, without going to higher order in the intruder-renormalized interaction.

In doing the partial summation, some folded terms should be included, in order

to bring the part of the interaction wholly inside the model space into the

energy denominators, and hence shift the pole to the correct point.

This prescription has been tested roughly in ^{18}O, where a semi-phenomenological deformed 4p2h state was used as an intermediate state to lowest order in perturbation theory, but without including the interaction within the model space in the energy denominator.[*] The resulting strong effect on the effective interaction was compatible with experimental observations and previous theoretical calculations.

The entire discussion to this point has assumed that some suitable single-particle basis has been defined and that all calculations are done in this basis. In an exact calculation, equivalent to a complete summation of the perturbation expansion, the choice of single-particle basis is irrelevant. The final results depend only on the total Hamiltonian and not on the way in which it is broken up into a diagonal one-body part and a perturbation. Convenience is then the only criterion for choosing a single-particle basis. In practice, of course, all computations are approximate, so that the choice of the single-particle potential can influence the results obtained. There is then a strong temptation to seek a criterion for choosing the single-particle potential within the structure of the theory itself. There result various theoretical definitions, most of them embodying some requirement of self-consistency of the single-particle potential[8]. The "correct" definition of this auxiliary quantity is thus still a matter of some debate, and it becomes significant to question how sensitive the results of the calculations are to single-particle assumptions.

The ideal situation would be that pertaining in infinite nuclear matter, as discussed by Mahaux[9]. There, provided the computation is carried to sufficiently high order (including three-body clusters, in practice) and provided the single-particle part of the perturbation is taken into account, the final results are quite insensitive to wide variations in the input single-particle potential. This was checked very carefully by Mahaux for the occupied-state potential. Much rougher calculations[**] applied to changes in the potential for low-lying unoccupied states confirm the lack of sensitivity even to large changes. This is very

---

[*] The calculation was performed by J. Rajewski, and will form part of his Ph.D. thesis.
[**] This refers to unpublished work of Y. Starkand.

satisfactory - the results of an adequate calculation should be insensitive to fairly wide changes in auxiliary quantities.

A similar check for the effective interaction calculation has not been performed. Generally, corrections associated with the single-particle potential have not even been evaluated. Different single-particle spectra are used in computing G-matrix elements and in computing perturbation theory contributions using these G-matrix elements. The reaction matrix shows sensitivity to the single-particle basis used, as do the perturbation theory calculations. As a striking example, the use of an approximate Hartree-Fock basis instead of a harmonic oscillator basis completely wipes out the collectivity of vibrations in the ^{16}O core[10]. This cannot be regarded as a satisfactory state of affairs. What is needed is either a compelling theoretical criterion for selecting the single-particle basis, followed by a careful calculation keeping track of single-particle corrections; or (perhaps "also") convincing evidence that calculations can be carried to a point where the results are reasonably insensitive to the choice of single-particle basis. Also required is some clear understanding of whether it is necessary to take explicit account of the fact that single-particle orbitals in an interacting system are not one hundred percent occupied or unoccupied. The inclusion of occupation probability factors, which has had such significant effects in the renormalized Brueckner-Hartree-Fock calculations[11], could dramatically affect perturbation calculations of the effective interaction.

As a final point of interest, it is amusing to contemplate the possibility, or otherwise, of a genuine first-principles calculation of nuclear spectra. This would involve starting with eighteen nucleons and the free nucleon-nucleon interaction (and how sensitive are the calculations to this auxiliary entity?), and ending with the binding energy and low-energy spectrum of, say, ^{18}F. Using current ideas, a possible plan of action is the following.

Some convenient single-particle basis, almost certainly a harmonic oscillator basis, is chosen and divided into high-energy and low-energy orbitals. A possible division might classify the s, p, sd, pf and sdg levels as low-energy levels (fig. 10). The high-energy renormalizations would then be taken care of by

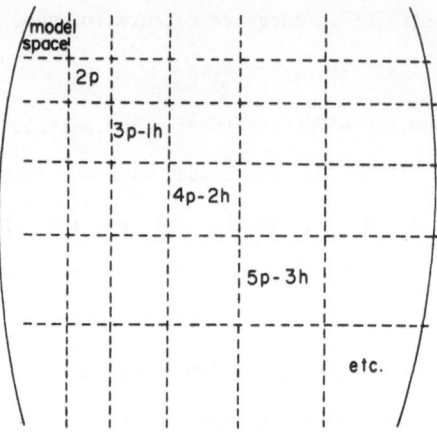

Fig. 11. Partition of the truncated Hamiltonian (in the space spanned by the
low-energy single-particle configurations, and with G replacing the
nucleon-nucleon interaction) into np,(n-2)h states

replacing the bare nucleon-nucleon interaction by a G-matrix,the excluded intermediate states in the two-particle sum being those with a particle above the sdg
shell. Three-body cluster effects would be assumed small. The ^{18}F states within
the truncated set of single-particle orbitals are then divided into model-space
states (filled s and p shells, two nucleons in the sd shell) and intermediate
states. The intermediate states are further classified as 2p, 3p1h, 4p2h, etc.
states (fig. 11). It is now required to compute the effective interaction in the
model space by renormalizations which include the effects of the virtual intermediate states. A barely tractable procedure would involve first diagonalizing
the G-matrix in each sub-space of the intermediate states (for instance, diag-
onalize in the 4p2h space) and then utilizing, to lowest order in perturbation
theory, the set of lowest-energy intermediate states. This procedure would take
into account collective effects and would avoid intruder-state inconsistencies.
It might be possible to obtain a more complete spectrum by enlarging the model
space to include the low-energy states found in the partial diagonalizations. In
principle, such a calculation is feasible at the present time.

To conclude, the theory of the nuclear effective interaction, though apparently quite well understood, still contains weak points, especially in connection with the single-particle aspects of the problem and with the treatment of high-energy intermediate states, but a reasonably coherent and consistent calculation appears possible.

REFERENCES

1. Barrett, B.R.: in The Two-Body Force in Nuclei (S.M. Austin and G.M. Crawley, eds.) (Plenum Press, N.Y., 1972); Kirson, M.W.: in Theory of Nuclei (G. Ripka and L. Fonda, eds.) (I.A.E.A., Vienna, 1972); Kirson, M.W.: in Proceedings of the Rome Conference on the Nuclear Many-Body Problem, 1972, to be published; Barrett, B.R., Kirson, M.W.: in Advances in Nuclear Physics, Vol. VI (M. Baranger and E Vogt, eds.) (Plenum Press, N.Y., 1973).

2. Brandow, B.H.: Revs. Mod. Phys. 39, 771 (1967).

3. Rajaraman, R., Bethe, H.A.: Revs. Mod. Phys. 39, 745 (1967).

4. Schucan, T.H., Weidenmüller, H.A., Ann. Phys. (N.Y.) 73, 168 (1972), and preprint.

5. Barrett, B.R., Kirson, M.W.: Nucl. Phys. A148, 145 (1970) [and A196, 638(1972)].

6. Kirson, M.W.: Ann. Phys. (N.Y.) 66, 624 (1971) [and 68, 556 (1971)], and to be submitted for publication.

7. Federman, P.: Nucl. Phys. A95, 443 (1967).

8. Kirson, M.W.: Nucl. Phys. A115, 49 (1968) and A139, 57 (1969); Brueckner, K.A., Meldner, H.W., Perez, J.D.: Phys. Rev. C7, 537 (1973).

9. Mahaux, C.: Nucl. Phys. A163, 299 (1971).

10. Ellis, P.J., Osnes, E.: Phys. Lett. 41B, 97 (1972) and preprint.

11. Becker, R.L.: in Proceedings of the Rome Conference on the Nuclear Many-Body Problem, 1972, to be published.

PROTON, GAMMA ANGULAR CORRELATIONS NEAR

ANALOGUE RESONANCES IN THE MASS 90 REGION

Zeev Vager

Department of Nuclear Physics, Weizmann Institute of Science, Rehovot, Israel

Several years ago, a group at the Weizmann Institute[1] had shown that it was possible to measure detailed information on some nuclear wave functions by the study of pp'γ angular correlations via analogue resonances. Encouraged by the success of the method a new project had been started[2]. The aim was the study of

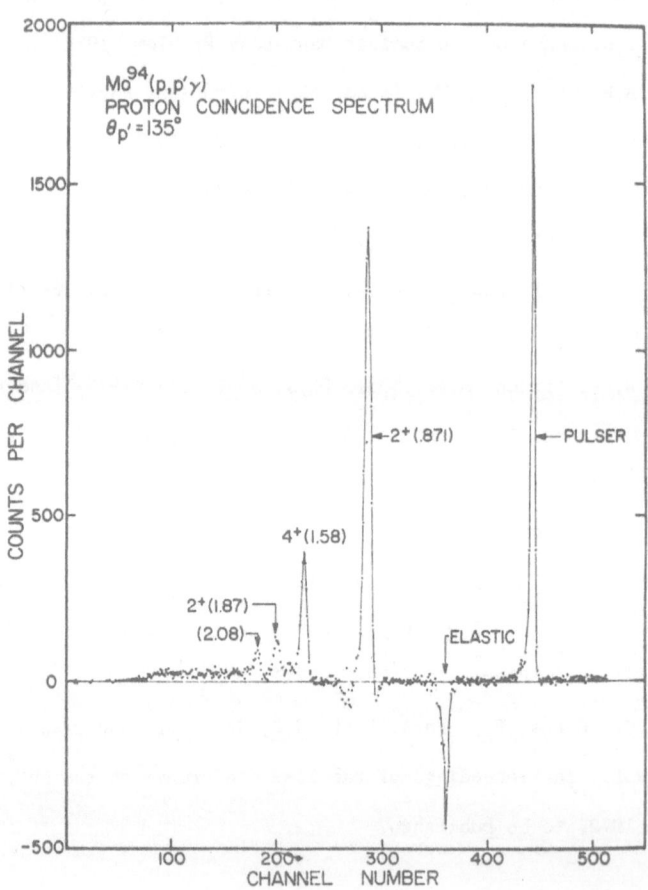

Fig. 1. Proton coincidence spectrum on the $5/2^+$ g.s. analogue resonance

nuclear wave functions of low lying states in the mass 90 region. The motivation

for this work, some experimental results and discussion of the interpretation diff-

iculties will be reported here together with some preliminary results of analyses

and its implications.

The motivation for this work can be illustrated with the aid of figure 1.

A proton spectrum in coincidence with $2^+ \rightarrow 0^+$ gamma rays from the ^{94}Mo(p,p'γ)

reaction is shown. The incoming proton energy is adjusted to excite the analogue

resonance of the $5/2^+$ ground state of ^{95}Mo. The 2^+, 4^+, 2^+ and (4^+) first excited

$$MO^{94} \; 2^+ \; E_x = .871 \, MeV$$

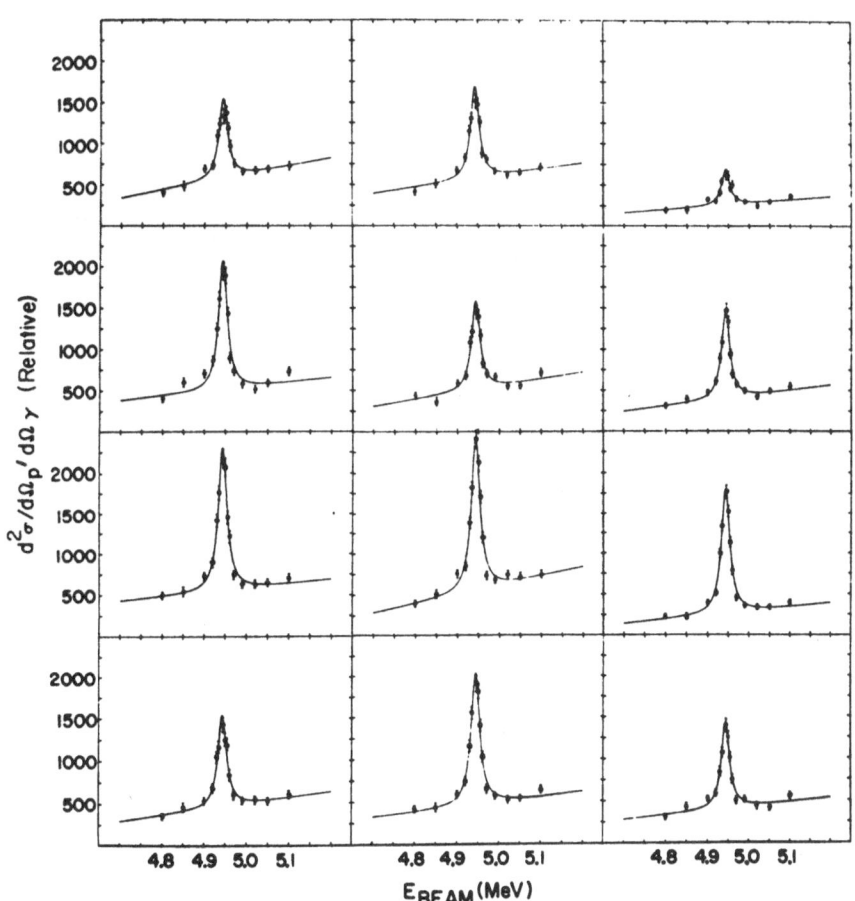

Fig. 2. Excitation functions of the pp'γ angular correlation through the 2^+(.871

MeV) state of ^{94}Mo, in the vicinity of the $5/2^+$ g.s. analogue states of ^{94}Mo are excited relatively strongly, as seen by the four proton groups in fig. 1. The resonance behaviour of the inelastic groups (figs. 2, 3, 4 and 5) is an indication that the $5/2^+$ g.s. wave function of ^{95}Mo has single particle components built on the four first excited states of ^{94}Mo. Furthermore, such components can be decomposed to several partial waves which are possible by angular momentum and parity conservation. It is clear that the specific decomposition of partial waves would display itself in the shape of the complex pp'γ angular correlation in each of the proton groups. The study of the pp'γ angular correlation as a

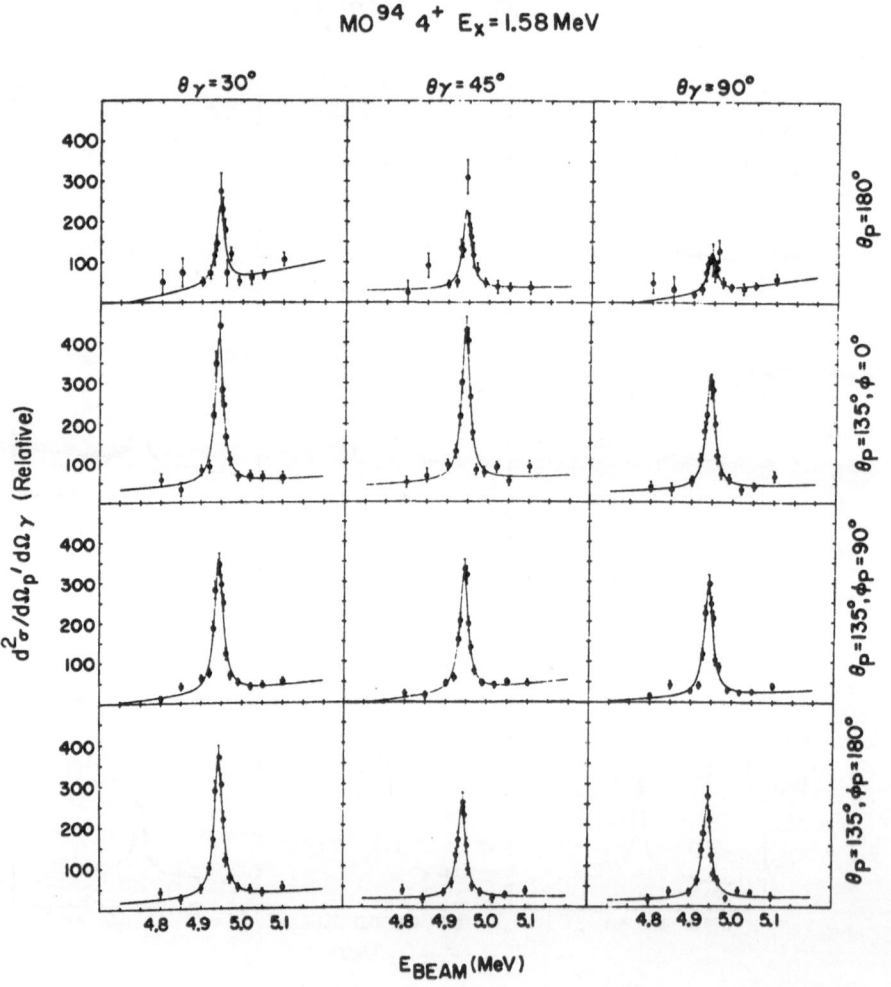

Fig. 3. Excitation functions of the pp'γ angular correlation through the 4^+(1.582 MeV) state of ^{94}Mo, in the vicinity of the $5/2^+$ g.s. analogue

function of the proton energy should result in all these partial waves and thus reveal an important part of the 5/2$^+$ g.s. wave functions. This is of special interest in the mass 90 region where the results can be compared with several shell model calculations.

Details on the experimental setup and its reliability are given elsewhere[2].

The interpretation would have been straightforward if there were no compound background below the measured resonances (see figs. 2,3,4 and 5). Unfortunately,

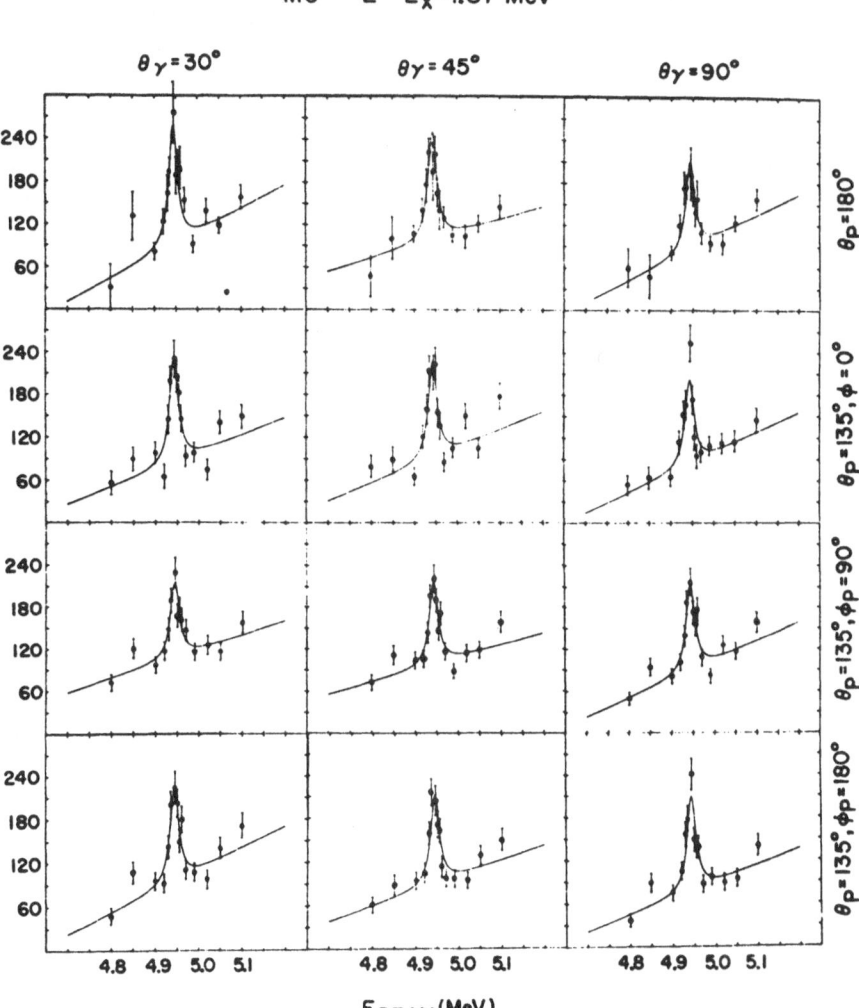

Fig. 4. Excitation functions of the pp'γ angular correlation through the 2$^+$(1.868 MeV) state of ^{94}Mo, in the vicinity of the 5/2$^+$ g.s. analogue

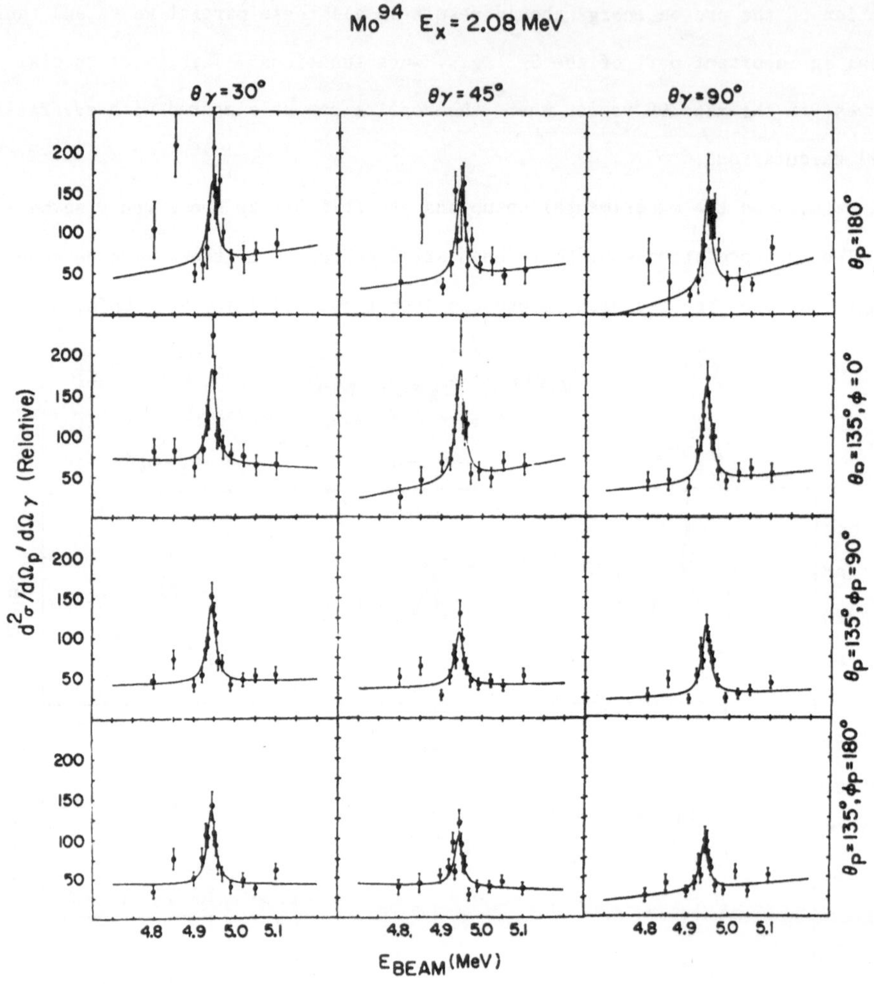

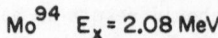

Fig. 5. Excitation functions of the pp'γ angular correlation through the (2.08 MeV) state of ^{94}Mo, in the vicinity of the $5/2^+$ g.s. analogue.

the partial widths of the analogue resonances determine only a fraction of the observed resonance anomalies. The rest is due to compound enhancement which is very important especially in the mass 90 region. The origin of the compound enhacement and the difficulties in the theoretical estimation is our next subject.

Consider an analogue resonance which can decay only by one proton channel. The average S matrix can be written in the form

$$<S> = <S^0> - i \frac{\Gamma_p}{E-E_R+i\Gamma/2} \qquad (1)$$

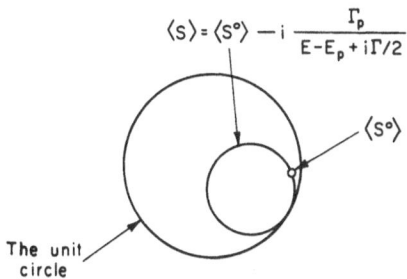

$$\langle S \rangle = \langle S^{\circ} \rangle - i \frac{\Gamma_p}{E - E_p + i\Gamma/2}$$

The unit circle

$\langle S^{\circ} \rangle$

Fig. 6.   The average S matrix contour at different energies near an analogue

resonance (eq. 1)

where S° is the background S matrix in the limit where the width of the analogue

resonance Γ tends towards zero, E_R is the apparent resonance energy and Γ_p is a

complex number, such that $|\Gamma_p| < \Gamma$.   The behaviour of $\langle S \rangle$ as a function of the

energy E is illustrated in fig. 6.   For energies which are from E_R, $\langle S \rangle$ is

about $\langle S^{\circ} \rangle$ - the optical model background.   At a certain energy E_o, $\langle S \rangle$ becomes

(almost) unitary, but there are other energies where $\langle S \rangle$ is appreciably smaller

than unity.   Conservation of flux arguments tells us that there must be a strong

energy dependent compound cross-section.   For the single channel case, unitarity

alone is enough for the calculation of the compound elastic cross-section

$$F = S - \langle S \rangle$$

$$\langle |F|^2 \rangle = 1 - |\langle S \rangle|^2 \qquad (2)$$

But when more than one channel is involved, the unitary alone is not enough and,

unless a model is given, the average of the compound cross section for each channel

are not known.   The only existing prescription which relates the average S matrix

to the average compound cross section is the Hauser-Feshbach formula.   The

intuitive feeling that the different decay modes should be proportional to the

corresponding transmission coefficients, can be shown to be equivalent to the

Hauser-Feshbach prescription.   On the other hand, it has been shown by Engelbracht

and Weidenmuller[3]   that the Hauser-Feshbach formula is justified only in the limit

where all the transmission coefficients involved are very small compared with unity.

In view of fig. 6, this condition definitely does not hold in the vicinity of analogue resonances. Conversely, Weidenmuller and Engelbracht[3] show that the ignored quantity in the Hauser-Feshbach formula is that which includes the level-level correlations. But analogue resonances induce strong correlations among the compound levels by mixing into those levels. It is concluded that the use of the Hauser-Feshbach formula for the estimates of compound cross section near analogue resonances is unjustified. At this point it is worthwhile mentioning the study of ref. (4) in which formulas for the compound nuclear reactions near analogue resonances are derived without ignoring the level-level correlations due to the mixing of the analogue. Such a formalism will be very useful for the reduction of the pp'γ data. Unfortunately, this study does not include yet the general pp'γ case.

In view of the above, we attempted to analyze the pp'γ data in the following semi-empirical approach. The very complex angular correlation averaged over the target thickness is a linear function of the energy dependent quantities

$$<S_{\beta\alpha}^* \, S_{\beta'\alpha}> \tag{3}$$

where α is the incoming channel and β, β' are any two possible outgoing channels. On resonance, the quantities given by (3) are assumed to be of the form

$$<S_{\beta\alpha}^* \, S_{\beta'\alpha}> - \text{Bckgrnd} \propto X_\beta X_{\beta'} + B_\alpha T_\beta \, \delta_{\beta\beta'} \tag{4}$$

The coefficient of proportionality, which has to include both the incoming partial amplitude and some enhancement effects, is taken as a free parameter. The first term is a multiplication of two decay amplitudes X_β ($\Gamma_\beta = |X_\beta|^2$) and represents the part of the angular distribution which displays interference between the analogue partial amplitudes. The average interference between an analogue partial amplitude and the background amplitudes is assumed to be zero. The last term is due to the decay of the analogue resonance via the background compound resonances and therefore is assumed to have negligible average partial wave interference. This term is assumed to be proportional to the corresponding outgoing wave transmission coefficient.

The formula was tested experimentally for one special case (ref. 5). A comparison between the ^{95}Mo$(d,t)^{94}$Mo$(0^+,2^+,4^+)$ reaction cross sections and the ^{94}Mo$(p,p'\gamma)^{94}$Mo$(2^+,4^+)$ reaction cross sections were made with the aid of the DWBA for the former reaction and eq. (4) for the latter reaction. The agreement is satisfactory but more experimental verifications are needed. Nevertheless, a preliminary analysis of many of our p,p'γ data were made using eq. (4) and are presented below.

The analyses presented include analogue resonances of $5/2^+$ ground state of the ^{87}Sr, ^{89}Sr, ^{91}Zr, ^{93}Zr, ^{95}Zr, ^{93}Mo, ^{95}Mo and ^{97}Mo nuclei. Only the decay to the first 2^+ and 4^+ states were analyzed and results are given as the ratios among the following five partial waves.

$$|(s_{1/2} \times 2^+)5/2^+>$$

$$|(d_{3/2} \times 2^+)5/2^+>$$

$$|(d_{5/2} \times 2^+)5/2^+>$$

$$|(d_{3/2} \times 4^+)5/2^+>$$

$$|(d_{5/2} \times 4^+)5/2^+>$$

The g-waves were ignored because of penetrability arguments. For convenience, the $|(d_{5/2} \times 2^+)5/2^+>$ component was normalized to one. The results are summarized in Table I and figs. 7, 8 and 9. A blank in the table means a statistically unobservable resonant cross section above background. According to a simple shell model argument, the measured states could be splitted into three categories.

TABLE I

Results: Relative Partial Amplitudes

	^{87}Sr	^{89}Sr	^{91}Zr	^{93}Zr	^{92}Zr	^{93}Mo	^{95}Mo	^{97}Mo
$\|(s_{1/2}\times 2^+)5/2^+>$	.75	1.12	-	-.25	-.083	1.12	-.086	-.26
$\|(d_{3/2}\times 2^+)5/2^+>$	-.20	.06	-	-.08	.003	.12	.34	-.34
$\|(d_{5/2}\times 2^+)5/2^+>$	1.	1.	-	1.	1.	1.	1.	1.
$\|(d_{3/2}\times 4^+)5/2^+>$	-	-	-	-.43	.12	-	.35	-
$\|(d_{5/2}\times 4^+)5/2^+>$	-	-	-	1.54	2.54	-	1.62	-

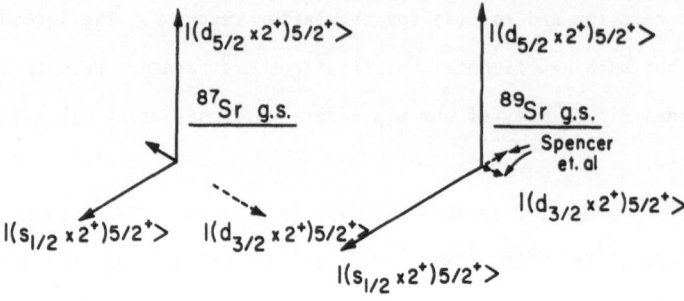

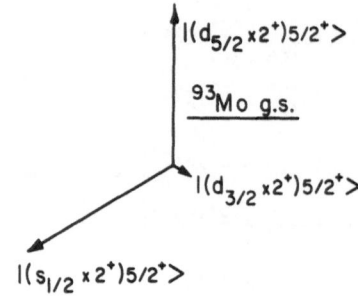

Fig. 7.   Vector diagrams for the measured wave function components of the $5/2^+$

g.s. of ^{87}Sr, ^{89}Sr and ^{93}Mo. (See table I)

1. Almost pure $d_{5/2}$ single particle states above the 0^+; e.g. ^{87}Sr, ^{89}Sr, ^{91}Zr
    and ^{93}Mo.

2. Both the $5/2^+$ and the 2^+, 4^+ final states are of $d_{5/2}^n$ nature; e.g. ^{93}Zr and
    ^{95}Zr.

3. The same as (2), but more mixing with other components due to the presence of
    two extra protons above a closed shell; e.g. ^{95}Mo and ^{97}Mo.

    The components of wave functions of nuclei in category (1) are illustrated in

fig. 7.   They all show a very similar pattern of almost equal amounts of $s_{1/2}$ and

$d_{5/2}$ components with the same relative phase and other negligible components.   An

exception is the ^{91}Zr g.s. where no corresponding resonance was observed, although

from the point of view of reaction mechanism and energies this case is not very

different than the ^{89}Sr g.s. or ^{93}Mo g.s. cases.   A possible explanation to this

is given by a comparison of the explicit shell model calculation for these cases

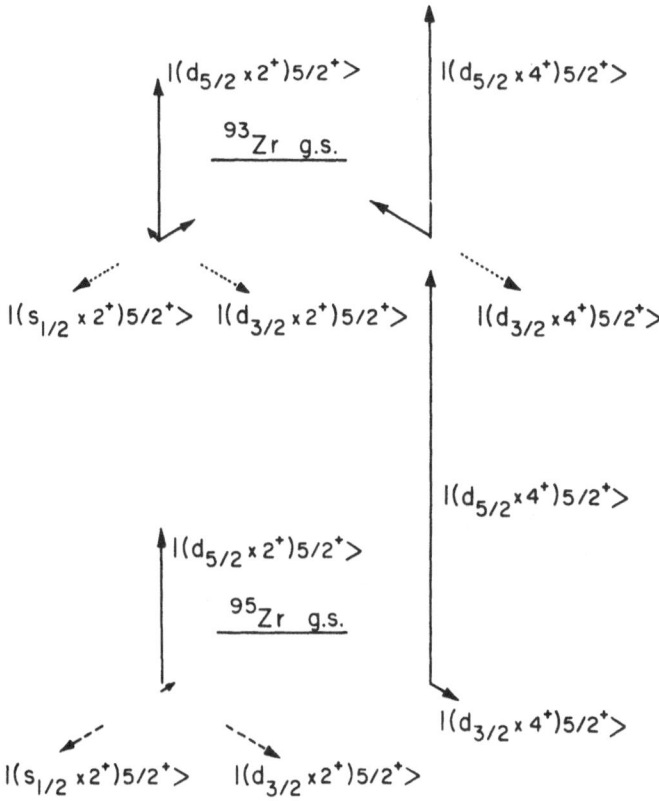

Fig. 8. Vector diagrams for the measured wave function components of the $5/2^+$ g.s. of ^{93}Zr and ^{95}Zr. (See table I)

(Table II). The square of the largest partial amplitude on the 2^+ of ^{89}Sr is larger by at least a factor of 20 than the corresponding one in ^{91}Zr. On the other hand the Spencer et al.[6] calculation predicts much smaller $s_{1/2}$ components and with an opposite phase relative to the $d_{5/2}$ component as compared with the experimental results.

The remarkable feature in fig. 8, the category 2 nuclei, is the purity of the $d_{5/2}^n$ configurations of both the $5/2^+$ states and the 2^+, 4^+ final states. The ratio of the $d_{5/2}$ components on the 4^+ and the 2^+ are not far from those expected by pure $d_{5/2}^n$ configurations. The partial waves of the category 3 nuclei are presented in fig. 9. The features resemble very much those of fig. 8, as expected, except that the configuration mixing with $d_{3/2}$ are more pronounced, probably due to the two

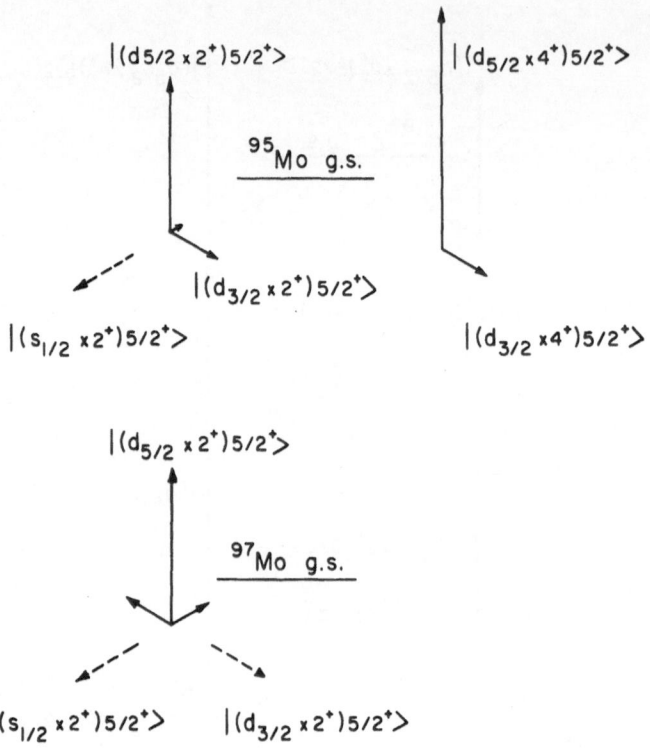

Fig. 9. Vector diagrams for the measured wave function components of the $5/2^+$ g.s. of ^{95}Mo and ^{97}Mo. (See table I)

TABLE II

Theoretical Calculations

I. Talmi and R. Gross [7] $5/2^+$ G.S. of ^{91}Zr overlap on ^{90}Zr states		J.E. Spencer, E.R. Cosman, H.A. Enge and A.K. Kerman [6] $5/2^+$ G.S. of ^{89}Sr overlap on ^{88}Sr states	
$(d_{5/2} \times 0_1^+)5/2^+$	.9957	$(d_{5/2} \times 0^+)5/2^+$	.9047
$(d_{5/2} \times 2^+)5/2^+$	-.0891	$(d_{5/2} \times 2^+)5/2^+$	.4040
$(d_{5/2} \times 0_2^+)5/2^+$	.0026	$(d_{3/2} \times 2^+)5/2^+$	.0649
$(d_{5/2} \times 4^+)5/2^+$	-.0241	$(s_{1/2} \times 2^+)5/2^+$	-.0983
		$(g_{7/2} \times 2^+)5/2^+$	-.0663

extra protons above the closed shell. It is not obvious why such two protons would not cause much more mixing both into the $5/2^+$ states and the 2^+, 4^+ final states.

In conclusion, the above preliminary analysis of the pp'γ data on analogue resonances in the mass 90 region showed that a great deal of information is gained about the structure of these states. The systematics of such results is an important tool in nuclear spectroscopy which is not yet employed.

REFERENCES

1. Abramson, E., et al.: Nucl. Phys. A144, 321 (1970).

2. Cue, N., et al.: Nucl. Phys. A192, 581 (1972).

3. Engelbrecht, C.A., Weidenmüller, H.H.: Hauser-Feshbach theory and Ericson fluctuations in the presence of direct reaction, preprint.

4. Agassi, D., Vager, Z.: Compound contributions to nuclear reactions involving the analogue resonance, preprint.

5. Ebramson, E., et al.: Phys. Lett. 38B, 70 (1972).

6. Spencer, T.E., et al.: Phys. Rev.

7. Talmi, I., Gross, R.: private communication.

ANALOGUE RESONANCES - SOME ASPECTS OF REACTION MECHANISM
AND SPECTROSCOPY REVISITED

J.P. Wurm

Max-Planck-Institut für Kernphysik, Heidelberg

In this brief talk I shall be concerned with isobaric analogue resonances (IAR) which have played a central role in nuclear spectroscopy during almost the last ten years.

What motivates revisiting IAR today are mainly two points: First, early promises that wave functions of low-lying states of (core + n) nuclei may be quantitatively decomposed have been given experimental foundation only quite recently[1]. This will be covered by the following talk of Zeev Vager. Second, the reaction mechanism - that is, all that complicates spectroscopy - has found increasing interest by itself. The reason, I believe, is that IAR in a unique way, studies the mutual influence of reaction modes of vastly different time scales. In inelastic proton scattering, those may be present as direct excitation of the target modes, the intermediate-resonance scattering induced by the analogue state, and the compound nuclear (CN) scattering.

In the following I shall give you a subjective choice of what I think are points of current interest, with emphasis on reaction mechanism.

1. BROADENING

The closest view into the compound nucleus certainly is provided by high-resolution experiments as they have been done at Duke[2]. Here, we are interested in average quantities rather than the properties of thousands of levels as appropriate to the more usual experimental situation

$$\Gamma^{CN}, D \ll \Delta E \ll \Gamma^{IAR}$$

where the beam-plus-target resolution ΔE is large compared to both Γ^{CN} and D of the compound levels, but much smaller than the total width of the IAR.

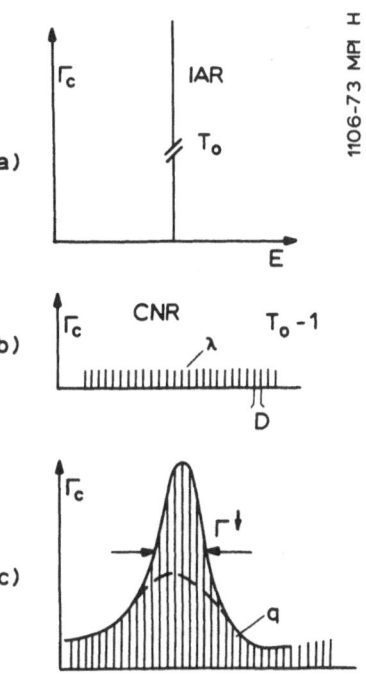

Fig. 1. Schematic view of partial-widths distributions for an isolated resonance

   (a), for compound-nucleus levels in the absence of an IAR (b), and for

   the giant resonance situation (c) where the original decay strength of

   the IAR is shared by levels q within the spreading width $\Gamma^{\downarrow}$

Fig. 1 is to remind you in a very schematic way of the situation in the com-

pound nucleus. The (large) decay with Γ_c^{IAR} for proton emission into channel c for

an isolated IAR is shown on top. The (small) decay width Γ_c^{λ} of possible compound

nuclear levels λ are indicated below on the assumption of equal level spacing D.

Their average is connected to the transmission coefficient T_c that can be found in

Blatt-Weisskopf

$$2\pi \; <\Gamma_c^{\lambda}> \; / \; <D> \; = \; T_c \; = \; 1 \; - \; e^{-4\eta_c}$$

and is expressed through the absorptive phase shift η_c of the optical model. This

relation is the well-known consequence of the fact that the (average) optical model

S-matrix deviates from unitarity by an amount just given by T_c:

$$T_c \; = \; 1 \; - \; \left| <S_{cc}> \right|^2$$

Line broadening occurs if both types of levels are allowed to interact (via some residual force V) as depicted at the bottom of fig. 1. The most obvious effect of the broadening is described by the spreading width $\Gamma^\downarrow$. In the picket-fence model we have the simplified relation $\Gamma^\downarrow \simeq 2\pi|V|^2\rho_\lambda$ reminiscent of Fermi's Golden Rule, $\rho_\lambda = 1/D$ [3].

The broadening has two consequences, First, the energy averaged S-matrix,

$$<S>_{cc} = e^{2i\delta}c \left(1 - i \frac{\Gamma_c}{E - E_{res} + i\Gamma/2}\right)$$

is not unitary, since $\Gamma^\uparrow = \Sigma_c \Gamma_c$ is smaller than the total width $\Gamma = \Gamma^\uparrow + \Gamma^\downarrow$. Second, and as a consequence, the CN absorption will be large on resonance and displays a Lorentzian

$$T_c \simeq \frac{\Gamma_c \Gamma^\downarrow}{(E - E_{res}) + \Gamma^2/4}$$

or a more complicated, asymmetric resonance shape[4,5]. The "Robson enhancement" of the compound nucleus scattering considerably complicates the spectroscopic analysis of IAR especially in the region of neutron deficient nuclei near A = 100 where many resonances occur below the neutron threshold. The conventional analysis uses the Hauser-Feshbach formula to calculate the compound scattering due to

$$S^{fl} = S - <S>$$

which contributes to the energy averaged cross section besides $<S>$:

$$<\sigma> = |<S>|^2 + <|S^{fl}|^2> = \sigma^{dir} + \sigma^{CN} \ .$$

Until quite recently, low-resolution experiments did not provide much information on the topics under discussion, e.g. "what is the amount of compound-elastic scattering?" or "how big is the spreading width?" Polarised-beam experiments have improved this situation[6]. The idea is that the analyzing power A, as derived from left-right asymmetry in the scattering of polarized particles as well as polarization P is caused only by bilinear amplitudes:

$$\begin{array}{l} A \ d\sigma/d\Omega \\ \qquad\qquad \simeq \ \Sigma_{\alpha\beta} \ Im \ <S_\alpha \ S_\beta^*> \ , \quad \alpha \neq \beta \ \ only. \\ P \ d\sigma/d\Omega \end{array}$$

In contrast, the cross section contains also quadratic terms

$$d\sigma/d\Omega \simeq \Sigma_{\alpha\beta} \; \langle S_\alpha S_\beta^* \rangle$$

α, β stand for physically different processes like potential or resonance scattering which can experimentally not be distinguished; or for different in- or outgoing channel numbers.

The most transparent case is that of elastic scattering from spin-0 nuclei since there is only one ingoing and one outgoing channel. S^{fl} enters only linearly into the expressions for σA, σP and therefore does not contribute to the energy average (as $\langle S^{fl} \rangle = 0$) which is not true for inelastic scattering; S^{fl} does of course contribute to the average differential cross section as it enters quadratically.

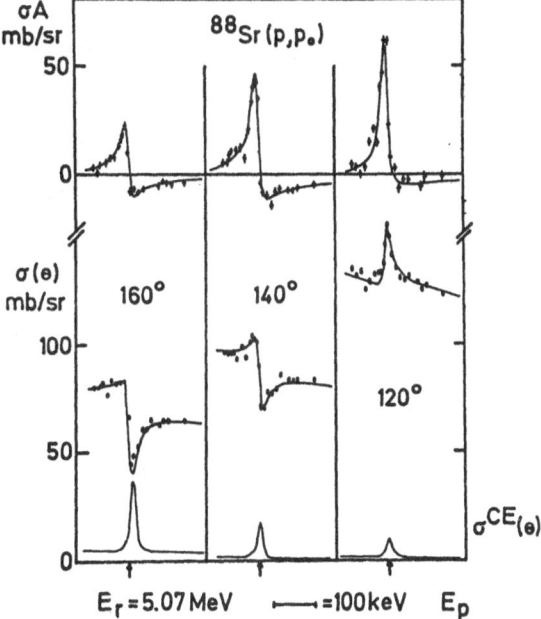

Fig. 2. Excitation functions of σA and σ for ^{88}Sr(p,p_o) near the $d_{5/2}$ g.s. IAR measured by the Erlangen group[7]. The A data were used to determine $\langle S_{cc} \rangle$. The calculation of σ using those parameters differs from the measured σ by the CN cross section σ^{CE} plotted at the bottom

Fig. 2 shows the data of the Erlangen group for elastic scattering from ^{88}Sr [7]. Excitation functions for $d\sigma/d\Omega \cdot A$ and $d\sigma/d\Omega$ are displayed in the vicinity of the 5.07 MeV IAR for three angles. As we have seen, $<S>$ should be taken to explain the $\sigma \cdot A$ data. This procedure - given an optical model - fixes the IAR parameters. Subtracting the calculated cross section $d\sigma/d\Omega \simeq |<S>|^2$ from the measured one leaves a surplus which we recognize as the compound-elastic cross section, exhibiting a sizeable enhancement on resonance (fig. 2, bottom). This is despite the fact that two neutron channels are open.

Having a measurement of the CN cross section we are in a position to determine the spreading width entering into the expression given above for T_c [8]. This possibility is important for medium-weight nuclei because of the large enhanced compound-nuclear scattering: a naive analysis of experimental data using $<S>$ only - as it is met frequently in literature - yields too small values of $\Gamma^\downarrow = \Gamma - \Gamma^\uparrow$ (because of too big values of $\Gamma^\uparrow$). Thus far, spreading widths have been determined that way for single channel cases only via $\Gamma^\downarrow = \Gamma - \Gamma_{elastic}$. The results[7] for several $d_{5/2-}$ g.s. IAR around Zr reveal a contribution of the line broadening $\Gamma^\downarrow$ to the total widths of 70-85% absolute values ranging between 10 and 20 keV. The method, however, should be applicable also to cases in which many proton (and other) channels are open. It derives some elegance from the fact that information is obtained entirely from an elastic scattering experiment without all the elaborate analysis which is unavoidable when using the $\Sigma\Gamma_c$-approach to $\Gamma^\downarrow$.

Much of the physical interest in the spreading width itself is due to the part played by the nuclear residual interaction as has been emphasized by Mekijan[9] and Auerbach et al.[10] In the region of $A \simeq 90$, the monopole excitation, according to these authors, is responsible for the dominant part in $\Gamma^\downarrow$ with 10-15 keV contribution.

2. CHANNEL CORRELATIONS

There is not much confidence that the Hauser Feshbach formula describes CN scattering in the vicinity of IAR correctly. IAR in general constitute doorways for different proton channels; let us assume two channels c and c'. Upon mixing,

decay amplitudes into these channels are added to the original compound-nucleus
decay amplitudes. Clearly, the resulting $\Gamma_{qc}^{1/2}$ and $\Gamma_{qc'}^{1/2}$ will be correlated so that
$<S_{ac}^{fl} \cdot S_{ac'}^{fl*}> \neq <S_{ac}^{fl}> <S_{ac'}^{fl*}> = 0$ does not any longer vanish.

The physical significance of channel correlations has been stressed in articles
by A.M. Lane[11] and C. Mahaux[12]. Both in high and low-resolution experiments
cases of clear evidence are rare.

I want to discuss an experiment we did in Heidelberg since it seems to be the
only one dealing with energy averaged IAR[13]. As suggested by the arguments just
given, the measurement of the <u>polarization</u> of inelastically scattered protons could
be a sensitive test for channel-channel correlations[37].

In the Hauser-Feshbach approach the compound-inelastic scattering may show
enhancement of the cross section but no polarization. In the case of correlated
channels $\sigma P \neq 0$ should be expected due to interfering decay channels in S^{fl}. It
should be stressed that the more convenient measurement of σA does not yield the
information required as it is insensitive to exit-channel interference. This has
been shown by Harney[14].

The experiment imposed several requirements on the IAR in order that a measured polarization - or a sizeable fraction of that can be conclusively ascribed to
compound-nuclear polarization $P^{CN} \simeq Im <S_{ac'}^{fl*} \cdot S_{ac'}^{fl*}>$ as arising from IAR-induced
correlations between channels c and c':

   (i)   large enhancement of the CN scattering;

   (ii)   decay via two different channels c, c';

   (iii)   greatly differing phase shifts in channels c and c', that is $\ell_c \neq \ell_{c'}$,
       in order that the imaginary part of the interference expression be
       large[15];

   (iv)   small polarization from $<S>$ due to direct scattering or nearby IAR.

We choose the $J = 3/2^+$ IAR at 7.61 MeV showing up strongly in inelastic scattering
$^{90}Zr(p,p')^{90}Zr$ to the 2.18 MeV 2^+ state[16]. The resonance is just at neutron
threshold and situated on a large compound-nuclear background.

The double-scattering experiment was performed using a carbon polarimeter
inside a scattering chamber (fig. 3) positioned at 135° left or right of the beam.

Two double-scattering spectra displaying asymmetries are shown in fig. 4, detail:
of the experiment and data analysis are described elsewhere[13].

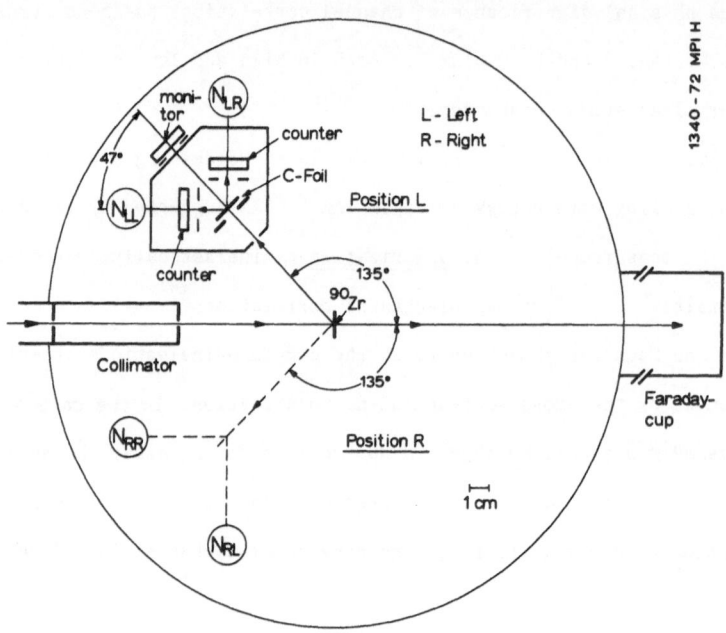

Fig. 3. Double-scattering arrangement for ^{90}Zr(p,p') using a carbon polarimeter

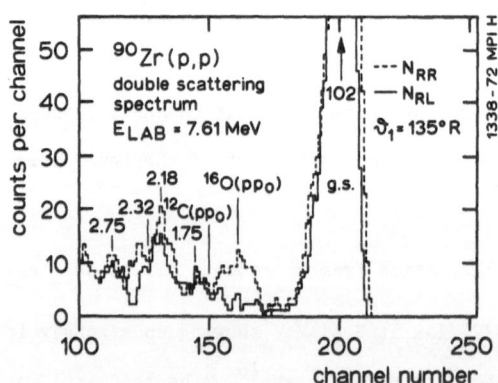

Fig. 4. Two double scattering spectra for left and right scattering from the
10 mg/cm^2 carbon foil

The measured polarization is (19 ± 5)%. Calculations of direct inelastic scattering using DWUCK[17] give less than 3% polarization. Calculation of P arising from the resonating part of $<S>$ requires knowledge of the "mixture" of various partial width amplitudes $\Gamma_{c'}^{1/2}$ for the decay to the 2^+ state. (p,p'γ)-experiments yielding that information are currently being analysed[18]. However, the following arguments are given to show that the size of the measured polarization is extremely difficult to explain on the basis of $<S>$. Let us denote the background cross section (assumed purely CN) on resonances by σ^{back}, by σ^{res} the resonance cross section above background and by σ^{dir} that part of σ^{res} which is due to $<S>$. Then in the absence of non-statistical effects, e.g. $P^{fl} = 0$, an effective direct polarization

$$P_{eff}^{dir} = P^{dir} \cdot \sigma^{dir} / (\sigma^{res} + \sigma^{back})$$

will be observed. To make a numerical estimate, we take the observed ratio $\sigma^{res}/\sigma^{back} \simeq 1$ and assume $\sigma^{dir}/\sigma^{res} \simeq 30$% and $P^{dir} \simeq 50$% [19]. This yields $P_{eff}^{dir} \simeq 7.5$% and is positive evidence for a significant contribution of P^{fl} to the measured value.

Very recently, the topic of average cross sections in the presence of direct reactions has found theoretical revival by Kawai et al.[20] and by Hans Weidenmüller and C. Engelbrecht[21]. The new theories can be used to calculate P^{fl}. One serious problem encountered when dealing with IAR as emphasized in the talk of Zeev Vager are the level-level correlations. In any case, the issue seems to be important enough to ask for more experiments.

A possibility to avoid the difficult and time consuimg double-scattering experiments has been suggested by Boyd et al.[22] combining polarized-beam with angular correlation p'-γ-experiments. In fig. 5 the measurable quantities using a polarized beam are listed in terms of basic cross sections $\sigma(m,m')$ specifying the proton spin components along an axis perpendicular to the reaction plane. $S(\theta)$ denotes the spin-flip probability and $\Delta S(\theta)$ the spin-flip asymmetry.

The equations can of course be solved for P yielding

$$P = A - 2S (\Delta S/S)$$

Polarization, Analyzing Power, etc.

1060-73 MPI H

$$2\, d\sigma/d\Omega \,(\Theta) = \sum_{m,m'} \sigma(m,m'; \Theta) = \sigma(\Theta)$$

$$\sigma(\Theta) = \sigma(\uparrow\downarrow,\uparrow\downarrow)$$
$$P_Z(\Theta)\,\sigma(\Theta) = \sigma(\uparrow\downarrow,\uparrow) - \sigma(\uparrow\downarrow,\downarrow)$$
$$A(\Theta)\,\sigma(\Theta) = \sigma(\uparrow,\uparrow\downarrow) - \sigma(\downarrow,\uparrow\downarrow)$$
$$S(\Theta)\,\sigma(\Theta) = \sigma(\uparrow,\downarrow) + \sigma(\downarrow,\uparrow)$$
$$\Delta S(\Theta)\,\sigma(\Theta) = \sigma(\uparrow,\downarrow) - \sigma(\downarrow,\uparrow)$$

Fig. 5. Expressions for polarization P, analysing power A, spin-flip probability

S and spin-flip asymmetry ΔS in terms of basic cross section $\sigma(m,m')$;

the proton spin components along k x k' are denoted by arrows, summation

$\sum_{m,m'}$ by double arrows

The way this relation is written is appropriate for the experiment. The spin-flip

asymmetry ΔS/S can be measured as sketched in Fig. 6 by detecting the gamma rays

from the $2^+ \to 0^+$ transition perpendicular to the scattering plane in coincidence,

and reversing the spin direction of the beam.

$$P_Z(\Theta)\,\sigma(\Theta) = A(\Theta) - 2\,S\,(\Delta S/S)$$

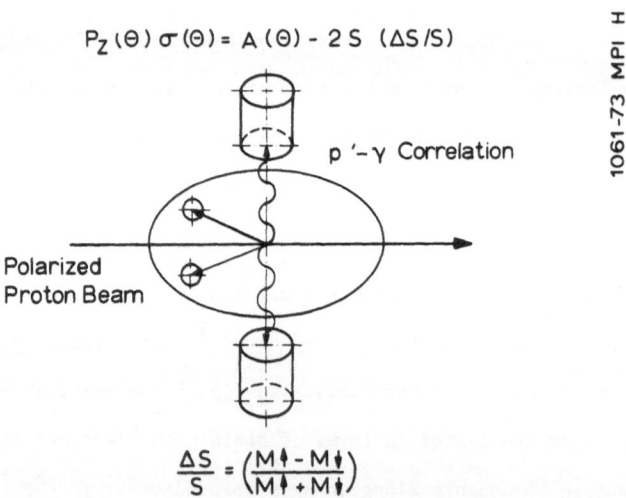

p'-γ Correlation

1061-73 MPI H

Polarized
Proton Beam

$$\frac{\Delta S}{S} = \left(\frac{M\uparrow - M\downarrow}{M\uparrow + M\downarrow}\right)$$

Fig. 6. Schematic view of a p'- correlation experiment with polarized beam to

measure the relative spin-flip asymmetry S/S. M , M are the coincident

rates for either spin direction of the beam

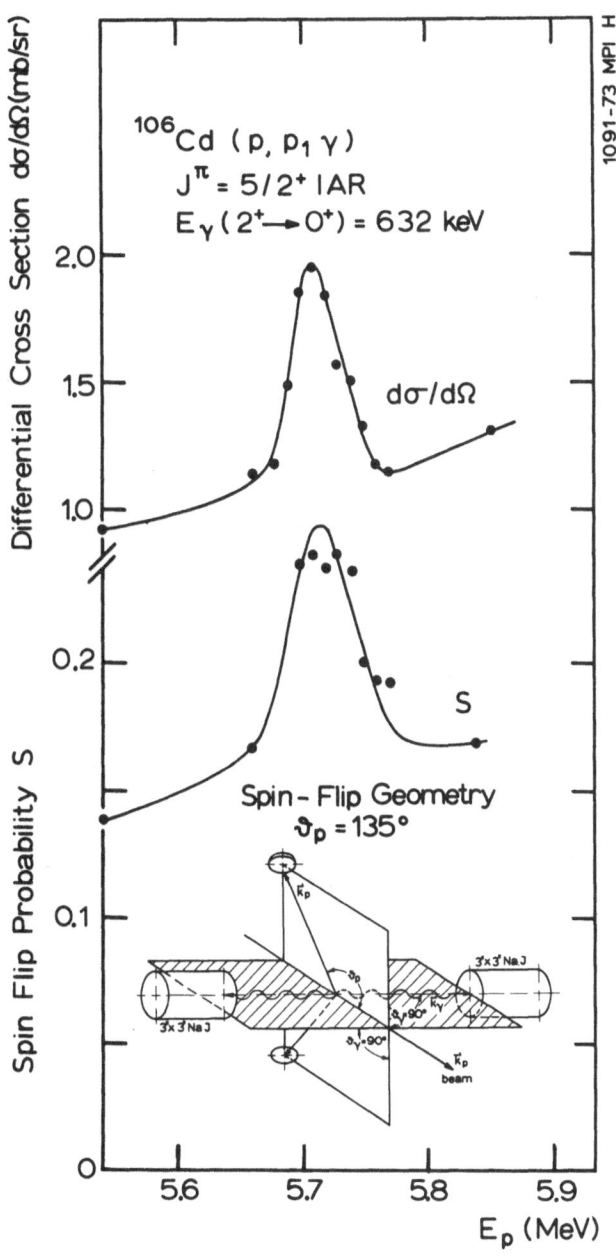

Fig. 7. Excitation function for dσ/dΩ and S measured by (p,p'γ) with NaI-
detectors perpendicular to the scattering plane. The partial widths of
this IAR for 2⁺-decay are known from (p,p'γ) using a different geometry
(ref. 1)

One point measured on resonance in principle suffices since $\Delta S/S$ vanishes for the off-resonance CN background. The determination of S with an unpolarized beam needs the measurement of an excitation function with absolute gamma-ray detection efficiency. Such data is displayed in Fig. 7 together with the (p,p'γ) setup. We have measured spin-flip excitation functions for a number of resonances which seem to be good candidates for the question of CN channel correlations.

3. ENHANCEMENT OF DIRECT BACKGROUND

Many open neutron channels and large $B(E\lambda)$ values for direct excitation tend to produce a large direct background. IAR then may provide a very interesting example of what may be termed "enhancement" of the direct scattering or "dynamical interference"[23]. As in the case of CN-enhancement, the disregard of it may prove quite hazardous for spectroscopy, since all partial widths $\Gamma_{c'}$, $c'\neq c$, may contribute to and enhance the cross section in channel c.

A good candidate seems to be the octupole excitation of ^{208}Pb (Fig. 8). In the region of the lowest IAR, the (p,p) and (p,p') excitation functions are very similar[24]. An extreme enhancement situation is encountered in the case that the inelastic resonance structure is entirely due to the elastic one combined with elastic-to inelastic direct coupling.

A graphic presentation is given in Fig. 9. Amplitudes (1) and (2) describe the resonant and direct scattering - each one in the absence of the other. For some IAR a coherent superposition of (1) and (2) has been treated numerically[1,25,26]. In addition, we consider the two-step amplitudes (3) and (4) where resonant and direct scattering follow each other in either sequence.

A <u>very</u> rough estimate of the size of (3) and (4) can be obtained as follows

$$\sigma_{cc'}^{(3)} \simeq \sigma_{cc}^{res}/(\sigma_{cc}^{res} + \sigma_{cc}^{back}) \cdot \sigma_{cc'}^{dir}$$

$$\sigma_{cc'}^{(4)} \simeq \sigma_{cc'}^{dir} \cdot \sigma_{c'c'}^{res}/(\sigma_{c'c'}^{res} + \sigma_{c'c'}^{back})$$

Using the cross sections of Fig. 8 on the g.s. IAR we obtain[27]

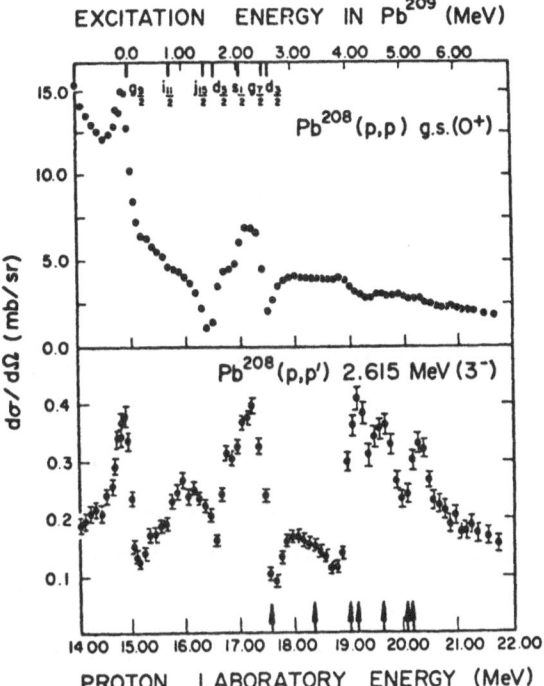

Fig. 8. Elastic and inelastic proton scattering leading to the 2.615 MeV octupole vibration state of ^{208}Pb. From ref. 24

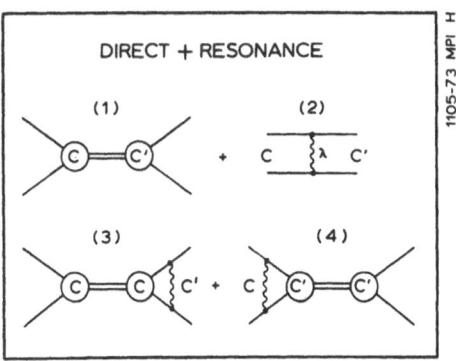

Fig. 9. Two-step amplitudes (3) and (4) in addition to the resonance amplitude (1) proportional to $\Gamma_c^{1/2} \Gamma_c^{1/2}$, and the direct (DWBA) amplitude (2) exchanging a phonon of multipolarity λ

$$\sigma^{(3)} \approx 0.1 \text{ mb}$$

$$\sigma^{(4)} \approx 5 \cdot 10^{-4} \text{mb}$$

which is to be compared with $\sigma^{(1)} \approx \sigma^{(2)} \approx 0.2$ mb.

We conclude that process (3) may be very important whereas (4) is here entirely negligible. It should be noted, that the importance of these terms is not necessarily connected to the accuracy to which (2) can be described in DWBA.

The formal development of the "direct enhancement" is an application of Feshbach's unified theory and has been specialized to IAR[28,10] and discussed in shell-model theories[5,29]. Numerical calculations have not been done to my knowledge as the formulas are not very practical for that purpose. Though final results are not yet available, it may be worthwhile to give a few relations to outline the method (the diagrams of Fig. 9).

Since we start from a two-potential formula we have two alternatives[30]:

(i) The modification of the partial width amplitudes

$$\Gamma_{c'}^{1/2} = <X_{c'} |U_1| \phi_n> \quad ;$$

$X_{c'}$ contains the direct transitions in Born approximation

$$X_{c'}(r) = \int G_{c'}(r,r') V_{c'c}^I(r') \, X_c^o(r') dr'$$

$$G_{c'}(r,r') = X_{c'}^o(r_<) \, 0_{c'}(r_>)$$

(ii) The modification of the direct reaction (DWBA) S matrix:

$$S_{cc'}^I = \int X_{c'} \, V_{c'c}^I \, X_c^{res} \, dr$$

by putting a one-level resonance into channel c (say):

$$X_c^{res}(r) = X_c^o(r) - \Lambda(E) \int G_c(r,r') \, U_1(r')\phi_n(r') \, dr'.$$

$\Lambda(E)$ resonates and contains Γ_c of the unmodified IAR.

X_c^o, $X_{c'}^o$ are the regular solutions to the homogeneous Schroedinger equation in channel c,c' . 0_c is the irregular outgoing-wave solution.

4. INTERMEDIATE STRUCTURE (DIFFERENT FROM IAR)

A case of intermediate structure not being related to isospin conservation has been proposed by Mittig et al. at Saclay[31]. Clustering of resonances was observed in ^{40}Ca + p around 7.2 MeV excitation energy in ^{41}Sc in the inelastic decay to the 3.73-MeV 3$^-$ state (Fig. 10). The authors suggested a doorway configuration (3$^-$ x p$_{1/2}$)J = 5/2$^+$ obtained by weak particle-core coupling and report a spreading width of 200 keV.

Whether the observed gross structure is due to a common doorway should be possible to decide by looking at common properties of the fine-structure resonances: (i) the mixture of partial decay amplitudes present in the doorway configuration should be reproduced on the average by the individual resonances. (ii) mixing should occur to J = 5/2$^+$ states only. These criteria should apply to isolated resonances. The fact that the calculated spacing D $\approx$ 30 keV [32] of J = 5/2$^+$ levels is not more than a factor 2 - 3 smaller than that of the observed resonances is in favour of the assumption made in Ref. 31 that states with 2p-1h structure are picked out; $\Gamma_q < D_q$ is therefore probably justified.

A (p,p'γ) angular correlation experiment is well suited to answer (i) since the angular correlation displays more complexity than the (p,p') angular distribution[33]. In addition, using the geometry of Goldfarb and Seyler[34] to measure the azimuthal dependence of the gamma-ray angular distribution, the spin values of the resonances may be determined. On top of the prominent resonances marked in Fig. 10 the coincidence rate was measured as a function of ϕ_γ (Fig. 11) [35]. The individual angular distributions lie within a fairly narrow band. This may be taken as evidence for the intermediate-structure interpretation that a single configuration is spread over many actual nuclear levels. Within the experimental errors, J = 5/2 is found for all except one or two weaker resonances. A full excitation function using a different geometry and poor beam resolution is currently being analysed. Since the interferences of nearby resonances are expected to cancel out upon averaging, the average correlation data - for any geometry chosen - should not change over the range of the gross structure. A similar observation has been made long ago for (p,γ) angular distributions being constant over the range of E1 giant resonances in light nuclei[36].

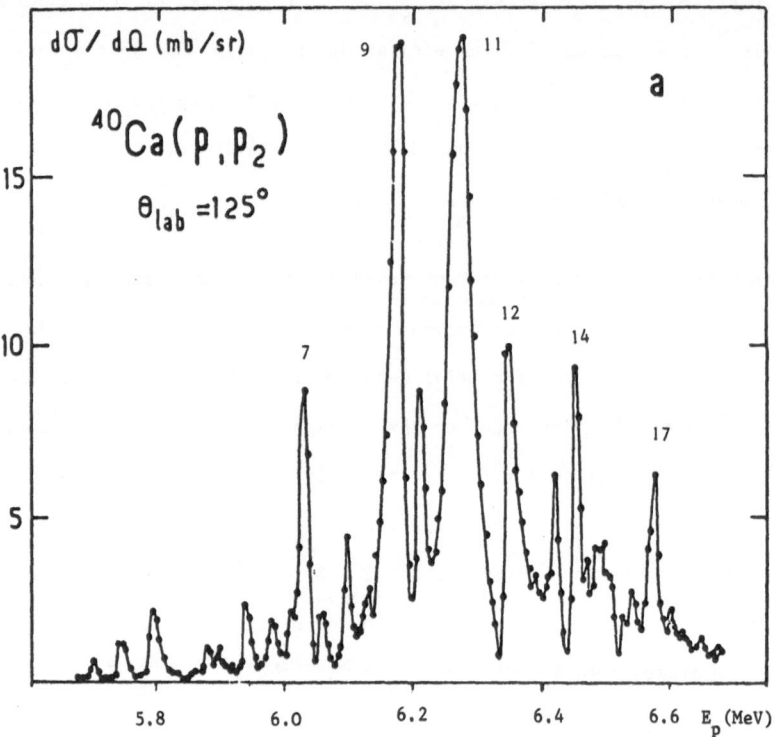

Fig. 10. Intermediate-resonance structure observed by Mitting et al. (ref. 27) in inelastic proton scattering to the lowest 3⁻ state in ^{40}Ca. Numbers indicate resonances on which (p,p'γ) measurements have been performed

5.   CONCLUDING REMARKS

I want to make two comments.

(i)   The detailed studies of the interplay of various reaction modes may have some important applications elsewhere. One field where CN scattering, intermediate structure and strong direct scattering may be present simultaneously is that of heavy ion reactions. The consideration of both CN and direct enhancement is evidently important when searching and interpreting structures in excitation functions as quasi-molecules etc.

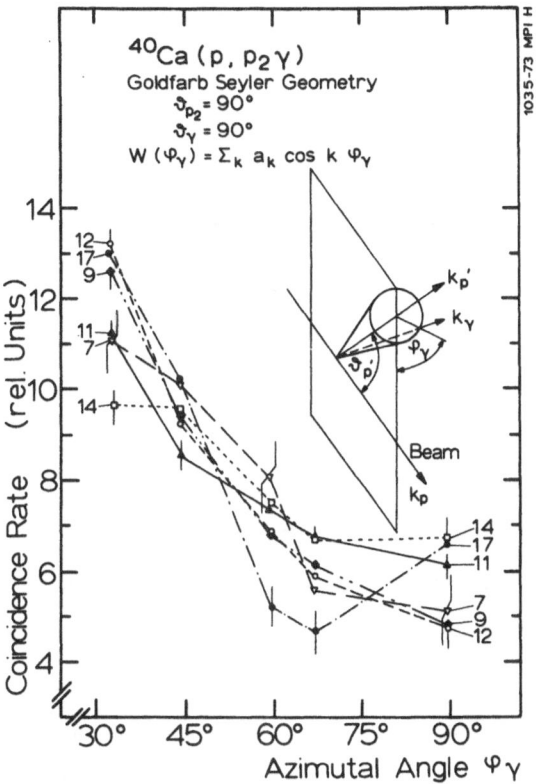

Fig. 11. p'-γ angular distributions in Goldfarb-Seyler geometry for ^{40}Ca(p,p$_2$γ). Numbers refer to the resonances marked in Fig. 10

(ii) The second remark relates to the methodical approach to intermediate structure phenomena. One way is the laborious level-by-level spectroscopy producing much more data than needed to learn the average quantities of interest. Alternatively, we may take energy-averaged data. But we take special care to prepare or resolve individual m-states: either of the projectiles by using polarized beams and measuring A; or of the emitted light particles by measuring their polarization; or of the residual nuclei by measuring angular correlations etc. The examples given show that sharp m- resolution may at least compensate for better energy resolution in order to measure CN scattering, determine spreading widths, detect induced channel-channel correlations or find out the simple configurations underlying intermediate structure.

I gratefully acknowledge the collaboration with R. Albrecht, K. Mudersbach and V. Zoran. I also want to thank Hans Weidenmüller and Zeev Vager for interesting discussions.

REFERENCES

1. Abramson, E., Eisenstein, R.A., Plesser, I., Vager, Z., Wurm, J.P.: Nucl. Phys. A144, 321 (1970.

2. For latest reference see article of E.G. Bilpuch in "Statistical Properties of Nuclei", ed. J.B. Carg, Plenum Press, New York - London 1972.

3. Lemmer, R.H. in "Intermediate Structure in Nuclear Reactions", ed. H.P. Kennedy, Univ. of Kentucky Press, Lexington 1968.

4. Feshbach, H., Kerman, A.K., Lemmer, R.H.: Ann. Phys. (N. Y.) 41, 230 (1967).

5. The asymmetry observed in (p,n)-reactions and explained first by Robson was very important for the understanding of IAR; see H.A. Weidenmüller and C. Mahaux in "Shell-Model Approach to Nuclear Reactions", North-Holland Publishing Co., Amsterdam 1969; many resonances display no asymmetries.

6. Graw, G., in "Polarization Phenomena in Nuclear Reactions", eds. H.H. Barshall and W. Haeberli, The University of Wisconsin Press, Madison 1970.

7. Kretschmer, W., Ph.D. Thesis, University of Erlangen 1972, Kretschmer, W., Graw, G.: Phys. Rev. Lett. 27, 1294 (1971).

8. Graw, G., Wurm, J.P.: Verhandlungen der Deutschen Physik. Gesellschaft 3, 209 (1971).

9. Mekjian, A.Z.: Phys. Rev. Lett. 25, 888 (1970).

10. Auerbach, N., Hüfner, J., Kerman, A.K., Shakin, C.M.: Rev. Mod. Phys. 44, 48 (1972).

11. Lane, A.M.: Ann. Phys. (N. Y.) 63, 171 (1971).

12. Mahaux, C., in Conference on Nuclear Structure Study with Neutrons, Hungary 1972.

13. Mudersbach, K., Albrecht, R., Wurm, J.P., Zoran, V.: MPI Heidelberg Annual Report 1971 and 1972 (unpublished) and to be published.

14. Harney, H.L.: Phys. Lett. 28B, 249 (1968).

15. Unitarity relates P^{CN} to S in the following way:

$$P^{CN} = IM \, \Sigma_a < S^{fl}_{ac} \, S^{fl*}_{ac'} > = - IM \, \Sigma_a < S_{ac} > < S^*_{ac'} >$$

In order that the r.h.s. be large, $\sin(\delta_c - \delta_{c'})$ should be large; δ_c are the optical model plus resonance mixing phases. Close to the Coulomb barrier δ_c is essentially the Coulomb phase so that $\ell_c \neq \ell_c'$ is a necessary condition for large P^{CN}.

16. Lieb, K.P., Kent, J., Moore, C.F.: Phys. Rev. 175, 1482 (1968).

17. Kunz, P.D.: DWUCK - a computer code for DWBA calculations, University of Colorado, 1969 (unpublished).

18. Abramson, E., Albrecht, R.: private communications.

19. A ratio of $\sigma^{dir}/\sigma^{res} = 5\,\%$ has been found from an Hauser-Feshbach analysis for the 5.73 MeV, $J = 3/2^+$ resonance in the $^{106}Cd(p, p\gamma)^{106}Cd(2^+)$ reaction, E. Abramson, I. Plesser, Z. Vager, J.P. Wurm, Contributions to the International Conference on Properties of Nuclear States, Montreal 1969, Les Presses de l'Universite de Montreal 1969, p. 369.

20. Kawai, M., Kerman, A.K., McVoy, K.W.: Ann. Phys. (N. Y.) 75, 156 (1973).

21. Weidenmüller, H.A.: Phys. Lett. 42B, 304 (1972) and Engelbrecht, C.A., Weidenmüller, H.A.: Max-Planck-Institut Heidelberg preprint MPI H-1973-V3, to be published.

22. Boyd, R., Davis, S., Glashausser, C., Haynes, C.F.: Phys. Rev. Lett. 27, 1590 (1971) and Phys. Rev. Lett. 29, 955 (1972).

23. Ratcliff, K.F., Austern, N.: Ann. Phys. (N. Y.) 42, 185 (1967).

24. Stein, N., in "Nuclear Isospin", ed. J.D. Anderson, S.D. Bloom, J. Cerny, and W.W. True, Academic, New York, 1969, p. 481.

25. Arking, R., Boyd, R.N., Lombardi, J.C., Robbins, A.B., Yoshida, S.: Phys. Rev. Lett. 27, 1396 (1971).

26. Ayoub, E.E., Asciutto, R.J., Bromley, D.A.: Phys. Rev. Lett. 29, 18¥ (1972).

27. We read off the values (c for elastic c' for 3^- channels)

$$\sigma^{res}_{cc} \simeq \sigma^{back}_{cc} \simeq 5 \text{ mb}$$

$$\sigma^{dir}_{cc} \simeq 0.2 \text{ mb, assume}$$

$$\sigma_{c'c'}^{res} \simeq (\sigma_{cc'}^{res})^2/\sigma_{cc}^{res} \simeq 0.05 \text{ mb},$$

and extrapolate $\sigma_{c'c'}^{back} \simeq 5 \text{ mb } (15 \text{ MeV})^2/(15 \text{ MeV} - 2.6 \text{ MeV})^2$

$$\simeq 20 \text{ mb}$$

assuming Rutherford scattering.

28. Stephen, R.O.: Nucl. Phys. A94, 192 (1967).

29. Ginocchio, J.N., Schucan, T.H.: Nucl. Phys. A156, 629 (1970).

30. The notation for the DI process is that of N.K. Glendenning, in Varenna Lectures Course 40, ed. M. Jean and R.A. Ricci, Academic, New York 1969, p. 332 and for the resonance scattering that of C. Mahus and H.A. Weidenmüller, Nucl. Phys. 89, 33 (1966).

31. Mittig, W.: Thesis, Orsay (1971) and Mittig, W., Cassagnou, Cindro, N., Papineau, L., Seth, K.K., in "The Structure of $1f_{7/2}$ Nuclei", ed. R.A. Ricci, Editrice Compositori, Bologna 1971.

32. Bohr, A., Mottelson, B.R.: "Nuclear Structure:, Vol. I., Benjamin 1969, p. 155, using a=6/MeV and the rigid-body moment of inertia.

33. Pure $p_{1/2}$-decay would yield an isotropic p' angular distribution; the p'-γ angular correlation could still measure three different angle functions; the complexity is essential when different $\Gamma_{c'}$ mix.

34. Goldfarb, L.J.B., Seyler, R.G.: Phys. Lett. 28B, 15 (1968).

35. Albrecht, R., Zoran, V.: to be published.

36. Allas, R.G., Hanna, S.S., Meyer-Schützmeister, L., Segel, R.E., Singh, P.P., Vager, Z.: Phys. Rev. Lett. 13, 628 (1964).

37. Ruh, A., Marmier, P.: Nucl. Phys. A151, 479 (1970).

THE SHAPE OF THE FISSION BARRIER

Hans J. Specht

Sektion Physik, University of Munich, Germany

1. INTRODUCTION

In the collective model of the nucleus, the nuclear shape can appropriately be described by a few dynamical variables β. We shall concentrate our interest here to the unusual shapes of heavy fissioning nuclei and the dependence of the nuclear potential energy surface on the most important of these variables describing the shape - the quadrupole (elongation) and the octupole (left-right asymmetry) degree of freedom.

Numerous calculations done over the last five years (all employing the so-called Strutinski theorem[1,2]) qualitatively agree in the well known picture of the "double humped fission barrier": Shell corrections, superimposed on top of the liquid drop deformation energy, lead to two minima in the deformation energy, the ground-state minimum and a second minimum at very large quadrupole deformations, located 2-3 MeV above the first and apparently stable against all other types of deformations investigated. They also agree in the prediction of a pronounced left-right asymmetry at the outer barrier. The detailed results of these calculations (covering statics and dynamics) as well as the many problems involved have been discussed extensively in recent reviews[2,3].

Rather than presenting series of "funny hills", we shall therefore restrict ourselves to a short survey of new experimental results related more or less directly to

a) the quadrupole degree of freedom

   (rotational levels and the moment of inertia of fission
   isomers, attempts at higher lying bands and quantum numbers
   of isomeric states, possibilities for measuring the quadrupole moment)

b) the octupole degree of freedom

   (barrier heights for symmetric and asymmetric fission)

As a short introduction, one example[4] of the numerous fission barrier calculations (fig. 1) demonstrates for a number of nuclei between ^{94}Pu and ^{114}X the smooth
behaviour predicted on the basis of the liquid drop model (dashed lines) and the
double-humped character arising from the addition of shell corrections (using in
this case as the average potential of the shell model a simple harmonic oscillator).
For a nucleus with its proton and neutron numbers away from magic numbers (magic
for a spherical shape!), the ground-state occurs at $\beta_2 \approx \epsilon = 0.2 - 0.25$ and the
secondary minimum at $\beta_2 = 0.6 - 0.7$. Mainly due to a decrease of the liquid drop
barrier and a shift to smaller deformations with increasing atomic number Z, there
is a "switch-over" in the relative heights of the two barriers from the first
barrier being the lower one to the opposite - a trend which, in fact, is qualitativ-

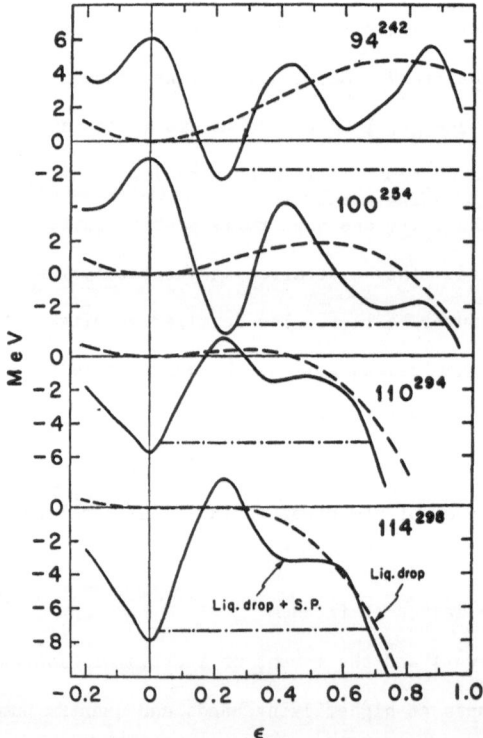

Fig. 1. Example of fission barrier calculations (from Nilsson[4]). The results
of the liquid drop model are shown by the dashed lines, those including
shell correction energies are shown by the full lines

ely (although not in detail[5]) supported by measured barrier heights[6-8] and the region

of occurance of fission isomers (see below).

Following arguments first proposed by Strutinsky[1,2], the remarkable oscill-

atory character of the shell corrections as a function of the quadrupole deforma-

tion - in the same way as in function of the mass number - is caused by variation

of the single-particle level density in the vicinity of the Fermi energy. As is

illustrated (fig. 2) in a schematic single particle level scheme for deformed

nuclei[1,2], single particle energies arrange themselves in bunches or "shells".

Nuclei with a filled shell, i.e. with a level density at the Fermi energy smaller

than the average have an increased binding energy compared to the average, because

the nucleons occupy deeper and more bound states; conversely, larger level density

is associated with a decreased binding energy.

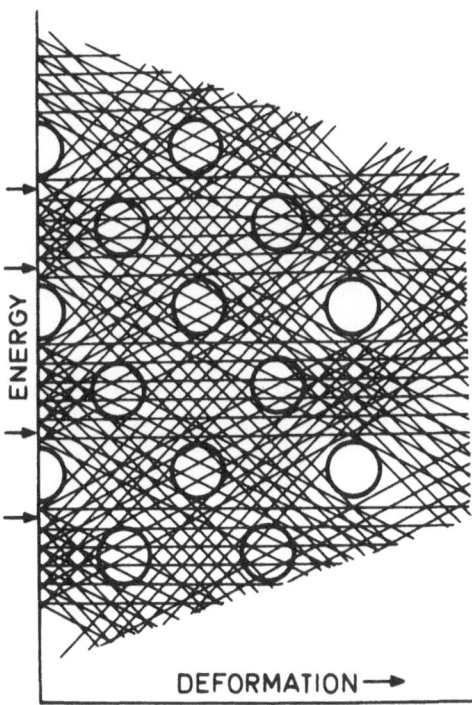

Fig. 2. Qualitative picture of the distribution of single-particle states in a

deformed nucleus. The regions of low level density are indicated by

circles (from Strutinsky[1])

Up to a year ago, experimental evidence supporting the double humped character of the fission barrier has only been indirect - in the sense of providing
information about the energies of the stationary points of the potential energy
surface, but not being sensitive at all to their deformation. This evidence is
based on two major categories of data:

a) the appearance of intermediate structure in fission cross sections[9-11]

   (transmission resonances, mainly observed in direct reaction induced

   fission[5,8,12], grouping of neutron resonances[13] connected with compound

   states in the second well),

b) the existence of spontaneously fissioning isomers, first detected in

   1962[14] and now interpreted as nuclei caught in the second minimum and

   tunneling through the outer barrier.

In this short introduction, we shall only discuss the latter. Following any
nuclear reaction exciting an actinide nucleus to energies above the fission barrier,
the probabilities of prompt fission, decay back into the first minimum and trapping
in the region of the second minimum are of the order of 0.9, 0.1 and 10^{-5} - 10^{-4}
resp.. A detailed investigation of the properties of fission isomers therefore
meets tremendous experimental difficulties, restricting our present knowledge mostly
to identification, half-lives and (in some cases) excitation energy above the ground
state. According to the latest survey[15], the actinide region of the chart of nuclei
(fig. 3) shows 33 isomers between ^{92}U and ^{97}Bk with half-lives ranging from 10^{-11}
to 10^{-2} sec - smaller by 20-30 orders of magnitude than the half-lives for spontaneous fission from the ground state because of the increase in tunneling probability.
Attempts[7,16] to understand the systematics of these half-lives in terms of the
barrier parameters have been moderately successful for the odd and doubly odd nuclei.
The puzzle of the rather constant half-lives for the even-even nuclei has, however,
only recently been solved (at least for the Pu isotopes): Due to improved techniques
for measuring subnanosecond half-lives[17], additional isomeric states with a lower
half-life have been found[17,18] in the same nuclei (two entries in fig. 3). They are
believed to represent the lowest state in the scond minimum and do, in fact, show a
systematic behaviour[18]. The longer-lived states are considered as two-quasiparticle

excitations just above the pairing gap in the second well with some hindrance both towards fission and γ-decay[11,17,18]. This interpretation has now been strengthened by measurements of fragment angular distributions (see below).

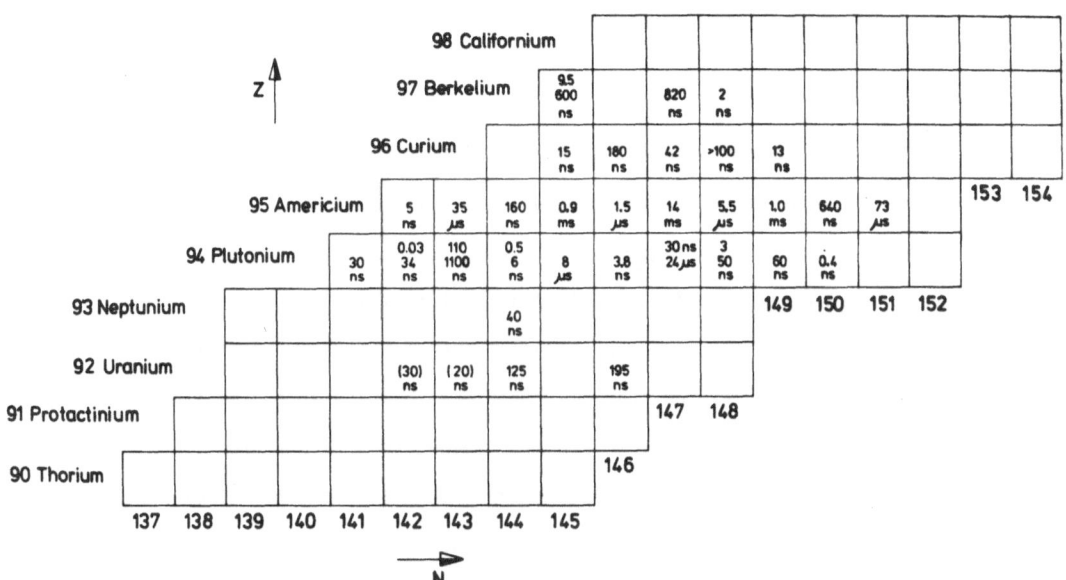

Fig. 3. Actinide region of the chart of nuclei showing the half-lives of all fission isomers known at present. Two numbers in the same field demonstrate the existence of spin-isomers within shape isomers

The sole occurance of fission isomers in this region of the chart of nuclei is in full agreement with the predicted behaviour of the relative barrier heights mentioned above: For Z > 97, the half-lives become too short to be observable with current techniques, for Z < 92 (and all the Np isotopes) the shape isomers presumably decay by γ-ray emission back into the first potential well. Unfortunately however, the investigation of these delayed electromagnetic transitions is even much more difficult than that of delayed fission; all attempts in this direction (including γ-rays[19] as well as e^--e^- delayed coincidences[20]) have so far not been very successful. The clean identification of these transitions remains a major challenge.

2. THE QUADRUPOLE DEFORMATION

Experimental information about the size of the quadrupole deformation associated with the second minimum can in principle be obtained by either identifying low-lying rotational bands in fission isomers (yielding, for example, moments of inertia), or - more directly - by determining the electric quadripole moment. We will discuss some results connected with the former, and only briefly mention the first attempts at the latter.

Following a nuclear reaction, any population of the secondary minimum will lead to electromagnetic transition preceding isomeric fission. Low-lying rotational bands can therefore be investigated by measuring either γ-ray transitions feeding these bands (which will be discussed later), or transitions within such bands. Specifically in an even-even nucleus (fig. 4), the final decay will proceed via E2 transitions within the "ground state" - rotational band built on the isomeric 0^+ level. These transitions should be completely converted (conversion coefficient $\alpha \gg 1$) at least up to the $8^+ \rightarrow 6^+$ transition, due to the smaller level spacing as compared to the normal ground state band. Furthermore, they should have lifetimes $\ll 0.1$ ns (see below), i.e. small compared to the spontaneous fission half-life of nearly all the isomers shown in fig. 3. Such a band can therefore be identified

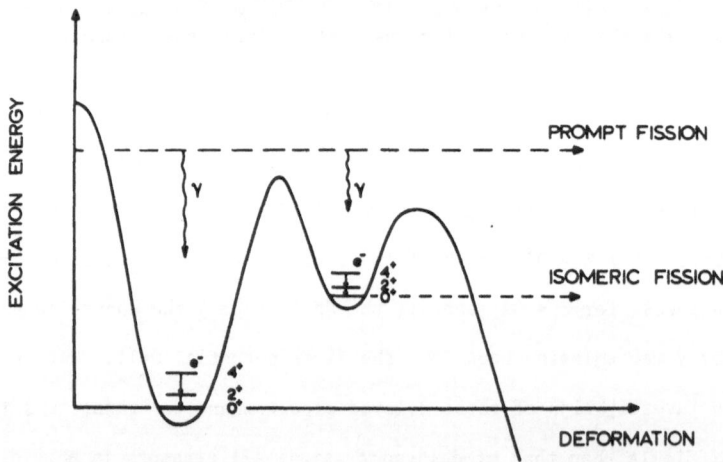

Fig. 4. Schematic drawing illustrating the occurence of electromagnetic transitions (γ-rays and/or conversion electrons) preceding isomeric fission

experimentally by measuring delayed coincidences between conversion electrons and fission fragments. We have performed two experiments of this type, using the reactions ^{235}U $(n_{th},\gamma)^{236m}U$ [21] and $^{238}U(\alpha,2n)^{240m}Pu$ [22,23]; two further experiments are in preparation[19,24].

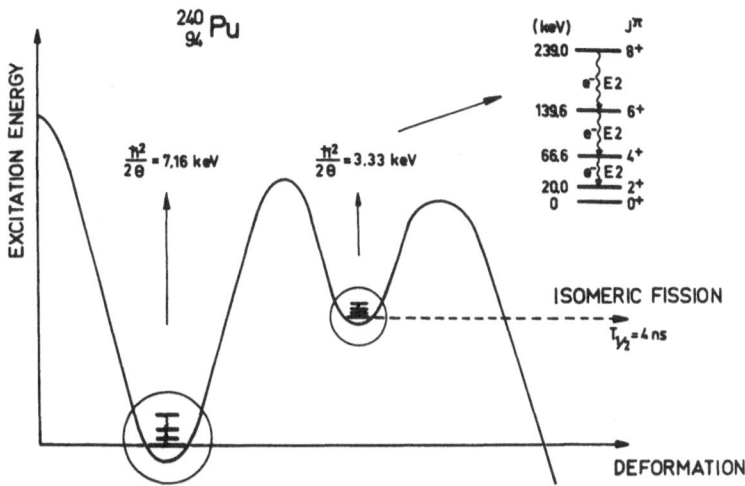

Fig. 5. Experimentally observed transitions within a rotational band built on the fission isomeric level in ^{240}Pu. The rotational constants result from a two-parameter fit to the transition energies (s. fig. 6)

The only rotational band positively identified so far is built on the 4 ns fission isomeric level in ^{240}Pu (fig. 5). Since the details of the experiment (performed at the tandem accelerator in Munich) have been described before[22,23], we will only briefly review the conclusions. The interpretation of the three transitions as a $K=0^+$ rotational sequence of E2 transitions is based on the correct relative energies and intensities of the electron conversion lines within each transition as well as on the absolute transition energies and intensities. The very large total intensities per isomer formed (0.7 for the $4^+ \rightarrow 2^+$ transition, for example) are, in fact, typical for "ground-state" rotational bands. A least squares fit of the transition energies (lower part of fig. 6) to the well known expansion of the rotational energies in powers of the quantity $I(I+1)$

$$E_I = A\ I(I+1) + B\ I^2\ (I+1)^2$$

yields a rotational constant A = 3.33 keV, which is much smaller than the value 7.156 keV[25] of the normal ground-state rotational band (upper part of fig. 6).

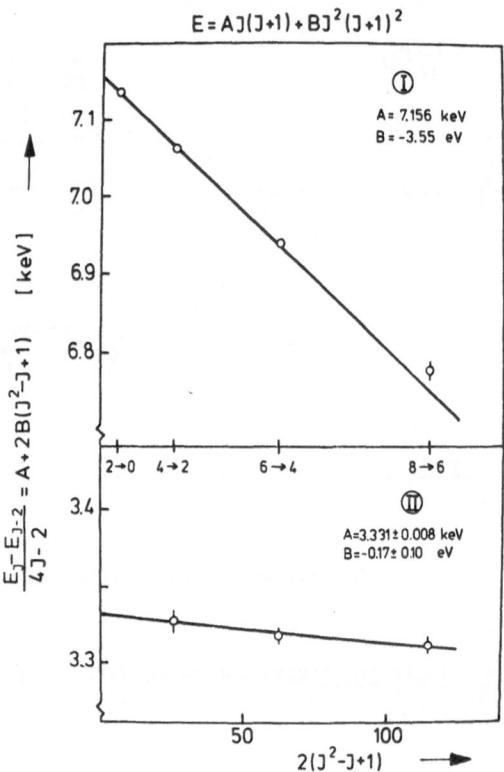

Fig. 6. Two parameter fit of the transition energies within the ground state rotational band (I) and the fission isomer rotational band (II) to the rotational energy dependence given on top of the figure (from ref. 22))

An even stronger decrease is found in the "non-adiabatic" parameter B = -0.17 eV compared to the value -3.55 eV for the ground state band. In a sense, the unusually low value of the rotational constant A which corresponds, to our knowledge, to the largest moment of inertia ever found in nuclei, presents the first direct evidence that isomeric fission is, in fact, connected with a nuclear deformation larger than the equilibrium ground state deformation. In fig. 7, the experimental moments of inertia are compared to their calculated deformation dependence according to the rigid rotor, the irrational fluid and the cranking model[11,26]. Assuming a quadru-pole deformation of 0.6 for the second minimum[2-4], the agreement with the values from the cranking model is excellent; further independent information about the size of the quadrupole deformation would, however, be needed to decide whether a pairing strength proportional to the nuclear surface area, $G \sim S$, should be preferred over a constant pairing strength, G = const. In any case, the conclusion of a change in deformation appears to be rather model-independent, since the large experimental value for the moment of inertia associated with the fission isomer even exceeds the rigid body limit at the ground-state deformation (fig. 7).

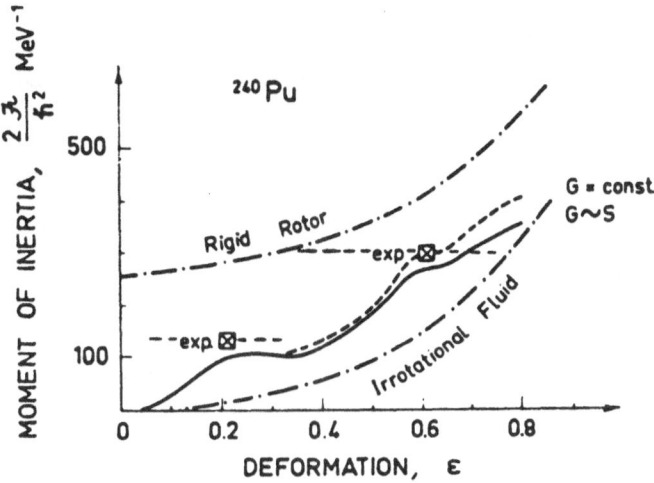

Fig. 7. Comparison of experimental and theoretical values of the moment of inertia associated with the lowest configuration of ^{240}Pu (from Bjørnholm[11,26])

The interpretation of the rotational term B presents an interesting problem in itself. On the basis of the cranking model, the rotational energy (and the moment of inertia) may also be expanded in powers of the rotational frequency ω rather than angular momentum[27]; the superiority of this expansion has, in fact, recently been demonstrated[28]. Using again only two parameters

$$E = \alpha \, \omega^2 + \beta \, \omega^4$$

(or, equivalently, the description by the two-parameter VMI-model[29]), an inter-
relation between the different constants of the type

$$B \sim - \beta \cdot A^4$$

can be obtained. Interestingly enough, the strong decrease found in the parameter
B is - within the limits of the statistical accuracy - compatible with such a fourth-
power dependence on A; the value of β , in other words, seems to be the same for
the two rotational bands. It remains unclear, however, whether this constancy of
β (or C in the VMI-model) can simply be interpreted as the rotation-vibration interaction ("centrifugal stretching") being the same in the two potential wells, - just
as the superiority of the ω^2-expansion itself has not yet been fully understood[28].

A possible identification of further low-lying rotational bands in fission
isomers could - in addition to moments of inertia - provide information about other
collective degrees of freedom in the region of the second minimum, for example about
γ - or octupole - vibrations. Unfortunately, however, the measurement of γ-rays
preceding isomeric fission (fig. 4) is even much more difficult than that of conversion electrons from the lowest band; previous attempts[30,31] have, in fact, been
unsuccessful.

Using the ^{238}U $(\alpha,2n)^{240}$Pu-reaction, we have - in parallel to the electron
experiment and with the same recoil technique[22,23] - tried to investigate these

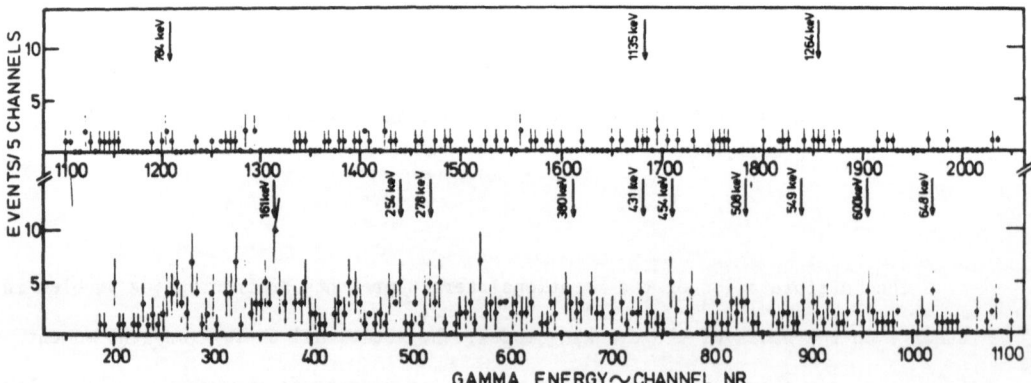

Fig. 8. Energy spectrum of γ-rays preceding isomeric fission from the
^{238}U$(\alpha,2n)^{240m}$Pu reaction

γ-rays by additional installation of a Ge(Li) detector[23,32]. The resulting γ-energy

spectrum (fig. 8) contains a number of structures (marked with arrows), whose summed

time spectrum of coincidences is correctly (i.e. as expected from the 4 ns isomeric

half-life) shifted relative to the prompt coincidence spectrum. However, the weak

relative intensities of at most a few percent/isomer at these positions - although

consuming a running time of nearly two weeks - might well be caused by statistical

fluctuations of a near - continuum of many weak and overlapping γ-transitions; this

interpretation is supported by simultaneous measurements[23] of coincidences between

γ-rays and conversion electrons from the ground-state band in the first potential

well.

As compared to the (α,2n)-reaction, the situation might be more favourable for

a γ-spectrum resulting from thermal neutron capture. We have therefore also invest-

igated delayed coincidences between γ-rays and fission fragments from the

$^{235}U(n_{th},\gamma)^{236}U$-reaction[33], using a 40 cm^3 Ge(Li)-detector at a neutron guide tube

of the Munich reactor to decrease the γ background down to the fundamental limits

caused by γ-rays from fragments. The total γ-energy spectrum (from a 10 days run)

corresponding to a time window of 32 - 190 ns before fission is shown in the upper

part of fig. 9, the spectrum corrected for the (dominant) contribution from chance

coincidence events in the lower part. Each position marked by brackets contains

approx. 250 ± 80 evemts; the summed time spectrum of coincidences (lower part of

fig. 10) shows a decrease consistent with the expected[15] isomeric half-life of

115 ns. Nevertheless, it is much too early to speculate about the significance

of the structures around 400 - 440 keV (whose energy separation is ≈ 20 keV, i.e.

equal to the $2^+/0^+$ splitting). The preparations for an improved set-up have just

been finished, and the experiment is now rescheduled for a running time of several

months.

As mentioned above, the quantity most directly related to the quadrupole

deformation of a nucleus is its intrinsic quadrupole moment. Any experiments in

this direction have to deal with such enormous difficulties, that no success has

been reported yet. We will shortly discuss two of the possibilities which we have

investigated so far - a direct measurement of the life-time of rotational transitions

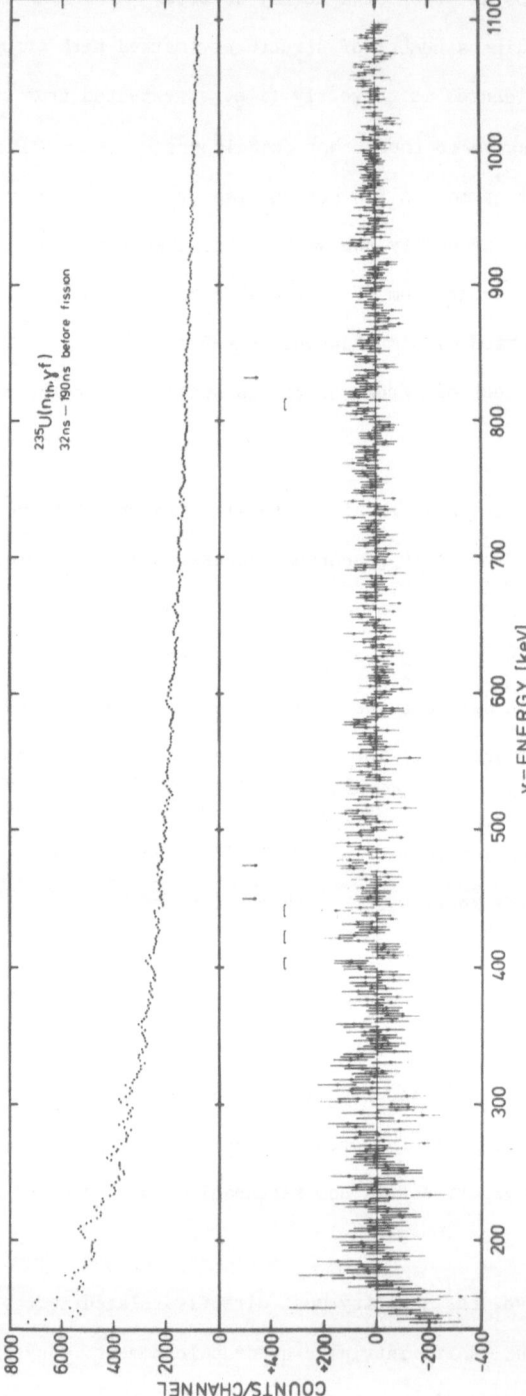

Fig. 9. Energy spectrum of γ-rays preceding isomeric fission from the ^{235}U(n_{th},γ) ^{236m}U reaction. Upper part: all events, lower part: chance coincidence events subtracted

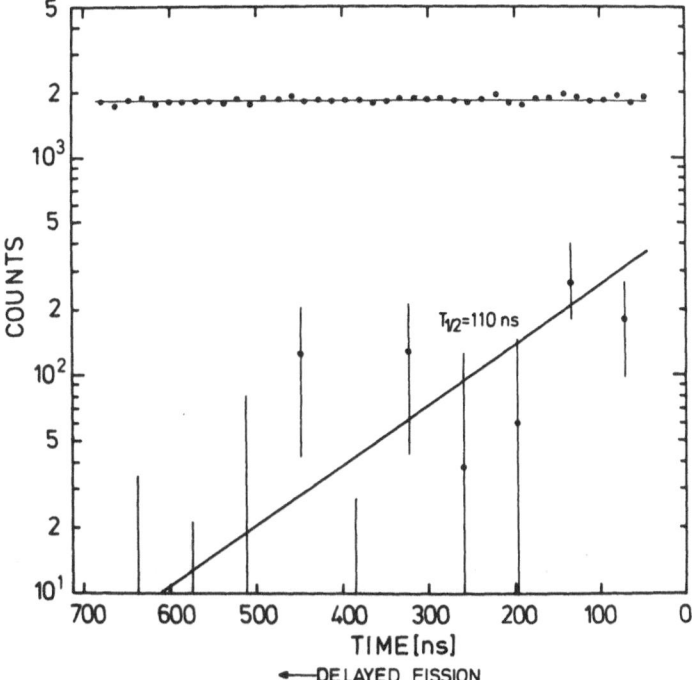

Fig. 10. Time spectrum of the sum of all events corresponding to the brackets
indicated in fig. 9. Upper part: all events, lower part: chance
coincidences subtracted

within the isomeric "ground state" band giving the B(E2)-value, and the application
of perturbed fragment angular correlation techniques.

Life-times estimates for the rotational transitions - based on an expected
increase of the quadrupole moment by a factor of 4 compared to the first minimum[34]
with a corresponding E2 enhancement factor of roughly 4000 ($\approx$ 250 for the ground
state band) - yield values between 6 ps ($8^+ \rightarrow 6^+$) and 19 ps ($2^+ \rightarrow 0^+$, no L-conversion).
Half-lives of this order can in principle be measured by a modified "recoil-distance"
method provided, the recoil velocity is sufficiently high. Unfortunately, only
heavy ion reactions populating the fission isomer (not the (α,2n)-reaction) lead to
such velocities. We have therefore started to systematically investigate prompt
fission and isomer production following transfer reactions with heavy ions on
actinide targets[35]. From that it appears, however, that both cross sections and

beam currents attainable at the moment are too low to perform such an experiment
in any reasonable length of time; we will probably have to await the new generation
of heavy ion accelerators.

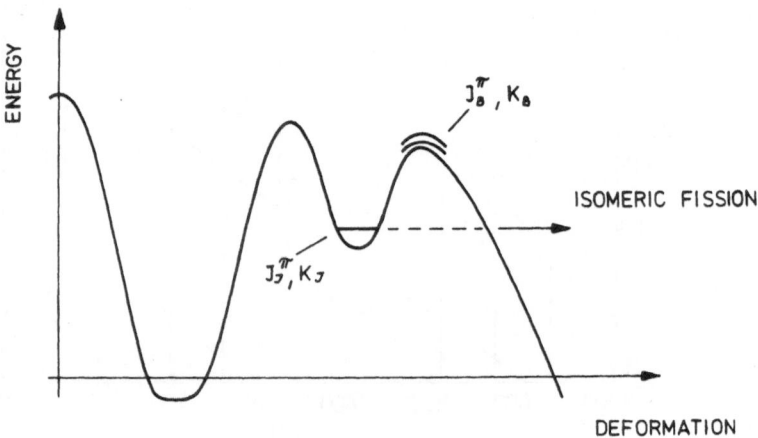

Fig. 11.   Illustration of the relevant quantum numbers of the isomeric state and
the lowest saddle state influencing the fragment angular distribution
in isomeric fission

The second alternative is based on the measurement of fragment angular
distributions in fission from isomeric states, aligned by preceding nuclear
reactions.   Investigations of this type (fig. 11) would allow to determine the
spin I_I of the isomeric state - either the "ground state" of an odd nucleus or a
spin-isomeric state in an even nucleus - if some independent information (from
induced fission) is known about the quantum number K_B of the lowest band at the
second barrier containing the appropriate spin and parity $I_B = I_I$ (it remains unclear at the moment whether K also should be conserved, i.e. $K_B = K_I$).   Apart from
the spectroscopic interest in this knowledge in itself, a perturbation of such
angular distributions could, at least in principle, provide information about the
g-factor or perhaps the quadrupole moment of the isomeric state (using an external
magnetic field or the interaction between the quadrupole moment and the electric
field gradient in an anisotropic lattice, resp.).

As a first step in this direction, we have tried to measure the 0°/90° anisotropy in isomeric fission of Pu-isotopes following the (α,2n) reaction[36]. The recoiling nuclei are stopped in Pb-absorbers of a multilayer U/Pb sandwich target in order to preserve the alignment up into the ns-region. The delayed fragments (fig. 12) are detected with an annular detector at 180° (relative to the α-beam) and two additional semiconductor detectors at 90° in coincidence using the pulsed beam from the Munich tandem accelerator. The results obtained so far for the σ (0°)/σ(90°) anisotropy in the c.m. system are the following:

^{235}U (α,2n) ^{237m}Pu (110 ns)   0.98 ± 0.03

^{236}U (α,2n) ^{238}Pu (6 ns)   0.68 ± 0.18

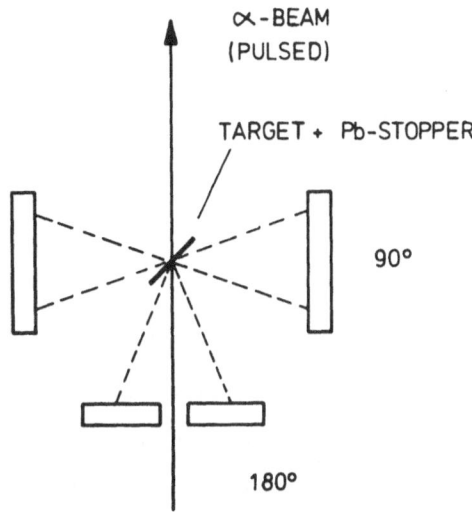

Fig. 12. Sketch of the experimental arrangement for measuring fragment angular distributions in isomeric fission

The complete isotropy in the case of ^{237}Pu could be caused either by a deorientation of the nuclear alignment within the 110 ns life-time or a spin $I_I = 1/2$ of the isomeric state. In the case of ^{238}Pu, the observed anisotropy <1 can be explained only by $I_I = K_B$. The bare existence of an anisotropy in this ee nucleus supports the hypothesis of K-isomerism within the second minimum mentioned in the

introduction (see above) and, in addition, allows the conclusion that the dominant
branch of the isomeric fission decay in this case proceeds via fission from the
isomeric state itself, not from the "ground state" of the second minimum after a
preceding γ-transition.

Using an improved set-up with a two-dimensional cylindrical multiwire counter
for the fragments, further reactions are presently being investigated. In view of
the tremendous experimental difficulties, however, the prospect for obtaining
quadrupole moments in the way described above cannot be judged with any optimism.
Fragment anisotropies have also recently been studied[37] with plastic foils as
fission track detectors; this technique is, of course, inherently not suitable for
time-differential perturbation measurements.

3. THE OCTUPOLE DEFORMATION

One of the historically most important problems in nuclear fission has been
to understand the character of the fragment mass distribution. As illustrated
schematically in fig. 13, the transition from a purely symmetrical distribution
(for nuclei with $Z < 85$) to the well known asymmetrical distribution (for the
heavier actinides) occurs somewhere between Ra and Pa. Measurements of, in fact,
triple-humped mass distributions in this region[38-43], systematic differences in
fragment total kinetic energies and their variances[39-43] and finally prompt neutron
emission[41] have been interpreted in terms of two different fission modes[44], symmetric and asymmetric, which apparently coexist in the same nucleus and compete
with relative probabilities depending on excitation energy (dotted lines in fig.13)
and fissioning nucleus. So far, however, it remained undecided experimentally
whether these modes are associated with different fission barriers.

As already mentioned in the introduction, the shell model calculations[2,3]
for the actinide nuclei all agree in the prediction of a pronounced left-right
asymmetry at the outer barrier with a potential energy dependence on the octupole
degree of freedom β_3 (fig. 14 [45], cut "Asym" along the dotted line) which looks
just like that of the familiar NH_3 molecule. The character of the fragment mass
split is then explained as a consequence of either this asymmetrically (pear-shape

FRAGMENT MASS DISTRIBUTIONS

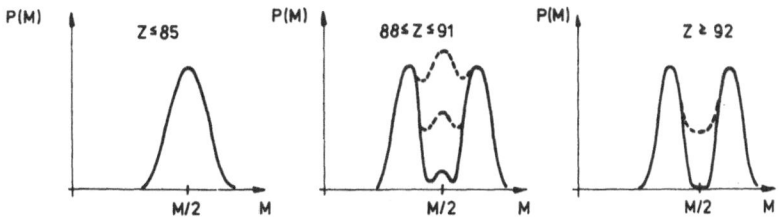

Fig. 13. Qualitative picture of fission fragment mass distributions for different

fissioning nuclei of atomic number Z

like) or a symmetrically (for Z < 85) distorted <u>outer</u> barrier, with slight hints

even for two different saddles in the same nucleus[3,46].

To obtain a crucial test for these calculations, we have measured the fission

probability and the fragment anisotropy (presumed to be determined at the barrier,

see above) close to the fission threshold, separately for the two mass components[47].

Such a study is not feasible for the heavier actinide nuclei both because of the

inner barrier being the higher one (see above) and the extremely low relative yield

($< 10^{-3}$) of the symmetric component close to the barrier; earlier attempts in this

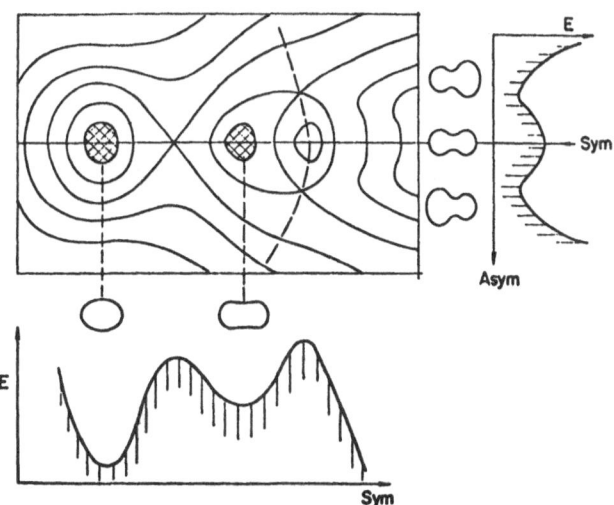

Fig. 14. Schematic drawing of the nuclear potential energy surface as a function

of symmetric and asymmetric deformation (from ref. 45))

direction[48-49] have, in fact, been unsuccessful. We have therefore selected Ac-isotopes, i.e. nuclei well within the triple-humped region, for which the outer barrier should be the higher one.

Sufficiently low excitation energies in the fissioning nuclei have been achieved by use of direct reactions induced in a ^{226}Ra target with a 23.5 MeV ^{3}He beam from the Munich accelerator. The outgoing light particles (p, d, t, α) from the direct reaction (fig. 15) were identified by a ΔE-E-telescope in coincidence with fission fragments, detected in two pairs of semiconductor detectors at 0° and 90° with respect to the recoil axis of the (^{3}He,d)-reaction. Further details about the set-up and the data analysis can be found elsewhere[47,50].

As an example of a fragment mass spectrum, fig. 16 exhibits the data for fission of ^{227}Ac at 0° corresponding to excitation energies between 7 and 13 MeV. They obviously show a triple humped distribution with clear minima between the two

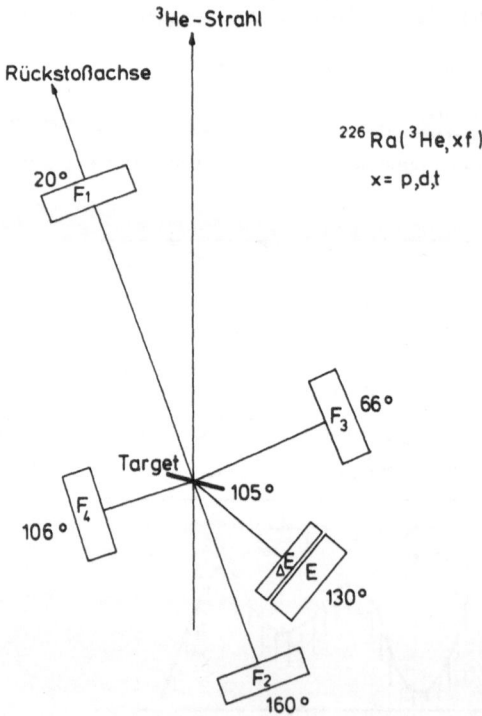

Fig. 15. Detector arrangement for measuring fragment mass distributions and angular anisotropies in direct reaction induced fission

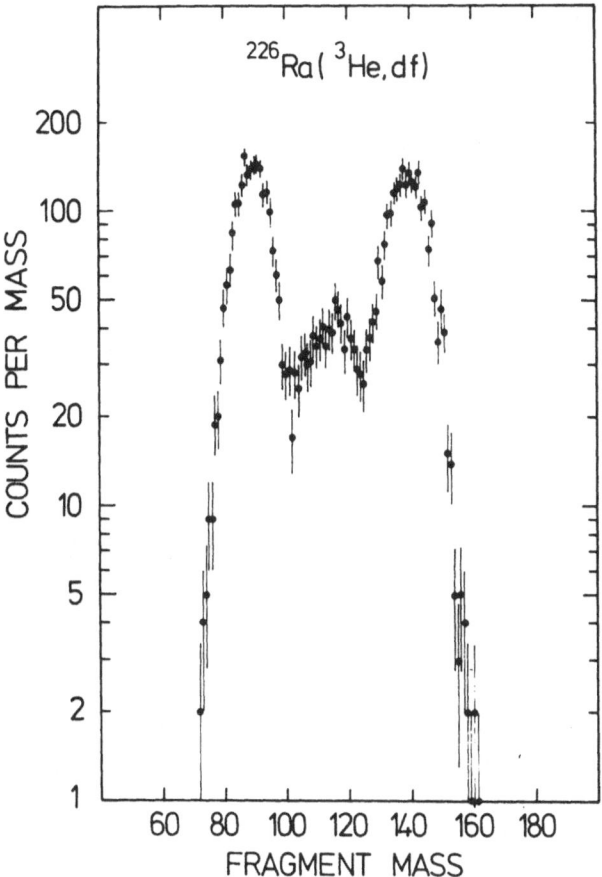

Fig. 16. Fragment mass distribution for fission of ^{227}Ac at excitation energies

between 7 and 13 MeV

mass components. Since second-chance fission is excluded here for energetical

reasons, we can conclude that both fission modes really occur in the same nucleus,

which had always remained unclear[42,45].

The fission probability, defined as

$$P_f = \frac{\sigma(^3He,xf)}{\sigma(^3He,x\ )} = \frac{\Gamma_f}{\Gamma_n} \quad (x = p,d;\ \Gamma_f << \Gamma_n)$$

and the fragment anisotropy $\sigma(0°)/\sigma(90°)$ are plotted in fig. 17 as a function of

the excitation energy in the final nuclei ^{227}Ac and ^{228}Ac, resp., separately for

the asymmetric and the symmetric mass component and corrected for chance coincidences, small instrumental tails from the asymmetric mass peaks in the symmetric region and break-up of the ^{3}He into p + p + n (influencing the (^{3}He,p) cross section). We will now discuss several interesting features directly visible in fig. 17.

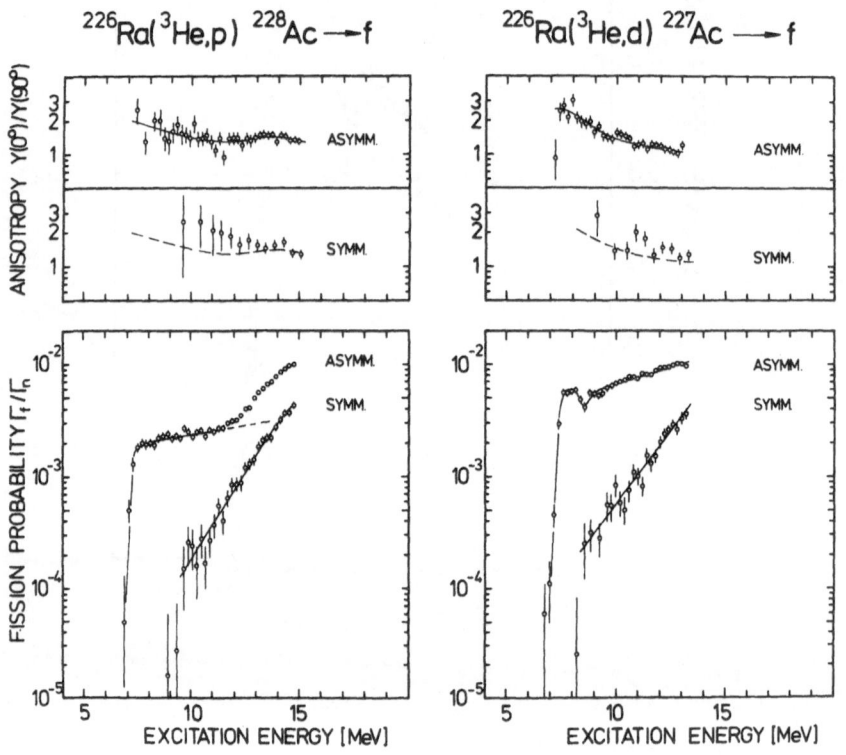

^{226}Ra(^{3}He,p) ^{228}Ac →f ^{226}Ra(^{3}He,d) ^{227}Ac →f

Fig. 17. Fission probabilities (bottom) and fragment anisotropies (top) as a function of excitation energy in the final nuclei ^{228}Ac (left) and ^{227}Ac (right), for the symmetric and the asymmetric mass component separately. The dashed lines are discussed in the text

Most important, symmetric and asymmetric fission appear, in fact, to be associated with different fission barriers; the symmetric barrier is higher than the asymmetric one by 1.2 and 2.0 MeV for ^{227}Ac and ^{228}Ac, respectively. The upper limit for a possible symmetric yield averaged over the region between the two thresholds relative to the symmetric yield just above the symmetric threshold is 6% (95% confidence limit). The fragment angular anisotropies (upper part of fig. 17) also appear to be different for the two mass components (the dashed line in the plots for symmetric fission marks the dependence for the asymmetric component). This difference further supports the interpretation of the different threshold behaviour as really being due to separate barriers.

Next, the fission probability for the asymmetric component increases - beyond the usual extremely rapid rise at the threshold - much more slowly with increasing excitation energy than should be expected on the basis of simple statistical model considerations[51] for the competition of fission and neutron emission (the increase above the dotted line for ^{228}Ac is caused by the onset of second chance fission). The fission probability for the symmetric component, on the other hand, increases very rapidly (exponentially) with increasing excitation energy, finally (at least for ^{228}Ac) even exceeding the asymmetric yield at an excitation energy as low as 5 MeV above the symmetric barrier. This surprising result can neither be due to second chance fission effects as has been speculated before[45], nor to a washing-out of shell effects with increasing excitation energy, which according to recent calculations[52] should be negligible at these low energies. It is not at all clear at the moment, how this "cross-over" as well as the rather general appearance of triple-humped mass distributions[39,40] (or at least their secondary effects) should be understood on the basis of the barrier shapes calculated so far.

Finally, the fission probability for the asymmetric component of ^{227}Ac shows a definite structure in the region of the fission threshold, similar to what has been observed in the ^{226}Ra (n,f) reaction[53]. Transmission resonances are not expected to occur in nuclei below Th [5,8] because of the switch-over in the relative barrier heights; their widths also should be much smaller. A possible relation of this structure to the onset of symmetric fission is also unclear.

We are presently pursuing further work in this direction, investigating other reactions as well as other target nuclei.

4. CONCLUSION

Identifying a rotational band, we have found definite experimental evidence for the large quadrupole deformation associated with spontaneously fissioning isomers, although the quadrupole moment itself remains to be measured. We have also found the existence of different barriers for symmetric and asymmetric fission, seemingly in accord with the expectations based on shell model calculations for the octupole deformation. The competition between the two mass components, however, remains to be clarified.

REFERENCES

1. Strutinsky, V.M.: Nucl. Phys. A95, 420 (1967); A122, 1 (1968).

2. Brack, M., Damgaard, J., Pauli, H.C., Jensen, A.S., Strutinsky, V.M., Wong, C.Y.: Rev. Mod. Phys. 44, 320 (1972).

3. Pauli, H.C.: Phys. Rep. 7, 35 (1973).

4. Tsang, C.F., Nilsson, S.G.: Nucl. Phys. A140, 275 (1970).

5. Back, B.B., Britt, H.C., Garrett, J.D., Hansen, O.: Phys. Rev. Lett. 28, 1707 (1972).

6. Bjørnholm, S., Lynn, E.: Rev. Mod. Phys. (to be published).

7. Britt, H.C., Bolsterli, M., Nix, J.R., Norton, J.L., Phys. Rev. C7, 801 (1973).

8. Back, B.B., Bondorf, J.P., Otroschenko, G.A., Pedersen, J., Rasmussen, B.: Nucl. Phys. A165, 449 (1971).

9. Bjørnholm, S., Strutinsky, V.M.: Nucl. Phys. A136, 1 (1969).

10. Lynn, J.E.: IAEA Sec. Symp. Phys. and Chem. of Fission, Vienna (1969) 249.

11. Bjørnholm, S.: EPS Conf. Nucl. Phys., Aix-en-Provence (1972).

12. Specht, H.J., Fraser, J.S., Milton J.C.D.: IAEA Sec. Symp. Phys. and Chem. of Fission, Vienna (1969) 363.

13. Migneco, E., Theobald, J.P.: Nucl. Phys. A112, 603 (1968).

14. Polikanov, S.M., Druin, V.A., Karnaukhov, V.A., Mikheev, V.L., Pleve, A.A., Skobelev, N.K., Subbotin, V.G., Ter-Akopjan, G.M., Fomichev, V.A.: Soviet Phys. JETP 15, 1016 (1962).

15. Britt, H.C.: Nucl. Data (1973, to be published).

16. Metag, V., Repnow, R., von Brentano, P.: Nucl. Phys. A165, 289 (1971).

17. Limkilde, P., Sletten, G.: Nucl. Phys. A199, 504 (1973).

18. Metag, V.: Contrib. to the IAEA Third Symp. Phys. and Chem. of Fission, Rochester (1973) and private communication.

19. Pedersen, J.: University of Washington, Seattle and Niels Bohr Institute, Copenhagen, private communication (1973).

20. Harrach, D.V., Khan, T.A., Konecny, E., Löbner, K.E.G., Osterman, P., Specht, H.J., Waldschmidt, M.: Jahresber. Beschleunigerlab. München (1972) 107.

21. Konecny, E., Specht, H.J., Weber, J., Weigmann, H., Ferguson, R.L., Osterman, P., Waldschmidt, M., Siegert, G.: Nucl. Phys. A187, 426 (1972).

22. Specht, H.J., Weber, J., Konecny, E., Heunemann, D.: Phys. Lett. 41B, 43 (1972).

23. Heunemann, D.: Technische Hochschule München, Dissertation (1972).

24, Christensen, J.: Niels Bohr Institute, Copenhagen, private communication (1972).

25. Schmorak, M., et al., Nucl. Phys. A178, 410 (1972).

26. Sobiczewski, A., Bjørnhold, S., Pomorski, K.: Nucl. Phys. A202, 274 (1973).

27. Harris, S.M.: Phys. Rev. 138, B509 (1965).

28. Johnson, A., Szymanski, S.: Phys. Rep. 7, No. 4, 181 (1973).

29. Mariscotti, M.A.I., Scharff-Goldhaber, G., Buck, B.: Phys. Rev. 178, 1864 (1969).

30. Browne, J.C., Bowman, C.D.: Phys. Rev. Lett. 28, 617 (1972).

31. Benson, D., Lederer, C.M., Cheifetz, E.: Nucl. Phys. A201, 445 (1973).

32. Heunemann, D., Konecny, E., Specht, H.J., Weber, J.: Jahresber. Beschleunigerlab. München, 103 (1972).

33. Horsch, F., Konecny, E., Löbner, K.E.G., Specht, H.J.: Jahresber. Beschleunigerlab. München, 104 (1972).

34. Sobiczewski, A.: Institute for Nuclear Research, Warszawa, preprint (1973).

35. Konecny, E., Kozhuharov, C., Specht, H.J., Weber, J.: Jahresber. Beschleunigerlab. München, 70 (1972).

36. Specht, H.J., Konecny, E., Weber, J., Kozhuharov, C.: Contrib. to the IAEA Third Symp. Phys. and Chem. of Fission, Rochester (1973).

37. Gangrsky, Yu. P., Khanh, Nguyen Cong, Pulatov, D.D., Hien, Pham Zuy:
     Dubna report P7-6466 (1972).

38. Jensen, R.C., Fairhall, A.W.: Phys. Rev. 109, 942 (1958); Phys. Rev. 118,
     771 (1960).

39. Britt, H.C., Wegner, H.E., Gursky, J.C.: Phys. Rev. 129, 2239 (1963).

40. Britt, H.C., Whetstone, S.L.: Phys. Rev. 133, B603 (1964).

41. Konecny, E., Schmitt, H.W.: Phys. Rev. 172, 1213 (1968); Phys. Rev. 172,
     1226 (1968).

42. Perry, D.G., Fairhall, A.W.: Phys. Rev. C4, 977 (1971).

43. Holubarsch, W., Pfeiffer, E., Gönnenwein, F.: Nucl. Phys. A171, 631 (1971).

44. Niday, J.B.: Phys. Rev. 121, 1471 (1961).

45. Tsang, C.F., Wilhelmy, J.B.: Nucl. Phys. A184, 417 (1972).

46. Möller, P.: Nucl. Phys. A192, 529 (1972).

47. Konecny, E., Specht, H.J., Weber, J.: Phys. Lett. (1973), to be published.

48. Kivikas, T., Forkman, B.: Nucl. Phys. 64, 420 (1965).

49. Vandenbosch, R., Unik, J.P., Huizenga, J.R.: IAEA First Symp. on the Phys.
     and Chem. of Fission, Salzburg I, (1965) p. 547.

50. Weber, J.: Universität München, Dissertation (1973), to be published.

51. Huizenga, J.R., Vandenbosch, R.: in Nuclear Reactions, edited by P.M. Endt
     and P.B. Smith (North Holland Publishing Co., Amsterdam, 1962), Vol. II.

52. Jensen, A.S., Damgaard, J.: Nucl. Phys. A203, 578 (1973).

53. Babenko, Yu.A., Ippolitov, V.T., Nemilov, Ju.A., Selitskii, Yu.A., Funshtein,
     V.B.: Soviet J. Nucl. Phys. 10, 133 (1969).

COMMON ASPECTS IN FISSION AND HEAVY ION REACTIONS

Klaus Dietrich

Physik Department of the Technische Universität Munchen

We would like to consider "common aspects" of nuclear fission (F) and reactions between medium-heavy and heavy nuclei ("heavy ion reactions", HIR). One of the first questions we might ask is: is there a connection between fission and a heavy ion reactions due to reciprocity?

From time-reversal invariance we have the well-known relation between S-matrix elements

$$S_{fi} = S_{\bar{i}\bar{f}} \qquad (1)$$

where i, f describe an initial and final channel, and $\bar{i}$, $\bar{f}$ are those asymptotic states which arise from i and f through reversal of time. If we consider only 2-body channels, this leads to the following relation between the cross-sections of unpolarized particles:

$$(2i_1^{(i)} + 1)(2i_2^{(i)} + 1) \; k_i^2 \; \sigma_{if} = (2i_1^{(f)} + 1)(2i_2^{(f)} + 1) \cdot \qquad (2)$$

$$\cdot \; k_f^2 \; \sigma_{\bar{f}\bar{i}}$$

$k_{i,f}$ = relative momenta in the incident (i) and final (f) channel $i_{1,2}^{(i)}$ = spin of the nuclei 1,2 in channel i; $i_{1,2}^{(f)}$ = spin of the nuclei in channel f. In low-energy fission, the nascent fragments have an average excitation energy of 10-20 MeV which they deliver by emission of neutrons ($\bar{\nu}_n$ = 2.1) and γ-rays ($\bar{\nu}_\gamma \approx$ 3-5). In a heavy ion reaction we can neither prepare this final state of some 5 particles and photons, nor can be produced an initial state of two so highly excited nuclei. Thus the reciprocity relations are not useful. Only for pairs of fragments which are produced in their groundstates already at scission can we practically use the reciprocity relation. We believe that this is not impossible for fission into 2 magic nuclei like

$$^{32}_{16}S_{16} + \gamma \rightarrow ^{32}_{16}S_{16}^{*} \rightarrow ^{16}_{8}O_{8} + ^{16}_{8}O_{8}$$

Let us now go through a number of topics common to fission and heavy ion reactions.

1. FISSION AND HEAVY ION REACTIONS AS GENERAL FORMS OF COLLECTIVE

MOTION-THEORY OF COLLECTIVE VARIABLES

F as well as HIR represent types of motion where many nucleons move in a similar way or where there is coherence between a large number of elementary excitations. Thus a theory of F or HIR implies a general theory of collective variables. On the other hand, we have here the testing ground for our theoretical understanding of collective motion. Let us shortly define the different methods and approximations used to describe collective motion and then point at the aspects of fission and HIR which might serve to probe them.

1A. The Semiclassical Approach

It treats the collective motion in the form of additional (superfluous) degrees of freedom $q_1(t)...q_f(t) \equiv q(t)$ of the system whose time-dependence is classically determined. It thus bases itself on a model Hamiltonian like

$$H = \sum_{\nu\mu} <\psi_\nu |T|\psi_\mu> c_\nu^\dagger c_\mu + \frac{1}{2} \sum_{\mu\nu\rho\kappa} <\psi_\mu\psi_\nu|v|\psi_\rho\psi_\kappa> \cdot c_\mu^\dagger c_\nu^\dagger c_\kappa c_\rho \tag{3}$$

where the single particle wave functions $\psi_\mu(x;q(t);\dot{q}(t))$, (x stands for space, spin) isospin variables) as well as the corresponding creation and annihilation operators $c_\mu^\dagger$, c_ν depend on time through the collective parameters q. The time-dependence of $q_i(t)$ is defined by an appropriately chosen classical Lagrangian L $(q(t), \dot{q}(t))$

$$\frac{d}{dt}\left(\frac{\partial L}{\partial \dot{q}_i}\right) = \frac{\partial L}{\partial q_i} \tag{4}$$

As an example, L could be constructed from parts of the fully contracted part of H^1 or it could be taken from an entirely classical description of the system. The problem is then to solve the time-dependent Schrödinger equation

$$H \psi = i \dot{\psi} \tag{5}$$

with appropriate boundary conditions. This is usually done by expanding ψ in terms of the stationary eigenfunctions ϕ_n of a simple part H_o of H

$$H_o \phi_n[q(t);\dot{q}] = \varepsilon_n[q;\dot{q}] \phi_n[q;\dot{q}] \tag{6}$$

$$\psi = \sum_{n} f_n(t)\ e^{-i \int^t d\tau\ \varepsilon_n[q(\tau)]}\ \phi_n \tag{7}$$

Let me note that it is desirable but not always possible to choose H_o such that

$$\frac{d\varepsilon_n}{dt} = 0$$

Examples:

i) Semi-classical treatment of multiple Coulomb excitation

and more generally of transfer-reactions between heavy ions[1]: $q(t) = r(t)$ = distance between the centres of the shell model potentials describing the ions

$$L = \frac{\mu}{2}\ \dot{r}^2 - \frac{Z_1 Z_2}{r}$$

(μ = rel. mass; $Z_{1,2}$ = charge of ions)

ψ_ν = single particle wave function in the two moving potentials[*,1].

ii) Fission

$q(t)$ = parameters describing the nuclear shape, possibly including $r(t)$. As an example, L could be given by the classical liquid drop; ψ_ν = eigenfunctions of a deformed average potential whose form is described by $q(t)$.

A theory of this form was applied by Boneh[3] for the migration from saddle to scission and by Kelson[3] to describe nuclear fission in the frame of simple models.

Remarks: we emphasize that the semi-classical picture does permit to describe intrinsic (non collective) excitations of the system. The intrinsic excitations are coupled to the collective motion. On the other hand, the classical equations

* If the ψ_ν are not orthogonal, one has to modify H by using the dual basis[2,1]

of motion in general do not contain an effect of intrinsic excitations. Attempts of including such effects are for instance the various rules for calculating recoil corrections in the semiclassical treatment of HIR. The method 1A can only be applied if the de Broglie wave lengths of the collective motion are small compared to typical dimensions of the system (HIR), or if the energy of the collective modes is much smaller than the energies of the intrinsic excitations.

1B. Cranking Theory and Time-dependent Hartree-Fock

The cranking theory is closely connected with the approach 1A: we use in (6) instead of H_o the full model Hamiltonian H. We then solve the system of equations for $f_n(t)$, which is obtained from (6) and (7) by treating terms proportional to the collective velocities or accelerations $(\dot{q}, \ddot{q})$ in lowest order of perturbation theory. We, furthermore, assume as a "boundary condition" that, for vanishing $\dot{q}$ and $\ddot{q}$, the solution ψ should coincide with the lowest "adiabatic state" ϕ_o:

$$\lim_{\dot{q}, \ddot{q} \to 0} f_n(t) = \delta_{no} \tag{8}$$

We thus obtain the expectation value $<\psi(t)|H|\psi(t)>$ as a quadratic form in q:

$$\tilde{H} \equiv <\psi|H|\psi> = \varepsilon_o + \frac{1}{2} \sum_{i,j=1}^{f} g_{ij} \dot{q}_i \dot{q}_j \equiv \varepsilon_o + \tau \tag{9}$$

$$g_{ij} = \sum_{n \neq o} \frac{<\phi_o|\frac{\partial}{\partial q_i}|\phi_n><\phi_n|\frac{\partial}{\partial q_j}|\phi_o>}{\varepsilon_n - \varepsilon_o} \tag{10}$$

We may interpret τ and ε_o as classical kin. and potential energies and quantize according to the prescription of Schrödinger. This leads to a Hamiltonian of purely collective motion:

$$H_{coll} = - \frac{1}{2}\Delta + \varepsilon_o \tag{11}$$

Δ = f-dim. Laplacian in curvilinear coordinates, $\hbar = 1$. We note that $\tilde{H}(t)$ of (9) could also serve to provide the classical time-dependence of q(t) in the theory given in 1A. A description which is closely related to 1B, is the so-called time-dependent HF or HB [4]. Since it is physically largely equivalent, I do not comment on it.

Contrary to 1A the cranking approach including the quantization (11) describes the collective motion quantum-mechanically. It should, however, be noted that this quantum-mechanical Hamiltonian H_{coll} is derived from a semi-classical picture and with a quantization rule which requires to be justified.

The cranking theory was extensively and very successfully used by Kumar and Baranger[4] and in a more phenomenological version by Gneuss[5] for the description of non-harmonic vibrations. Here, we mainly have in mind the applications of this theory in HIR and F:

(i) Heavy ion reactions

The picture of adiabatic energy surfaces ε_n has been used by Gneuss and Scheid[6] on the basis of a phenomenological energy functional and Fliessbach[7] in a more microscopic description of ^{16}O - ^{16}O scattering. In both methods one finds valleys in the potential $\varepsilon_o[r]$ at a distance r close to the sum of the nuclear radii. The question whether these valleys may give rise to quasimolecular states

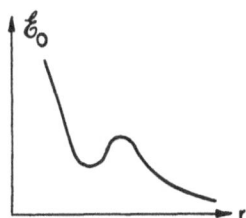

depends on the strength of the coupling to intrinsic excitations in this range of r. The possible existence of such resonances is one of the most exciting aspects of HIR. It is also tied to the general question of nuclear viscosity (see sect. 4). The quasi-molecular states in HIR would be the counterpart of the scission-configuration whose decay is slowed down by a scission barrier[8,9]. As far as I know, the relative motion of the ions has so far only been treated classically in the frame of this approach, i.e. the Hamiltonian (9) was not quantized.

(ii) Fission

Here method 1B was applied in connection with a phenomenological calculation of $\varepsilon_o[q]$ using Strutinsky's method. The manyfold successes of this method are so well-known that I need not digress on them. I just like to mention a perhaps less known recent success: if one calculates the penetrability in 1-dimensional WKB-approximation on a series of paths through the potential landscape using the variable inertia given by (10), variation of the paths and a slight readjustment of the fissility parameter, lead to the correct order of magnitude for the life-times of groundstates as well as the shape isomeric states of the actinides and to a preference for asymmetric fission with the correct average mass-ratio[10]. We feel that this result is rather encouraging. It challenges us to try for a better understanding of why Strutinsky's method works so well and of why the simple-minded cranking approach is seemingly better than·one might expect it to be.

1C. Generator Coordinate Method (GCM)

A well-known way to go beyond a semiclassical approach to collective variables is the GCM initiated by Hill and Wheeler[11] in 1953. It has since that time been investigated by many authors. In this method one expands the eigenstate $\tilde{\psi}$ of the time-independent Schrödinger equation

$$\hat{H}\,\tilde{\psi} = E\,\tilde{\psi} \tag{12}$$

in terms of basic states ϕ_n which depend not only on discrete quantum nrs (n) but also on one or a set of continuous parameters (the "generator coordinates"). $\hat{H}$ is the "true" many-body Hamiltonian, which depends only on the effective two-body forces, and the degrees of freedom of the nucleons, not on any external parameters. The basic states are usually generated by a model Hamiltonian like (1) or a simplified version of it, which depends on (time-independent) generator coordinates

$$\tilde{\psi} = \sum_n \int dq \; g_n(q_1 \ldots q_f) \; \phi_n(x; q_1 \ldots q_f) \tag{13}$$

As an example, ϕ_n could be given by

$$\phi_n = \prod_\nu \; (c_\nu^\dagger)^{n_\nu^{(n)}} \; |0> \tag{14}$$

$$n = \{n_1^{(n)} , n_2^{(n)} , \ldots, n_N^{(n)}\} \qquad\qquad N = \text{nr of s.p.s.}$$

The states ϕ_n are not orthogonal for different values of the GC. A system of integral-equations for the weight-functions $g_n(q)$ is obtained from (12), (13). The advantage of the method over the semiclassical treatments 1A and 1B is that the collective variables do not appear as superfluous degrees of freedom and that their dynamics is not treated classically or semiclassically. The disadvantage of the GCM is that it is practically impossible to carry out any practical calcul-ations for a set of 3-5 deformation parameters q_i needed to describe the family of shapes in F and that its practical application in HIR is certainly not simple either.

Basic to an application of the method in F and HIR is thus the question of how to reduce the number of GC to only one and, nevertheless, obtain a theory which is superior to the semiclassical methods 1A, 1B. I think that the success of the cranking theory combined with WKB in F [10] and of the semi-classical approach in HIR may suggest the following way:

(i) <u>HIR</u>

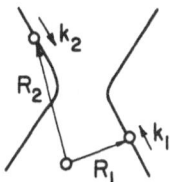

Let us assume that the classical motion of the nuclei on the Coulomb trajectories provides a good first approximation. This can be concluded from the comparison between experimental and theoretical work in Coulomb excitation and transfer reactions (below or at least not far above the Coulomb barrier). Let us define a variable s as GC such that $R_{1,2}(s)$ and $k_{1,2} = m \dot{R}_{1,2}(s)$ depend on s in the same way as they depend on time in the classical theory. As an example, we could use a set of basis functions ϕ_n of the type (14) where c_ν^+ creates a particle in a shell-model state

$$\psi_\nu(x) = \phi_\nu(x - R_{1,2}(s)) \cdot e^{ik_{1,2}(s)x} \tag{15}$$

and expand $\overset{\gamma}{\psi}$ in terms of them

$$\overset{\gamma}{\psi} = \sum_n \int dsg_n(s) \, \phi_n(R_{1,2}(s); \dot{R}_{1,2}(s)) \tag{16}$$

It can be shown[13] that the GCM, in the limit of vanishing overlap for different s, becomes equivalent to the cranking theory. Thus a GCM based on the "Ansatz" (16) is certainly superior to the semi-classical methods 1A, 1B. By using different trajectories in (16) for the incident and exit channel one would obtain "recoil corrections". The Ansatz (16) is, however, too restrictive to describe diffraction phenomena. But in this case, the semi-classical method is anyhow not useful as a first approximation.

(ii) Fission

For excitation energies E of the compound nucleus smaller than the classical threshold E_f we could be guided by the success of the WKB calculation of the lifetimes (see IB) and choose the trajectory $q_i(s)$ obtained for maximal penetrability for providing a suitable set of basis functions $\phi_n[q_1(s)...q_f(s)]$ in the "Ansatz" (16). It should be noted that in both cases (HIR, F) the choice of the trajectory depends on the energy E. The projection on eigenstates of angular momentum introduces no further GC and may be neglected if the deformation is large enough.

2. SHELL EFFECTS

Shell effects as a function of the density of shell model levels close to the Fermi energy are known to be of paramount importance for the deformability of the nucleus. In fission, they are responsible for the appearance of shape isomers, and of the asymmetric mass-division, and for the sawtooth form of the average excitation energy as a function of the neutron and proton nrs of the fragments. In HIR, they influence the change of the shapes (and thus of the Coulomb barriers) due to the interaction between the nuclei, and they will be of importance for the eventual existence of quasi-molecules. They might also determine the transfer of nucleons or nucleon groups by their influence on level densities (see section 3). Generally speaking, the interesting aspects of of F and HIR concerning shell effects are: (i) we can study nuclei with very strongly deformed shapes and consequently, the shell structure as a function of deformation, (ii) through the composite nature of both partners (fragments in F, target and projectile in HIR) the density of levels which depends on the shell structure may become a dominant feature more frequently than so far experienced.

3. STATISTICAL EFFECTS

This brings up the question of whether we may expect that F and HIR may also bring new aspects to the study of statistical properties. Of course, the statistical properties which lead to the concept of the compound nucleus are very much the

same in F and HIR than in other compound nuclear reactions. We just note in passing that nuclear fission, at least at low energies, is a very clean case of a
compound nuclear reaction.

What is, I think, of a greater interest is the eventual importance, in F and
HIR, of statistical properties within subgroups of states, which may a bit sloppily
be called "partial thermal equilibrium". I mention two possible cases of such a
phenomenon, both representing controversial, but amusing suggestions:

A) Fission

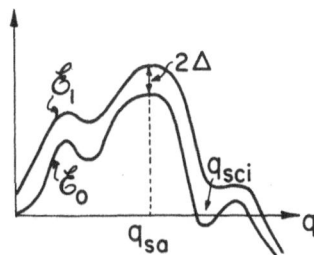

Let ε_0 and ε_1 be the lowest adiabatic surface and the lowest 2 quasi-particle
excitation of an even-even nucleus respectively (separation in energy = 2 x gap).
Let us assume that the energy E of the system is smaller than $\varepsilon_1[q_{sa}]$: $E < \varepsilon_1[q_{sa}]$.
Then we know that at the saddle, only the collective states built on the "ground-
state band" ε_0 are excited. On the other hand, at the scission point, the available excitation energy exceeds $\varepsilon_1[q_{sci}]$ and even the energy of many higher
adiabatic states. Two alternatives may happen on the passage from saddle to
scission: (i) the passage is slow enough that all configurations are populated
with equal probability which implies a complete thermal equilibrium at the scission
point[12], (ii) the passage is rapid enough so that only the strongly coupled collective substates of a given band are statistically populated[13] which implies a
partial thermal equilibrium at scission.

In the latter case, the ensemble of thermally occupied states at scission is
rather suddenly enlarged when the excitation energy E becomes larger than $\varepsilon_1[q_{sa}]$,
because in this case, also the collective states of the 2 quasi particle-band will
be populated. This has consequences for the distribution of excitation and kinetic
energies of the fragments (f.i. decrease of average kin-energy of fragments with
increasing excitation energy) apparently in agreement with experiment[13].

B) <u>Heavy ion reactions</u>

In order to explain recent transfer experiments performed in Dubna (projectiles like N, O on heavy targets), Bondorf <u>et al</u>.[14] introduced the following picture:

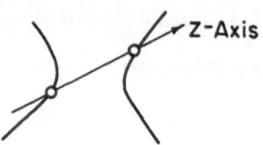

the two nuclei move essentially on classical trajectories. Close to the turning point, transfer of nucleons takes place in such a way that a thermal equilibrium is established for those degrees of freedom which correspond to a single particle motion with magnetic quantum number $\ell_z = 0$ (z-axis is in the direction of line joining the two centers of the nuclei). The idea is that for nucleons moving perpendicular to that line the matrix elements of the interaction H_{int} between the nuclei are smaller on the average, so that no equilibrium is achieved in the relatively short time of nuclear contact. It means that in the formula for the cross-section in a statistical theory, one has to replace the ordinary density of states by the density of states with $\ell_z = 0$ [14]. This picture is still debated on.

I cite it as an example of the type of statistical assumptions which may turn out to be relevant for HIR. The coupling between the different degrees of freedom of the system is also the essence of another domain of problems which we could circumscribe by "optical potentials" and nuclear viscosity.

4. OPTICAL POTENTIALS, NUCLEAR VISCOSITY

This is an important topic which is debated in all recent meetings on HIR or F and, so far, nobody knows anything precise on it. We only define in very short terms which theoretical steps we believe should be taken. As a tool we use the GCM.

(i) We define an "optical potential" which contains the virtual and real population of intrinsic excitations of the system and which occurs as an operator in the equation for g_o. This may be done by eliminating the functions $g_{n \neq o}$ from the set of Hill-Wheeler equations obtained from the "Ansatz" (13) or (16). The simplest way of achieving this, is to introduce the dual basis $\overset{\sim}{\phi}_n(s')$ defined by

$$<\overset{\sim}{\phi}_m(s')|\phi_n(s)> = \delta_{nm}\,\delta(s-s')$$

which may serve to define the Feshbach projection operators. Furthermore, it may be useful to choose the generator coordinate s as proposed in section 1C.

(ii) The resulting expression for the optical potential should be analyzed on the same lines as this has been done for the optical potential of nucleon scattering: random phase assumption for matrix-elements involving complicated intrinsic excitations, special attention to doorway state phenomena (quasimolecular states in HIR, saddle point and isomeric states in F).

(iii) In addition to that it would be interesting to investigate the WKB limit of the equation for $g_o(s)$ together with the limit of δ-overlap. A problem related to optical potentials is the concept of "viscosity" which is suggested by the semiclassical picture. It is tempting to introduce a viscosity term into the classical equations of motion for q(t) in order to take into account the average loss of translational kinetic energy due to the excitation of intrinsic degrees of freedom. This viscosity term must be related to the number of open channels and to the coupling to them.

5. MULTIPLE TRANSFER OF PARTICLES OR QUANTA

Let us consider two composite systems in a "weak" time-dependent contact. "Weak" should mean that the intrinsic wave functions of the two constituents are not drastically changed by the coupling. "Composite" should mean that the number of particles in each of the two parts is sufficiently large for collective excitations to occur in both systems. Nuclear examples of such systems are: (i) in HIR: two heavy or at least medium heavy nuclei close to the turning point of the trajectories at an energy below (but close to) the Coulomb barrier E_B, possibly also quasimolecular states. (ii) in F: the two nascent fragments close to the scission point. There are two physical phenomena which may lead to a "multiple transfer" of particles or quanta of the system. In some cases the two aspects are only different ways of expressing the same physical contents (f.i. Josephson effect between superconductors).

I.   Symmetry violation in both parts of the system

Let us assume that there is violation of the same continuous symmetry (ex: rotational invariance, gauge invariance) in both the constituents of the system. This means that the (unperturbed) states of part $1\overline{(2)}$ can be written as a projection operator $\hat{P}_n$ (n = ensemble of quantum nrs related to the symmetry operators) acting on an "intrinsic" state $\phi_{1(2)}$ which is not an eigenstate of the symmetry operators

$$\psi_{1(2)}^n = \hat{P}_{n_{1(2)}} \phi_{1(2)} \tag{20}$$

This "spontaneous" symmetry violation can be understood from the observation that there is a component V of the interaction which has large attractive matrix elements with the intrinsic state $\phi_{1(2)}$:

$$<\phi_{1(2)}|V|\phi_{1(2)}> = \text{large} \tag{21}$$

examples: $\phi_{1,2}$ =   Slater det. of Nilsson wave functions

   V   =   Q.Q force

or: $\phi_{1,2}$ =   BCS state

   V   =   pairing force

If the symmetry violation is sufficiently large (implying large particle numbers in 1 and 2), statement (21) holds also for matrix-elements between the "true" states $\psi_{1(2)}^n$:

$$< \psi_{1(2)}^{n'}|V|\psi_{1(2)}^n> = \text{large} \tag{21'}$$

Consequently, operators which contain the same constituents as the interaction V, will have large matrix-elements between the states $\psi_{1(2)}^n$. This can be made explicit, if we can use a group structure[15,16].

examples: large BE2-values (for n'≠n) between rotational states, large spectroscopic factors for 2n- or 2p-transfer to or from superconducting target nuclei.

It can be shown that in this case the same interaction V which is responsible for the symmetry violation in the parts 1 and 2 of the system leads also to a coherently enhanced coupling V_{12} between the systems, i.e.

$$\langle \psi_1^{n_1'} \psi_2^{n_2'} |V_{12}| \psi_1^{n_1} \psi_2^{n_2}\rangle = \text{large, same sign for all}$$
$$n_1, n_2, n_1', n_2' \tag{22}$$

In these respects, "large" means that the matrix elements in question are larger, in absolute value, than corresponding ones of the remaining part of the nuclear interaction.

The observation (22) may have the consequence, that the interaction Hamiltonian V_{12} between the parts 1 (described by a Hamiltonian H_1) and 2 (Hamiltonian H_2).

$$H = H_1 + H_2 + V_{12} = H_o + V_{12} \tag{23}$$

can no longer be treated as a perturbation but must be **diagonalized** in the subspace of the functions $\psi_1^{n_1} \cdot \psi_2^{n_2}$. Such a system may be called a "Josephson junction". The necessity of a diagonalization means physically the occurence of a **multiple** transfer of quanta Δn between the two parts of the system:

examples: $\Delta n =$ Cooper pairs for 1,2 = superconducting nuclei

$\Delta n =$ angular momentum quanta for 1,2 = deformed nuclei[16]

(multiple Coulomb excitation)

II. Degenerate bosons in both parts of the system

All what we said in I is also correct, if $\psi_{1(2)}^n$ or at least a substantial part of it may be written as a degenerate system of bosons (described by boson creation op. $A_{1(2)}^{\dagger}$)

$$\psi_{1(2)}^{n_{1(2)}} = (A_{1(2)}^{\dagger})^{n_{1(2)}} |0\rangle \tag{24}$$

Now $n_{1(2)}$ is the occupation number of the lowest boson state in 1(2). Consider now an interaction V_{12} of the type

$$V_{12} = M(A_1^{\dagger}A_2 + A_2^{\dagger}A_1) \tag{25}$$

where M is a strength factor. Then, obviously the matrix elements (22) of V_{12} will be large because they contain the factor $\sqrt{n_1 n_2}$ which is larger the higher the degree of condensation:

$$\langle \psi_1^{n'_1} \psi_2^{n'_2} | V_{12} | \psi_1^{n_1} \psi_2^{n_2} \rangle \approx \sqrt{n_1 n_2} \cdot M \cdot$$

$$\left(\delta_{n'_1, n_1+1} \, \delta_{n'_2, n_2-1} \; + \; \delta_{n'_1, n_1-1} \, \delta_{n'_2, n_2+1} \right) \tag{26}$$

The factor $\sqrt{n_1 n_2}$ shows explicitly how the enhancement of a matrix element may arise.

Examples: a) again the Josephson effect between two superconducting systems, since we may in this case approximately describe a pair of Fermions as a boson.

b) possibly α-particles or quartets as exchange quanta[17]. It should, however, be noted that M is smaller the larger the number of particles in the cluster, because a small penetration factor enters for each nucleon which is to be transferred.

c) according to a recent suggestion of Migdal[18], nuclear matter and possibly also large enough nuclei should contain a condensate of π-mesons. If this turns out to be correct, one should observe an enhanced exchange of pions between two heavy nuclei in a HIR or possibly also between the fragments at scission. Since these pions are off the mass-shell this enhanced exchange would only exhibit itself as a modification (probably increase) of the nuclear interaction between the two systems.[*]

Let me conclude this section by commenting on the kind of physical effects which we may expect:

As for the coupling between deformed nuclei this is the well-known multiple Coulomb excitation which is observed in HIR and which certainly also occurs in F while the deformed fragments fly apart. As for the coupling between "superfluid" nuclei, one should observe "resonances" in the cross section for transfer of a given number of pairs at a given scattering angle[15]. As far as F is concerned, a Josephson effect between the nascent fragments was considered to be the source of asymmetric mass division by Gaudin[15]. This is most probably outdated by the

---

[*] I am grateful to M. Danos for clarifying discussions on this matter.

strikingly consistent success of the theory of shell effects. We believe that in F the existence of a "Josephson effect" is equivalent to the simple statement that low energy fission is sufficiently adiabatic so that there is little breaking of pairs at scission. Considering the fission of an even-even nucleus this would mean that pairs of even-even fragments should be more frequent than pairs of odd-even fragments and those again more frequent than pairs of odd-odd fragments. I would like to note that odd-even effects of this kind have recently been observed by several authors[19].

6. ELECTRONIC TRANSITIONS

Finally, I would like to draw your attention to the study of electronic transitions following a HIR or F. Very beautiful work has been done on this field by P. Armbruster et al.[20].

The electromagnetic transitions between the electronic states of the out-going atoms are observed after the HIR or F took place. The energy of the emitted radiation permits interesting insight in the mechanism and the time scale of a HIR and possibly of the fission process.

Let us consider a HIR:

We may think of two physical extremes: (i) the "rapid process (RP)" occurs in the case that the time τ_0 of impact (τ_0 = diameter of el. orbit/relat. veloc. of nuclei) is much smaller than the period T of the electronic motion on the orbit considered: $\tau_0 \ll T$. The periods T of the orbital electronic motion vary over a wide range due to the different binding energies of electrons in outer and inner orbitals. Indeed, the condition $\tau_0 \ll T$ is fulfilled for the lightly bound electrons. It turns out that electrons with periods $T \gg \tau_0$ are all stripped on their way through the target. The qualitative explanation is that they do not follow the non-uniform motion of the nucleus plus the core electrons.

For the electrons in the lowest orbitals (K,L,M shell) the opposite limit is relatively well fulfilled. (ii) the "slow process (SP)" occurs in the opposite case $\tau_0 \gg T$. In this limit, the electrons follow adiabatically the change of the Coulomb field which is produced by the motion of the two nuclei. More precisely:

in each subspace of states with given conserved quantum numbers, the same number
of lowest states which was occupied at time t = -∞ would continue to be occupied
at each distance r(t) of the nuclear charges (t=0: turning point of trajectory).

A conserved quantum nr would be for instance the projection of the angular
momentum on the axis connecting the nuclear charge centres.

We assume that the electronic transition occur <u>after</u> the atoms have separated
and that there is no transfer of protons between the two nuclei (for the sake of
simplicity). Then, in spite of possible crossings of occupied and unoccupied
levels, no electronic transitions will be observed except in the case that an
electron is ejected into the continuum while the nuclei are close. This may be
expected to happen if one of the occupied levels is pushed into the continuum
(fig. 1) or into the region of orbitals with T >> τ_o where it is stripped instanta-
neously.

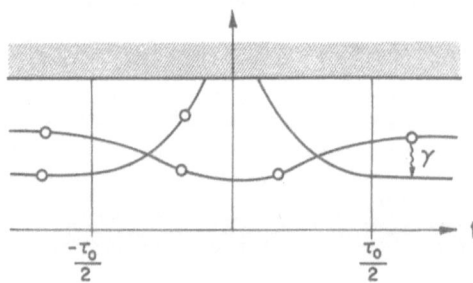

Fig. 1. Two occupied levels cross, one electron is ejected into the continuum,
        γ-transition after second crossing. Note: each time t corresponds to
        a distance r(t) of the nuclear charges. r(0)=distance of closest approach

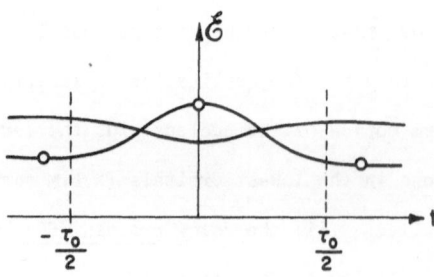

Fig. 2. Crossing of occupied and unoccupied level without emission of an electron

145

By measuring the γ-transition after the HIR one may thus infer from the
diagram of adiabatic levels, which levels had reached the continuum or the levels
with T >> τ_o during the HIR.

There is another physical aspect of the process which is due to the fact,
that it is not completely adiabatic: consider two levels which would cross if
they were not coupled by a small, but finite interaction, H_{cplg} (fig. 3).

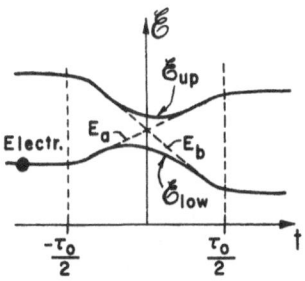

Fig. 3. Close approach of occupied and unoccupied interacting levels[21] (full line
with, broken line without interaction)

We furthermore, assume that prior to the collision only the lower level ε_{low} is
occupied. In the adiabatic limit ($\dot{r} \approx 0$), the electron will always remain on the
lower level and thus no γ-transition will be observed. To the extent that the
process is non-adiabatic, the electron will move to the upper orbit close to t = 0
and consequently, a γ-deexcitation may be observed later on. The probability W of
finding the electron in the upper level ε_{up} at $t >> \frac{\tau_o}{2}$ is given by

$$W = \exp \left\{ - \frac{2\pi |<\phi_a|H_{cplg}|\phi_b>|^2}{\hbar \left| \left(\frac{dE_a}{dt} - \frac{dE_b}{dt} \right)_{t=0} \right|} \right\} \tag{27}$$

where ϕ_a, ϕ_b are the states without the small coupling H_{cplg}. They belong to the
unperturbed energies E_a, E_b[21]. W is the probability with which a γ-deexcitation
will be observed for $t >> \frac{\tau_o}{2}$. The denominator $\left| \frac{dE_a}{dt} - \frac{dE_b}{dt} \right|$ depends on the speed
$\dot{r}(t)$ close to the turning point.

One thus measures a quantity which is directly related to the time-scale of
the HIR. Analogous phenomena can be studied in F. Very interesting aspects arise

if a <u>nuclear</u> quasi-molecule is formed together with a vacancy in the K-shell and if this atom, through its recoil, is scattered by an atom of the target. Dipole transitions are then so much enhanced that they may occur before the nuclear molecule has decayed[20] and thus provide an experimental proof for the formation of nuclear molecules.

The vast scope of the subject forced us to be very short on the different issues. Our main purpose was to show that F and HIR encompass a huge field of interesting problems. Some of these problems concern the understanding of many body systems, quite generally, and atomic nuclei provide only a convenient system to study them.

REFERENCES

1. Broglia, R.A., Winther, A.: Nucl. Phys. <u>A182</u>, 112 (1972); Physics Reports Vol. 4C, N°4, 153 (1972).
   Dietrich, K., Hara, K.: "On the semiclassical description of HIR" preprint (submitted to Nucl. Phys.) Jan. 1973, Phys. Dept. of TECHN. UNIV. MUNCHEN.

2. Moshinski, M., Seligman, T.H.: Annals of Phys. <u>66</u>, 311 (1971).

3. Boneh, Y.: Weizmann Institute, private communication.
   Brandt, A, Kelson, I.: Phys. Rev. <u>183</u>, 1025 (1969).

4. Kumar, K., Baranger, M.: Nucl. Phys. <u>A110</u>, 490 (1968) (ref. to earlier work).

5. Gneuss, G., Greiner, W.: Nucl. Phys. <u>A171</u>, 449 (1971). (ref. to earlier work).

6. Greiner, W., Scheid, W.: Journ. de Phys. C6, supp. au n°11-12, tome 32, 91 (1971) (ref. to earlier work).

7. Fliessbach, T.: Zeitschr. f. Physik <u>238</u>, 329 (1970); Z. Physik <u>242</u>, 287 (1971).

8. Schmitt, H.: Lyssekil conference 1969.

9. Nörenberg, W.: Phys. Rev. <u>C5</u>, 2020 (1972).

10. Pauli, H.C., Götz, U.: preprint, University of Basel, 1973.

11. Hill, D.L., Wheeler, J.A.: P.R. <u>89</u>, 1102 (1953); Mihailovic, M.V., Rosina, M.: "GCM for nuclear bound states and reactions", Proc. of a working sem. Aug./ Sept. 1972, Ljubljana (ref. to earlier work.)

12. Fong, P.: P.R. <u>89</u>, 332 (1953); P.R. <u>135</u> (1964), <u>B1338</u> (ref. to earlier work).

13. Nörenberg, W.: "Zur mikroskopischen Beschreibung der Kernspaltung",
    Habilitationsschrift, Univ. Heidelberg 1970; Conf. on Phys. and Chem. of
    Fission", Vienna 1969, p. 51.

14. Siemens, P.J., Bondorf, J.P., Gross, D.H.E., Dickmann, F.: Phys. Lett. 36B,
    24 (1971).
    Bondorf, J.P., Dickmann, F., Gross, D.H.E.m Siemens, P.J.: Journ. de Phys. C6,
    suppl. au N°11-12, tome 32, C6-145 (1971)

15. Gaudin, M.: Nucl. Phys. A144, 191 (1970).
    Kleber, M., Schmidt, H.: Zeitschr. f. Phys. 245, 68 (1971).
    Dietrich, K.: Ann. of Phys. 66, 480 (1971); Phys. Lett. 32B, 428 (1970).
    Dietrich, K., Hara, K., Weller, F.: Journ. de Phys. Co, suppl. au N°11-12,
    tome 32, C6-183 (1971); Phys. Lett. 35B, 202 (1971).

16. Dietrich, K., Hara, K.: Journ. de Phys. C6, suppl. au N°11-12, tome 32,
    C6-179 (1971).

17. Eichler, J.: Phys. Lett. 37B, 250 (1971).

18. Migdal, A.B.: preprint, sbumitted to ZhETF, July 1972.

19. Tracy, B.L., Chaumont, J., Klapisch, R., Nitschke, J.M., Poskanzer, A.M.
    Roeckl, E., Thibault, C.: P.R. C5, 222 (1972)
    Nifenecker, H., Girard, J., Matuszek, J., Ribrag, M.: suppl. Journ. de Phys.
    tome 33, C5-1972, 29
    Ehrenberg, B., Amiel, S.: P.R. 6c, 618 (1972),
    Amiel, S., Feldstein, H., Nucl. Chem. Dept., Soreq Nuclear Research Center,
    Yavne, Israel: "A systematic odd-even effect in the independent yield dis-
    tributions of nuclides from thermal neutron fission of ^{235}U", to be published
    in Fission Symposium, Rochester, Aug. 1973.

20. Armbruster, P., Mokler, P.H., Stein, H.J.: Phys. Rev. Lett. 27, 1623 (1971);
    ibid 29, 827 (1972).

21. Zener, C: Proc. Roy. Soc. A137, 696 (1932).

DETERMINATION OF NUCLEAR DEFORMATIONS FROM

INELASTIC SCATTERING OF HEAVY IONS NEAR THE COULOMB BARRIER

Dietrich Pelte

Max-Planck-Institut für Kernphysik, Heidelberg

and

Uzy Smilansky

Weizmann Institute of Science, Rehovot

1. INTRODUCTION

Inelastic scattering of heavy projectiles below the Coulomb barrier (Coulomb Excitation)[1] is a reliable and comfortable tool to measure nuclear properties. Its main advantages are:

a) The process is completely determined by the well understood electromagnetic interaction.

b) The nuclear structure information is obtained by measuring the matrix element of the charge multipole operators. The deduced quantities are thus strictly model independent. (In most cases it is convenient to parametrize the measured multipole moments matrix elements in terms of collective degrees of freedom[2]. This is certainly a model dependent procedure but it is necessary in order to correlate the Coulomb excitation results with data obtained from other experimental sources and to get a unified picture of the nuclear charge distribution).

c) The analysis of the experiments can in most cases be carried out in the framework of a well established semi-classical theory[3].

In view of the success of the Coulomb excitation technique in providing so much data concerning nuclear charge deformations, it seems reasonable to try and carry Coulomb excitation like studies to energies slightly above the Coulomb barrier. We might expect that some of the attractive features of the sub-Coulomb region will remain while a wider range of phenomena and nuclear properties could be studied. As long as we restrict ourselves to the energy region mentioned above, the nuclear and Coulomb effects have almost the same order of magnitude, so that

by studying the interference between them we might get a good means to compare
charge and nuclear mass deformations. Since the relation between nuclear mass and
nuclear potential is not well understood, one would need some reliable model to
provide the link between potential and mass distributions.

In this lecture we shall describe recent attempts to investigate the advantages and difficulties encountered in such studies. We shall start by describing
the problems encountered already at the level of elastic scattering. We shall then
turn to discuss inelastic reactions which can be characterized as one step process.
They provide a further check on the validity of the theories proposed to explain
the elastic scattering data, and to remove some ambiguities inherent to these
theories. Finally, multiple-steps inelastic reactions will be described with a
special attention to the deduction of Hexadecouple deformations. Some problems
connected with the measurement of the E4 matrix elements applying the pure Coulomb
excitation technique will be discussed, and a comparison between the charge and
mass moments will be presented.

2. ELASTIC SCATTERING

Typical results of elastic scattering experiments[4] near and above the Coulomb
barrier are shown in fig. 1. The onset of the nuclear interaction is clearly
demonstrated by the abrupt decrease of the elastic cross sections at the back
angles. At forward angles, deviations from the Rutherford cross sections appear
at higher bombarding energies. The typical decrease in the cross sections is
preceeded by an increase (or even a few oscillations) which indicate the diffract-
ive nature of the scattering process. In this lecture we shall be concerned with
phenomena that can be described in terms of pure potential scattering. The energy
regions considered are such that the elastic cross sections at back angles remain
larger than about one tenth of the Rutherford result.

In order to describe the elastic cross sections in a quantitative manner, we
have to introduce the nuclear interaction. We use a phenomenological optical
model, thus adding four parameters which should be determined by fitting the
experimental data. We use the same form factor for the real and imaginary parts.

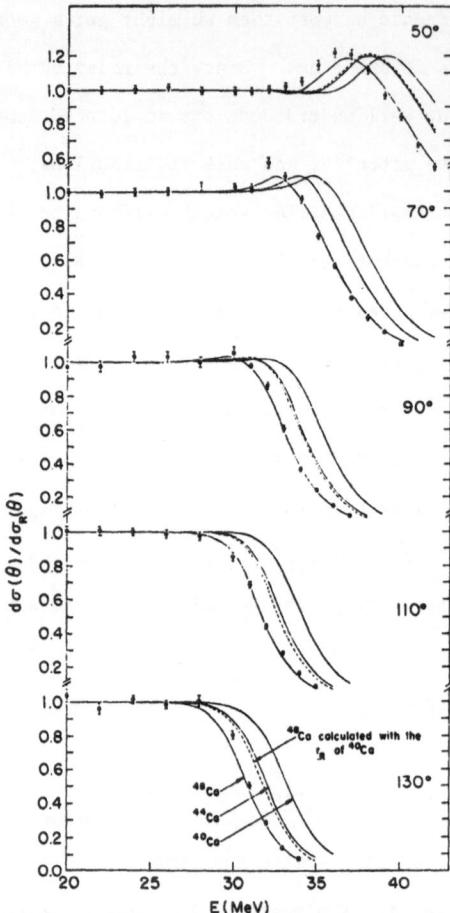

Fig. 1. Measurements of the ratios $d\sigma_{el}(\theta,E)/d\sigma_{Rutherford}(\theta,E)$ for ^{16}O scattered from ^{48}Ca. Optical model fits for this data as well as fits for smilar experiments on other Ca isotopes are included. The dashed lines represent calculations for ^{16}O on ^{48}Ca calculated with the optical parameters deduced for ^{40}Ca (ref. 4)

$$V_{NUC}(r) = -(V_o + iW_o)(1 + \exp(r - R_o)/a)^{-1} \qquad (1)$$

The real part of the total potential composed of the nuclear and Coulomb potential is displayed in fig. 2, together with the ratio $d\sigma_{el}(\theta=\pi)/d\sigma_{Rutherford}(\theta=\pi)$. It is seen that in the region of interest the nuclear potential is already about 50 times weaker than its maximum value. The Coulomb barrier peaks at a radius $r_R \stackrel{\sim}{=} R_0 + 5*a$. At such distances the nuclear potential can be approximated by

$$V_{NUC}(r) = - (V_0 + iW_0) \exp(R/a) \exp(-r/a) \qquad (2)$$

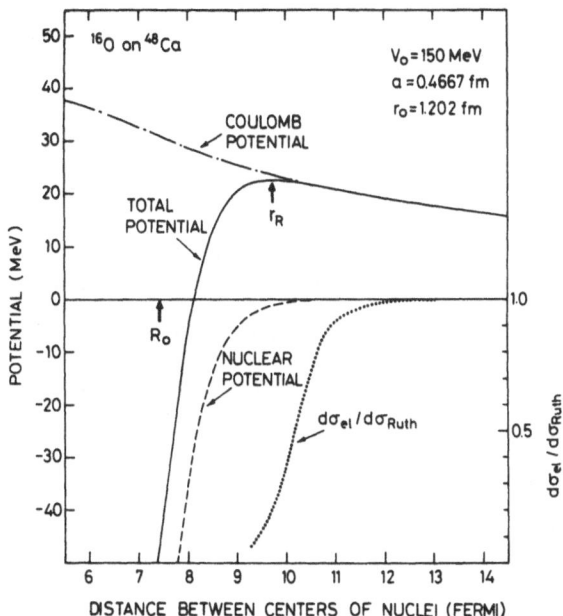

Fig. 2.   The various components which contribute to the interaction between ^{16}O and ^{48}Ca as a function of the distance between the centers of the colliding ions.   The ratio $d\sigma_{el}(\theta=180°)/d\sigma_{Rutherford}(\theta=180°)$ is also shown as a function of the distance of closest approach, deduced from the bombarding energy with the assumption of a Rutherford trajectory

Both V_0 and W_0 appear in the expressions only through the combination $V_0\exp(R/a)$ and $W_0\exp(R/a)$. This expresses the well-known ambiguity (introduced by Igo[5]) in the determination of optical model parameters from elastic scattering reactions characterized by a strong volume absorption. Under certain circumstances, however, these ambiguities can be circumvented by specifying the optical model potential in a more appropriate way. We can thus try a description in which the absorption is taken into account by assuming that all the flux which penetrated the Coulomb barrier is completely absorbed. For backward scattering it is especially easy to calculate the barrier penetrability P, and try to fit the measured $d\sigma_{el}/d\sigma_{Rutherford}$ by the expression:

$$d\sigma_{el}/d\sigma_{Rutherford} = 1-P \quad \text{for} \quad \theta=\pi . \tag{3}$$

Approximating the vicinity of the barrier maximum by a second order polinomial, one can get an analytical form for P so that[6]

$$d\sigma_{el}/d\sigma_{Rutherford} = [1 + \exp (E - V_B)/\omega]^{-1} \tag{4}$$

where V_B is the height of the barrier and ω is proportional to the second derivative of the potential at the maximum. The elastic cross sections for ^{16}O and ^{18}O scattered from ^{40}Ca and ^{52}Cr at $=180°$ are analyzed according to the eq. (4) and are shown in fig. 3. It is clearly demonstrated that the "never come back" approximation is quite successful for the case of ^{16}O ions, while deviations occur for the scattering of ^{18}O on the same targets. Thus, it seems that at least for very tight projectiles (like ^{16}O or ^{4}He) we should be able to use the picture presented above. It can be formulated more rigorously in terms of boundary conditions on the scattering wave functions at the inner side of the Coulomb barrier. At a certain $R_m < r_R$ the wave function should match to a function which has only an incoming part for smaller radii than R_m. In this way[6] (denoted by IWBC - Incoming Wave Boundary Condition) we can calculate the elastic cross sections using the real part of the optical potential. It was demonstrated by Eisen and Vager[7] that the results of such calculations are insensitive to the choice of R_m provided that it is sufficiently smaller than r_R. The nuclear potential takes the form

$$V_{NUC}(r) = V_o \, e^{R/a} \, e^{-r/a}$$
$$= \tilde{V}_o \, e^{r/a} . \tag{5}$$

so that we are left with only two optical parameters $\tilde{V}_o$ and a which can be determined from the data. r_R can be easily determined from V_o, a and the product $Z_1{}^*Z_2$. The cross sections measured for scattering of ^{18}O and ^{3}He indicates that surface

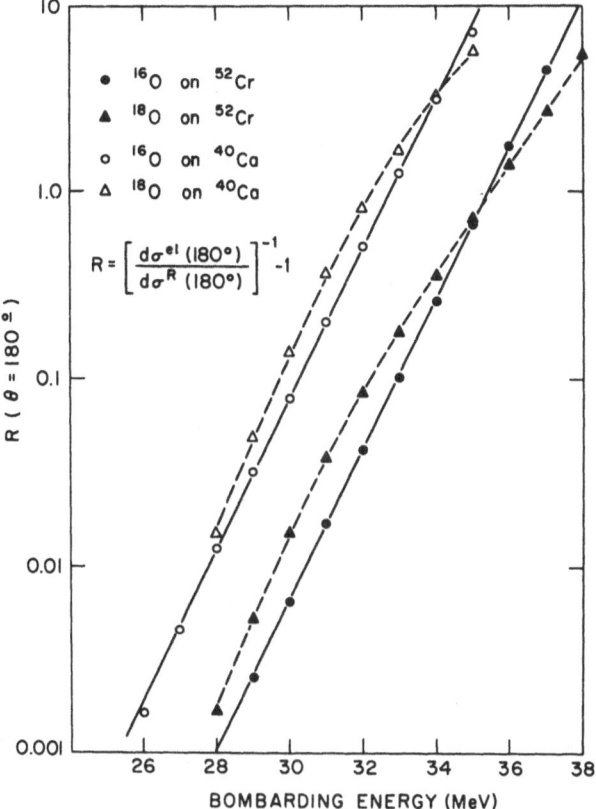

Fig. 3. Analysis of various elastic scattering experiments in terms of the extreme "never come back" approximation

absorption should be included on top of the IWB condition. This is done by writing the optical potential in the form

$$V_{NUC}(r) = V_{NUC}^{Real}(r) \ (1 + ic). \tag{6}$$

IWBC fits in which the three variables a, r_R and c are used as parameters are shown in fig. 4.

As is the case in multi-parameter fits, the deduced parameters are correlated, and their accuracy is not the same. This fact reflects the sensitivity of the assumed theory to variation in the parameters. The standard error analysis theory shows that one can define a set of parameters for which the correlation matrix is

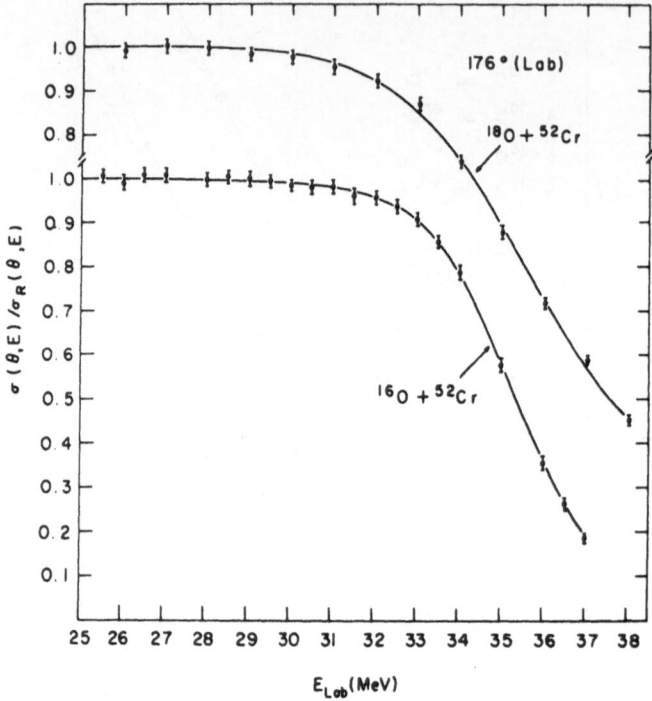

Fig. 4. IWBC analysis of the elastic scattering of $^{16,18}O$ on ^{52}Cr (ref. 8)

diagonal. In this representation a hierarchy of the parameters can be established, where the parameters with the larger eigenvalues have the smaller uncertainties.

When such an analysis is applied to our problem, one can see that the parameter r_R (the location of the Coulomb barrier) is an eigenparameter and it can be determimed in the most accurate way. The two other parameters correspond to two combinations of a and c, for which we cannot give a physical interpretation.

An attempt to study systematics of optical model parameters using back-scattered ^{16}O and ^{18}O was recently reported[8]. Figs. 5 and 6 summarize these experiments by showing the radius r_R and diffuseness parameter **a** for various targets. It is seen that the ^{16}O data can be well described by assuming a radius $r_R = 1.555 \times (Ap^{1/3} + A_T^{1/3})$ fm and a nearly constant diffuseness a = 0.48 fm. The corresponding values for the charge distributions as obtained from electron scattering are $r_c = 1.05 \times A_T^{1/3}$ fm and a = 0.55 fm. The ^{18}O data on the other hand, do not behave in the same regular way and the deduction of trends in r_R or

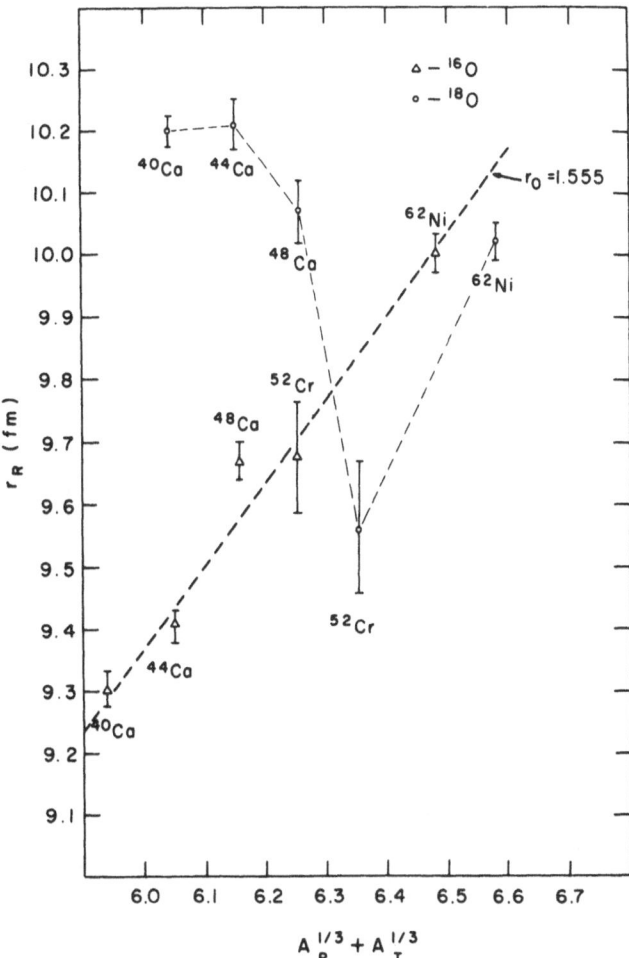

Fig. 5. The radius parameter r_R as a function of $A_P^{1/3} + A_T^{1/3}$ for the scattering of $^{16,18}O$ on various targets. The heavy dotted line is a best fit for the results of ^{16}O with $r_o = 1.555$ fm. The dotted line through the ^{18}O points serves only to guide the eye (ref. 8)

a is impossible. This comparison demonstrates that in general the parameters with which we describe the optical interaction are not simple additive properties of the two colliding ions but that they might depend on the properties of the composite system as a whole.

Very interesting theoretical and experimental studies were recently reported

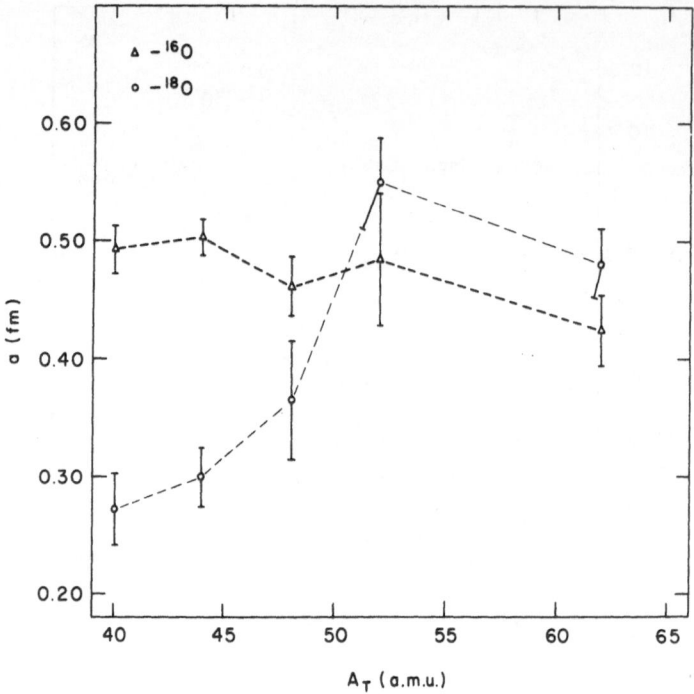

Fig. 6. The diffuseness parameter a as a function of A_T. The dotted lines are only to guide the eye (ref. 8)

from Copenhagen[9-12] in which the semi-classical aspects of the elastic (as well as inelastic) scattering were investigated. Elastic scattering experiments were conducted with [16]O projectiles and the results obtained for the lower energy domain and more backward angles are qualitatively the same as reported in previous experiments. At the higher energy range and at more forward angles, the interference effects before the sharp decrease in the cross section become more prominent as can be seen for example in fig. 7. Optical model analysis of these results indicate that fits could be somewhat improved if the range of the imaginary optical potential exceeded that of the real part while the diffuseness was decreased.

These same experiments were also analyzed in the framework of the semi-classical time dependent theory as was developed by R. Broglia and A. Winther[13]. In this theory, the trajectories are described clasically, while the phase information is not lost and the quantum mechanical rules of interference are observed.

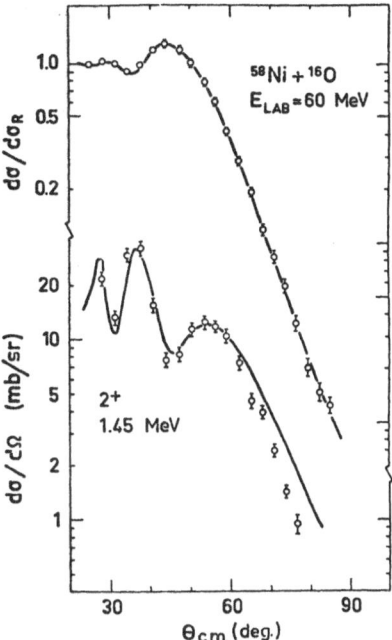

Fig. 7. Elastic and inelastic cross sections of ^{16}O on ^{58}Ni. The line through

the experimental points is theooptical model (and DWBA) fits to the data

(ref. 10)

The real part of the optical potential determines the trajectory and deflection

angle for every impact parameter b. A classical cross section is assigned to a

trajectory via the relation

$$[\frac{d\sigma}{d\Omega}]_{classical} = \frac{b}{\sin\theta} \mid \frac{db(\theta)}{d\theta} \mid \tag{7}$$

A complex time dependent amplitude C(t) is now defined along the trajectory

and the semi-classical scattering amplitude f(b,θ) is defined by

$$f(b,\theta) = C(t = +\infty) \; [\frac{d\sigma}{d\Omega}]^{1/2}_{classical} \tag{8}$$

$$C(t = -\infty) = 1.$$

The modulus of C(t) is mainly determined by the absorption induced by the ima-

ginary part of the optical potential. The importance of the phase of C(t) become apparent only when several impact parameters contribute to the scattering for a given angle. This happens when the attractive potential is sufficiently strong. In such cases the semi-classical scattering amplitudes are added coherently and the cross section is given by

$$\frac{d\sigma}{d\Omega} = \left| \sum_{b} \left[\frac{d\sigma}{d\Omega}(b,\theta) \right]^{1/2}_{class} C(t=+\infty) \right|^2 \qquad (8)$$

This expression must be modified when the contributing impact parameters are too close to the rainbow angles where $\frac{d\theta}{db} = 0$.[14]

The interference patterns observed at forward angles are very well reproduced as is shown in fig. 8.

We can conclude this section with the following remarks.

The various methods proposed to describe the elastic scattering near the

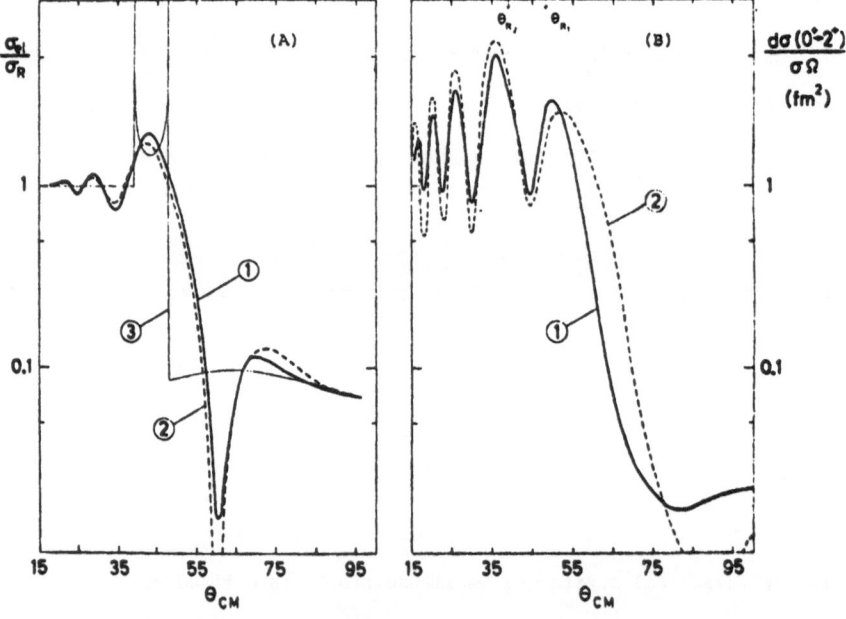

Fig. 8. Comparison between semi-classical and quantum mechanical calculations for elastic and inelastic scattering of ^{16}O from ^{58}Ni. Curve (1) gives the semi-classical result, curve (2) the quantum mechanical and (3) gives the classical cross sections. (A) refer to elastic scattering while (B) gives the inelastic data (ref. 11)

Coulomb barrier give good agreement with the experimental data. They all depend on optical model parameters which cannot be unambigously determined because these experiments are sensitive only to the nuclear surface. In the next section it will be shown that inelastic scattering provides another probe by which deeper parts of the nuclear surface are measured and consequently remove the ambiguities inherent to the elastic scattering.

3. INELASTIC SCATTERING, ONE STEP PROCESS

The main difficulties which arise in the description of inelastic scattering are concerned with the derivation of the form factor for excitation and relating its parameters to other nuclear properties. One may try a solution to these problems in the case of "collective" excitations by assuming that the optical potential deforms and vibrates together with the nuclear matter that induces it. Given an elastic spherical model $V(R^{(o)},r)$ one can deform it by replacing the radius parameter $R^{(o)}$ by a function $R^{(o)}(\Omega)$. Since $R^{(o)}$ is usually taken as the sum of radii of the colliding ions we can write:

$$R^{(o)} = R_P^{(o)} + R_T^{(o)} \rightarrow R_P^{(o)}(1+\Sigma\alpha_P^{\lambda\mu}Y_{\lambda\mu}(\Omega)) + R_T^{(o)}(1+\Sigma\alpha_T^{\lambda\mu}Y_{\lambda\mu}(\Omega)) \qquad (9)$$

where the suffixes P and T stand for the projectile and the target degrees of freedom, respectively. The deformation-parameters α can be related to the charge moments operators assuming a uniform charge distribution. To first order in α

$$m_T(E\lambda,\mu) = \frac{\lambda+1}{4\pi} Z_T^2 R_{CT} \cdot \alpha_T^{\lambda\mu} \qquad (R_{CT} \text{ is the charge radius}). \qquad (10)$$

With the help of eq. (9) one can express the deformed optical potential. To first order in the α parameters

$$V(R(\Omega),r) \cong V(R^{(o)},r) + \frac{\partial V}{\partial R^{(o)}}(R^{(o)},r)\ (R_P^{(o)}\ \Sigma\ \alpha_P^{\lambda\mu}Y_{\lambda\mu}(\Omega) +$$

$$R_T^{(o)}\ \Sigma\ \alpha_T^{\lambda\mu}Y_{\lambda\mu}(\Omega)) + \dots \qquad (11)$$

If we take for $V(R^{(o)},r)$ a Saxon-Wood function, we see that the inelastic form factor (which is proportional to $\frac{\partial V}{\partial R^{(0)}}$) is peaked at the nuclear surface.

A more satisfactory way to deduce the nuclear form factor would be to fold the nuclear mass distribution with the nucleon-alpha interaction. The form factors calculated in this way[15,16] have the same shape as the phenomenological form factors (11) but disagree in the normalization particularly for the higher multipolarities. Satchler[17] included α exchange effects in the derivation of the form factor and found that the disagreement mentioned above is reduced. Due to the exchange effects the interaction becomes non-local and actual calculations were not carried out but for a very few cases[17]. With the lack of conclusive method for constructing form factors from "first principles", and in view of the success of the phenomenological picture to interpret a variety of (α,α') experiments, we shall use the latter picture. The interaction is thus completely determined once the spherical optical potential is specified and the radius parameters R_p, R_T and R_{CT} determined. The relation (10) is used to supply the matrix elements of the operators. It now has to be seen to what extent the inclusion of inelastic scattering data reduces the ambiguities inherent in the optical model analysis of elastic scattering.

In this section we shall discuss transitions to excited states for which the DWBA or semi-classical first order approximations are applicable. Under these conditions the calculation of cross sections are rather simple and fast so that a detailed study of the dependence of calculated quantities on the optical parameters becomes practical. This is not the case when the full coupled-channels problem has to be solved.

The main difficulties in DWBA calculations at energies near the Coulomb barrier stem from the important role played by the Coulomb excitation process. The long range of the Coulomb interaction compel us to integrate the wave equations to larger distances and include also many partial waves. The later is not a handicap once it is realized that the higher partial waves do not interact at all with the nuclear fields. Thus only the Coulomb interaction contribute and recursion relations can be used to calculate the S matrix elements in a fast and accurate way[8]. Without the inclusion of some few hundreds partial waves in this way one cannot get the desired accuracy in the calculations.

The inelastic scattering amplitude is a sum of the nuclear and Coulomb amplitudes. The real part of these two interactions have opposite signs and therefore a destructive interference in the cross section is expected to be observed once the two amplitudes are of the same order of magnitude. The observed interference patterns are not as deep because of the inclusion of the imaginary potential both in the diagonal and non-diagonal elements of the interaction.

Typical excitation functions for α and ^{16}O inelastic scattering are shown in fig. 9 and fig. 10. The interference pattern stands out more strongly in the ^{16}O scattering[10]. The theoretical fits to the data show that the general features are well reproduced by the DWBA, especially for the (α,α') experiment where very good fits are obtained. It may well be that multiple effects be important in the $(^{16}O,^{16}O')$ case.

We shall use the (α,α') data to study the problems which were discussed above in relation with the optical model description and the inherent ambiguities[19]. The

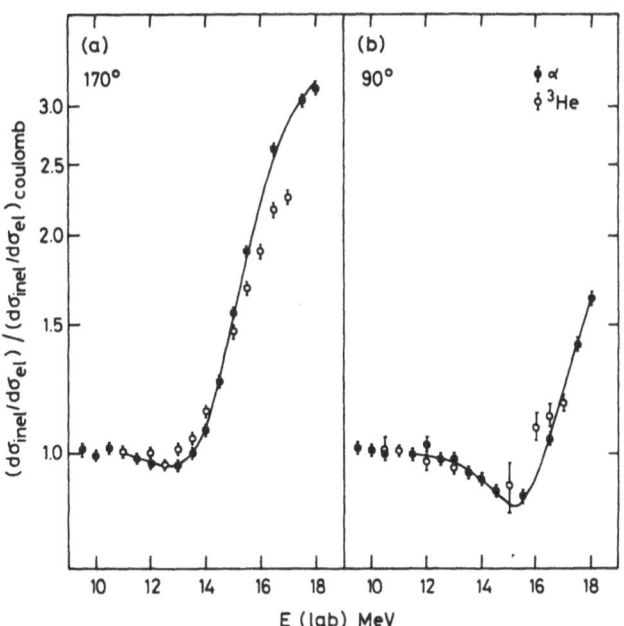

Fig. 9. Inelastic scattering of ^{4}He and ^{3}He from ^{128}Te at backward angles. The excitation probabilities: $d\sigma_{inel}/d\sigma_{el}$ are normalized by the results of pure Coulomb excitation calculations. DWBA fits are shown (ref. 19)

theory depends on five parameters: $R^{(o)}$, a, V_o, W_o and $<0||\alpha_T^2||2^+>$. Fitting the elastic data, we vary only the first four parameters and as is shown in fig. 11, there is a straight line in the $(V_o, R^{(o)})$ plane which the χ^2 is almost constant. The width of this line corresponds to one standard deviation in R_o. If now, for each set (a, V_o, W_o) determined by the elastic data we vary R_o to get the best fit to the inelastic data, we get the other straight line in fig. 11. The fact that χ^2 changes significantly along the line shows that the ambiguity in the elastic data is removed. The two lines intersect near the minimum of χ^2 on each of them. The analysis described above was carried out by restricting the matrix element of the α_T^2 by the relation (10). Preliminary studies indicate that removing the restric-

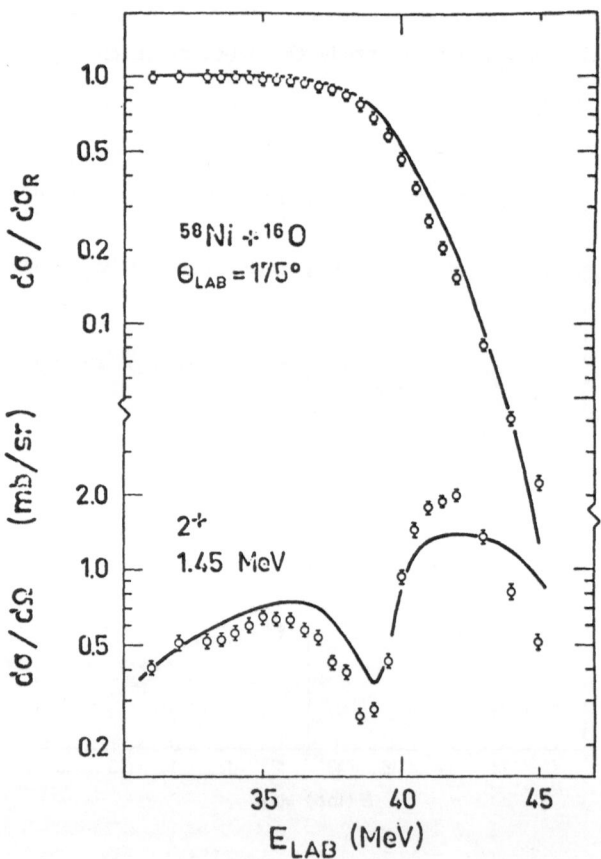

Fig. 10. Inelastic scattering of ^{16}O from ^{58}Ni at backward angle θ=175°. DWBA fit to the data are shown (ref. 10)

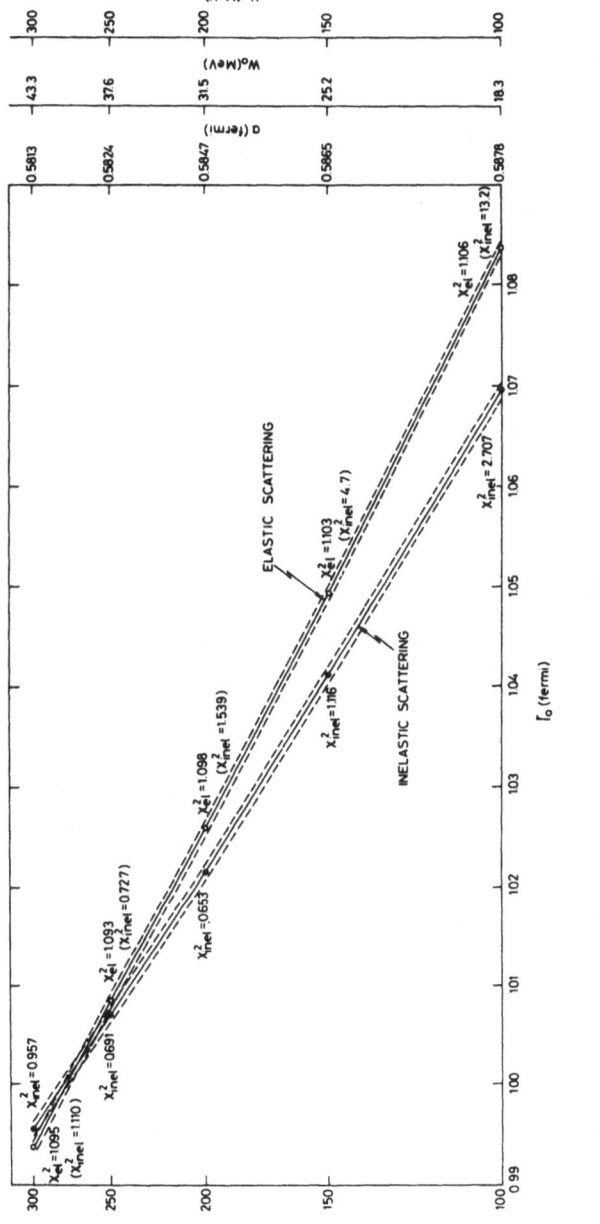

Fig. 11. χ^2 fits to the elastic and inelastic data for ^{4}He scattering on ^{128}Te. Lines of minimum χ^2 are shown in the $(r_o, R^{(o)})$ plane and the corresponding values of a and W_o are indicated on the left columns. The values of χ^2 (for elastic or inelastic data) are given for various points on the lines

tion (10) does not introduce a new ambiguity so that $<0||\alpha_T^2||2>$ can be determined
and compared with the analogous charge deformation matrix element.

The semi-classical theory for inelastic scattering was applied successfully
by the Copenhagen group[9-12] to the measurements of Christiansen et al[10]. The ideas
which underline this theory are similar to those discussed in the previous chapters.
Most of the features of the inelastic cross sections were reproduced, and definite
predictions about the relative phases between the oscillation of the elastic and
inelastic cross sections at forward angles were derived.

It is perhaps proper to conclude this section by emphasizing again the fact
that the success of either the DWBA or semi-classical calculations do not go beyond
our phenomenological picture of the reaction mechanism. Therefore, care should be
exercised when the derived quantities are compared to other spectroscopic data,
even though the agreement might be very convincing.

4. MULTISTEP PROCESSES IN ROTATIONAL NUCLEI

In the last sections it was shown that elastic scattering at bombarding
energies near the Coulomb barrier probes the interaction at distances outside the
nuclear radius and that the phenomenological optical potential is not uniquely
described by this process. However, by including one step processes in the
scattering mechanism the ambiguities in the parametrization of the potential seem
to be removed. This suggests that the study of inelastic scattering to various
excited states at bombarding energies slightly above the Coulomb barrier might
give information about the nuclear charge and mass distributions at the nuclear
surface. In order to demonstrate the effects that are observed in experiments of
such kind we show in fig. 12 some results[20] obtained with α scattering on rare
earth nuclei. The upper part of this figure shows the elastic scattering ratios
$d\sigma/d\sigma_R$ at a scattering angle θ_{Lab} = 170° for various energies of α particles
bombarded onto ^{152}Sm and ^{154}Sm. The inelastic cross ections for the first excited
2^+ and 4^+ states in these nuclei are displayed in the lower part of fig. 12. The
excitation probabilities defined as $d\sigma(J^+)/d\sigma(0^+)$ are normalized to the corresponding calculated results of pure Coulomb excitaiton. The general structure of the

excitation functions is the same for both nuclei, differences are observed, for
example, in the widths of the interference minima at $E \stackrel{\sim}{=} 15$ MeV and in the slopes
at higher bombarding energies. In the following we shall discuss the problems
that are encountered when the observed structures in the excitation functions are
used in order to deduce nuclear mass and charge deformations. In doing so we
first want to review some aspects of Coulomb excitation and shall point out where
the semi-classical approximation has to be replaced by more accurate procedures in
order to obtain reliable results.

Coulomb excitation into the lowest 2^+ states provides a simple method to
deduce the nuclear quadrupole charge deformation. The determination of higher
charge multipoles by Coulomb excitation is difficult. The reason is the electric

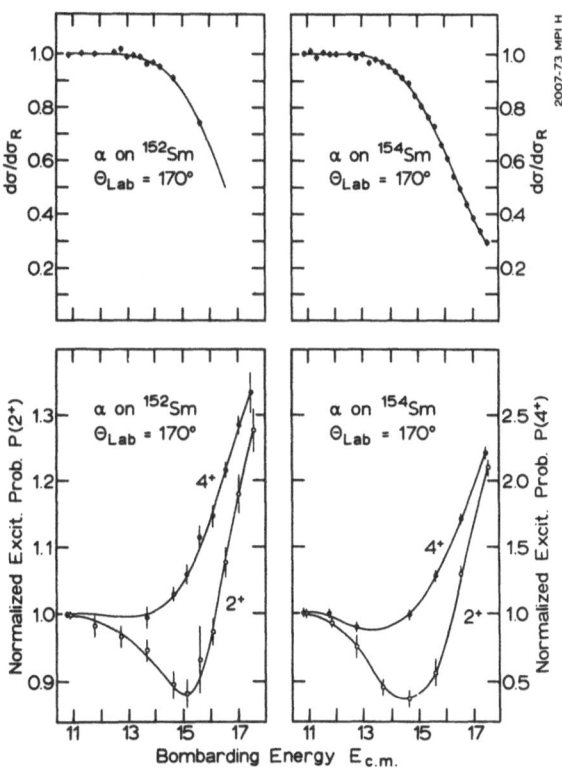

Fig. 12. Elastic scattering and excitation probabilities observed in α scattering
on ^{152}Sm and ^{154}Sm. The results are normalized to pure Coulomb excitation

form factor which is proportional to $1/r^{\lambda+1}$ and favours low multipolarities λ. However, with α particles the nucleus can be approached to such small distances that the influence of the direct hexadecupole excitation to the first excited 4^+ state of a rotational nucleus becomes observable. Figure 13 displays the percentage change in the excitation probabilities $P(2^+)$ and $P(4^+)$ of the lowest 2^+ and 4^+ states of ^{152}Sm due to the hexadecupole matrix element M_{04} which connects the ground state to the 4^+ state. These calculations were performed in the framework of the semi-classical approximation[3]. The change in $P(4^+)$ is around 20% for the largest values of M_{04} measured in rare earth nuclei. As can be also seen from fig. 13 the measurement of $P(4^+)$ does not provide a unique determination of M_{04}. The sign of M_{04} is generally inferred from theoretical predictions. The results of the semi-classical theory have to be corrected due to quantal effects that become important when the Sommerfeld parameter $\eta = Z_1 Z_2 e^2/hv$ is as small as

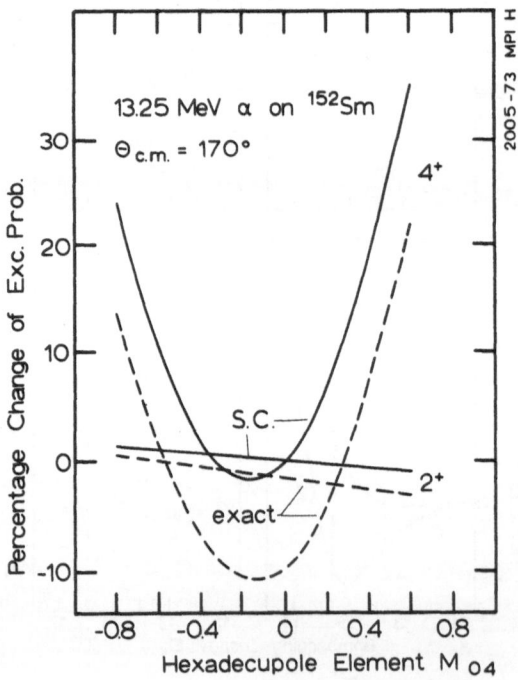

Fig. 13. Dependence of the 2^+ and 4^+ excitation probabilities on the hexadecupole matrix element M_{04}

$\eta \approx 10$ which is the case in Coulomb excitation with α particles. The assumption of classical trajectories overestimates the distances up to which the projectiles probe the electromagnetic interaction. The size of these corrections depend therefore on the slope of the form factor near the classical turning point. They increase with the order of the multiple E2 excitations that are necessary to populate high spin states. The reason is that multiple E2 excitations of order n can be considered as an effective E(2n) transition and the corresponding effective form factor has a stronger radial dependence than in the case of one step E2 excitations. The quantal correcitons to the 2^+ and 4^+ excitation probabilities are displayed in fig. 13 as dotted lines. The excitation probability of the 2^+ state is changed by about 1.5% whereas the correction amounts to 12% for the excitation probability of the 4^+ state. This might change the deduced hexadecupole matrix element M_{04} by as much as 30%. The quantal corrections are therefore generally taken into account when the calculated $P(4^+)$ is compared to experiments. The corresponding correction to $P(2^+)$ is considered as small enough to be neglected. It should be noted, however, that the 4^+ excitation probability is quite insen-

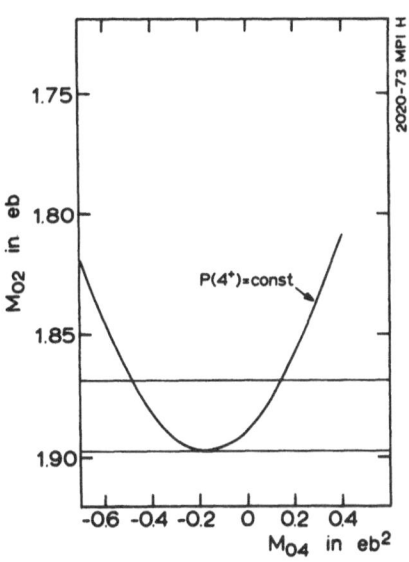

Fig. 14. M_{02} versus M_{04} curve for a fixed transition probability $P(4^+)$. The two horizontal lines correspond to an uncertainty of 1.5% in M_{02}

TABLE I

E2 and E4 Transition Matrix Elements and

Charge Deformations Deduced from them

	ref.	M_{02}[eb]	M_{04}[eb^2]	$\beta_2^{(c)}$	$\beta_4^{(c)}$
^{152}Sm	a	1.857 ± 0.016	0.374 ± 0.090	0.284	0.063
	b	1.835 ± 0.007	0.470 ± 0.070	0.273	0.092
	c	1.841 ± 0.008	0.350 ± 0.070	0.283	0.057
	d	1.830 ± 0.020	0.450 ± 0.090	0.274	0.087
	e	1.836 ± 0.019	0.368 ± 0.017	0.281	0.062
	f	1.852 ± 0.017	0.510 ± 0.080	0.272	0.103
^{154}Sm	a	2.116 ± 0.020	0.530 ± 0.070	0.309	0.091
	b	2.066 ± 0.008	0.653 ± 0.050	0.291	0.128
	c	2.074 ± 0.008	0.430 ± 0.080	0.311	0.065
	d	2.053 ± 0.023	0.670 ± 0.080	0.288	0.134
	f	2.065 ± 0.010	0.570 ± 0.090	0.298	0.104
^{158}Gd	b	2.225 ± 0.012	0.390 ± 0.090	0.324	0.044
	c	2.230 ± 0.030	0.340 ± 0.110	0.328	0.030
	f	2.228 ± 0.018	0.380 ± 0.170	0.325	0.041
^{160}Gd	b	2.276 ± 0.014	0.317 ± 0.100	0.335	0.021
	c	2.290 ± 0.020	0.360 ± 0.106	0.334	0.031
^{162}Dy	c	2.320 ± 0.010	0.270 ± 0.010	0.333	0.007
^{164}Dy	b	2.356 ± 0.018	0.220 ± 0.140	0.339	-0.008
	c	7.360 ± 0.010	0.280 ± 0.110	0.336	0.007
	f	2.365 ± 0.012	0.220 ± $^{0.190}_{0.230}$	0.341	-0.009

sitive to the values of M_{04} near the minimum which occurs at $M_{04} = -0.15$ eb^2 in the case considered here. Small errors in the quadrupole matrix element M_{02} are therefore enhanced when M_{04} is deduced from the measured excitation probability $P(4^+)$. This is demonstrated in fig. 14 which displays the relation between M_{02} and M_{04} for a constant excitation probability $P(4^+)$. In this calculation the E2 strengths between the $4^+ \rightarrow 2^+$ and the $2^+ \rightarrow 0^+$ states were assumed to follow the predictions of a rigid rotator. As is inferred from this figure a change of M_{02} by 1.5% would change M_{04} by 0.33 eb^2 in the worst case. This change is comparable to the one due to quantal corrections in $P(4^+)$. As pointed out already the quantal corrections become increasingly important for higher states in the rotational band which are populated by a large number of multiple E2 transitions. An estimate for the semi-classical nature of these processes is given by the condition $\eta/2n \gg 1$. In Coulomb excitation of ^{152}Sm with α particles the quantal corrections are in the order of 30% for the 6^+ state and 60% for the 8^+ state at backward angles[21].

Table 1 contains the results of Coulomb excitation studies in the rare earth and actinide regions using α particles as projectiles. The first two columns show the E2 and E4 transition matrix elements M_{02} and M_{04} leading from the ground state to the 2^+ and 4^+ states, respectively. These matrix elements were used to calculate the $\beta_2^{(c)}$ and $\beta_4^{(c)}$ charge deformations which are defined by

$$M_{02} = \int r^2 Y_{20}(\Omega)\rho^{(c)}(R(\Omega),r)dV, \qquad M_{04} = \int r^4 Y_{40}(\Omega)\rho^{(c)}(R(\Omega),r)dV$$

The model assumed for the charge density $\rho^{(c)}(R(\Omega),r)$ is a deformed fermi form

$$\rho^{(c)}(R(\Omega),r) = \rho_0^{(c)} / (1 + \exp(r - R(\Omega))/a)$$

with $\qquad R = R_0^{(c)} (1 + \beta_2^{(c)} Y_{20}(\Omega) + \beta_4^{(c)} Y_{40}(\Omega))$.

The charge radius $R_0^{(c)} = 1.1 A_2^{1/3}$fm and the diffuseness a = 0.6 fm were taken from the results of μ-mesic atoms[22]. Nuclei in these mass regions are characterized by strong β_2 deformations as is well known. Strong β_4 deformations are found for the actinides and in the beginning of the rare earth region. The small values of β_4 obtained in the middle of the rare earth region should be taken with caution

TABLE I (Cont'd.)

	ref.	$M_{02}[eb]$	$M_{04}[eb^2]$	$\beta_2^{(c)}$	$\beta_4^{(c)}$
^{166}Er	b	2.399 ± 0.012	0.300 ± 0.180	0.328	0.012
	c	2.400 ± 0.020	$0.060 \, ^{+\,0.120}_{-\,0.180}$	0.344	-0.048
	f	2.411 ± 0.009	$0.210 \, ^{+\,0.210}_{-\,0.410}$	0.336	-0.011
^{168}Er	b	2.435 ± 0.014	0.285 ± 0.200	0.332	0.006
	c	2.400 ± 0.020	$0.200 \, ^{+\,0.120}_{-\,0.180}$	0.333	-0.013
^{170}Er	c	2.410 ± 0.020	$0.240 \, ^{+\,0.120}_{-\,0.180}$	0.329	-0.003
^{174}Y	b	2.435 ± 0.012	0.230 ± 0.170	0.320	-0.005
^{230}Th	g	2.830 ± 0.020	1.049 ± 0.236	0.224	0.104
^{234}U	g	3.228 ± 0.031	1.761 ± 0.244	0.226	0.174
^{236}U	g	3.388 ± 0.022	1.300 ± 0.320	0.253	0.113
^{238}U	g	3.421 ± 0.022	1.122± 0.262	0.260	0.090
^{238}Pu	g	3.544 ± 0.050	1.449 ± 0.228	0.255	0.124
^{240}Pu	g	3.641 ± 0.021	1.179 ± 0.252	0.270	0.089

References

a) Brückner W., Pelte, D., Povh, B., Smilansky, U., Traxel, K.: in Proceedings of the Osaka Conference on Nuclear Moments and Nuclear Structure, to be published.

b) Greenberg, J.S., Shaw, A.H.: ibid.

c) Erb, K.A., Holden, J.E., Lee, J.Y., Saladin, J.X., Saylor, T.K.: Phys. Rev. Lett. 29, 1010 (1972).

d) Stephens, T.S., Diamond, R.M., de Boer, J.: Phys. Rev. Lett. 27, 1151 (1971).

e) Bertoozi, W., Cooper, T., Ensslin, N., Heisenberg, J., Kowalski, J., Mills,M., Turchinetz, C., Williamson, C., Fivoznisky, S.F., Lightlody, J.W., Penner, S.: Phys. Rev. Lett. 28, 1711 (1972).

f) Ebre, K.W.: University of Frankfurt, Germany, private communication.

g) Milner, W.T., McGowan, F.K., Bennis, C.F., Ford, J.L.C., Robinson, R.L., Stelson, P.H.: J. de Physique C5, 112 (1972).

because quantal corrections for $P(2^+)$ were generally not taken into account in the analysis.

At energies above the Coulomb barrier the cross section of the inelastic 2^+ scattering displays the known interference between Coulomb and nuclear excitation as was shown in fig. 12. The interference effects in the inelastic 4^+ scattering are much less pronounced at backward angles but become stronger at forward angles. For a systematic study of the different contributions that lead to the observed behaviour of the 4^+ excitation function, $R(4^+)$ was calculated[20] for different combinations of the potential deformation $\beta_4^{(P)}$ and the hexadecupole charge moment M_{04}. The absolute values of these two quantities were kept fixed whereas their signs were changed as is indicated in fig. 15 by the superscripts of R. The most important components in the excitation amplitude for the 4^+ state are: (a) two step processes due to either E2 Coulomb or $\lambda = 2$ nuclear excitations, (b) direct E4 excitation, and (c) direct $\lambda = 4$ nuclear excitation.

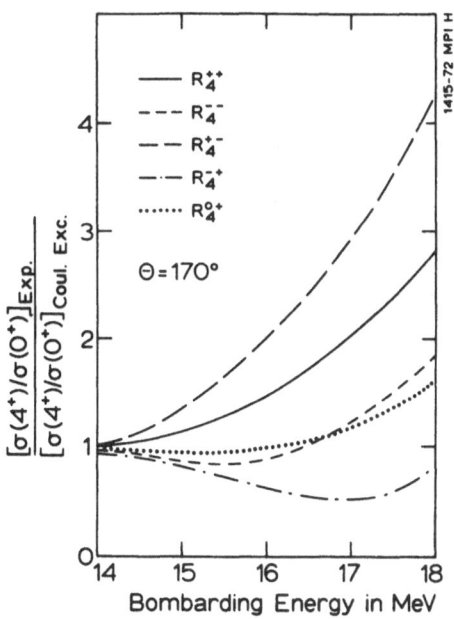

Fig. 15. Normalized 4^+ excitation probabilities for various sign combinations (see text) of M_{04} and $\beta_4^{(P)}$ (ref. 20)

$$f_{4+} = A + M_{04} B + \beta_4^{(P)} C$$

where A, B, and C correspond to the processes (a), (b), and (c), respectively. The cross section will be of the form

$$d\sigma_{4+} = |A|^2 + M_{04}(A \cdot B) + \beta_4^{(P)}(A \cdot C) + \dots$$

with the notation $(X \cdot Y) = 2\text{Re}(X \cdot Y^*)$. Thus the bulk of the cross section comes from the square of (a) while processes (b) and (c) contribute to the cross section mainly through their interference with (a). As can be shown this simple formulation describes the calculated cross sections to an accuracy of better than 10% in the entire energy region studied here. The magnitude and sign of the 4^+ interference is mainly determined by the values of $\beta_4^{(P)}$ and M_{04} and by the phase difference between the two interference terms. In situations like α scattering which are characterized by a weak imaginary component in the optical potential, the phase difference is around 160° at backward angles. Therefore the contribution of one term is constructive when the other term interferes destructively for equal signs of $\beta_4^{(P)}$ and M_{04}. This result indicates that the interference effect in the 4^+ excitation probability can be used in general to determine uniquely the sign of M_{04} which was not obtained in Coulomb excitaiton studies. This simple picture might change, when the imaginary part of the potential becomes comparable to the real part. In this case the relative phase changes in the direction of 90° and the interference effects between processes (a) and (c) might disappear altogether in special situations. This behaviour is quite similar to the behaviour of the interference in the 2^+ excitation probability and might be used to determine the magnitude of the imaginary component in the nuclear form factor[23].

Because of the strong coupling between the low lying states in a rotational nucleus the analysis of the elastic and inelastic scattering cross sections has to be performed by the method of coupled channels. The problems in deriving the nuclear form factors were discussed already in section 3.

We have for the phenomenological form factor in the case of rotational nuclei

$$U_L(r) = \int U(R(\Omega),r) \, Y_{LO}(\Omega) d\Omega$$

with the deformed Wood-Saxon potential $U(R(\Omega),r)$ that was used in section 3. In relating the potential deformations $\beta_\lambda^{(P)}$ to the mass deformations $\beta_\lambda^{(M)}$ one assumes the empirical relation

$$\beta_\lambda^{(M)} R_o^{(M)} = \beta_\lambda^{(P)} R_o^{(P)} \tag{12}$$

where $R_o^{(M)}$ and $R_o^{(P)}$ are the mass and the potential radii, respectively. This relation was used already to deduce the equation (11). By changing the potential radius $R_o^{(P)}$ to the radius $R_o^{(M)}$ of the nuclear mass distribution a mass deformation $\beta_\lambda^{(M)}$ can be deduced and compared to $\beta_\lambda^{(c)}$ obtained from Coulomb excitation. It is hoped that the sensitivity of the observed interference effects in elastic and inelastic scattering might allow one also to check these assumptions.

The complete analysis of the α scattering experiments is completed[20] only for

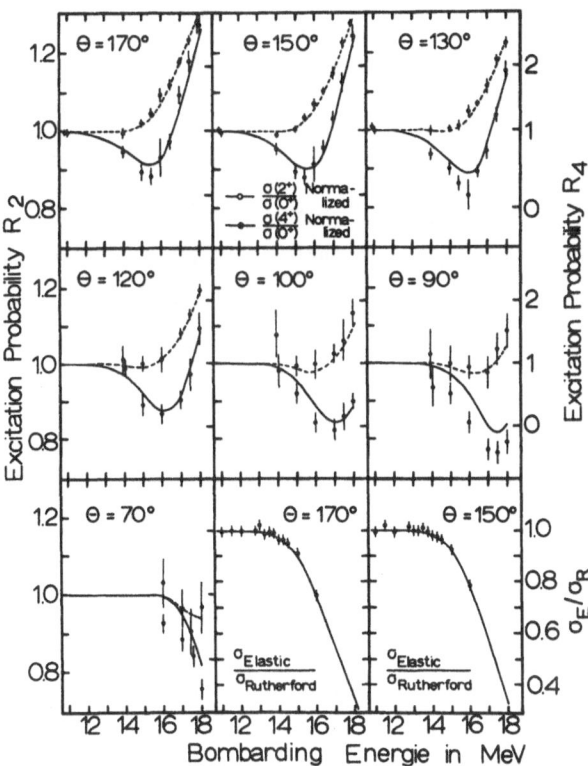

Fig. 16. Optical model analysis of elastic and inelastic α scattering on ^{152}Sm (ref. 20)

the case of ^{152}Sm. The nuclear form factors were calculated from the optical

potential including the weak imaginary part. Contrary to the calculation reported

in section 3 the potential deformations were treated as parameters to be determined

by the experimental data. Therefore six optical parameters are to be determined by

fitting the data, and the theoretical cross sections are results of coupled channels

calculations. Thus, a full fitting analysis could not be done within an acceptable

computer time limit. This difficulty was circumvented following the analysis of

Glendenning[24]. He showed that in the analysis of high energy scattering the collec-

tive modes of excitation in rare earth nuclei are the strongest in removing flux

from the entrance channel and that the optical potential is practically the same

for all nuclei in the rare earth region if these inelastic channels are directly

included in the calculation. We therefore took the geometrical properties of the

optical potential from the results[25] of α scattering experiments on rare earth

nuclei at a 30 MeV bombarding energy. The depth of the real and imaginary part of

the optical potential and the deformation parameters were then determined from the

experimental data. The results are displayed in fig. 16. The general behaviour

of the excitation probabilities is well reproduced with the exception of some

features like the width of the 2$^+$ interference at backward angles and the depth of

the interference minimum at forward angles. It is an unsolved problem whether or

TABLE II

Results of the Optical Model Analysis

of α scattering on ^{152}Sm

ref.	Bombarding energy (MeV)	V_o (MeV)	W_o (MeV)	R_o (fm)	a (fm)	$\beta_2^{(P)}$	$\beta_4^{(P)}$	$\beta_2^{(M)}$	$\beta_4^{(M)}$
20)	14-18	45.0	11.2	7.963	0.605	0.214 ± 0.007	0.038 ± 0.007	0.279 ± 0.009	0.050 ± 0.009
25)	27.5-32.5	50.0	11.2	7.963	0.605				
26)	50	65.9	27.3	7.685	0.637	0.205	0.040	0.248	0.048

not these discrepancies reflect, for example, details of the nuclear form factors
that are not included in the phenomenological derivation. The values of the
extracted parameters are quoted in table II together with the corresponding results
of α scattering at higher bombarding energies. If potential deformation parameters
are scaled down to a radius 6.09 fm according to relation (12) the resulting values
are as shown in the last two columns. They agree within the limits of the experimental errors with the charge deformations.

This agreement, however, cannot be considered as representative for the nuclei
of the rare earth region. Extensive studies[25,26] of α scattering on these nuclei
have been carried out at 30 and 50 MeV. The results of these studies are displayed
in fig. 17. This figure also contains the mean values of $\beta_2^{(c)}$ and $\beta_4^{(c)}$ obtained

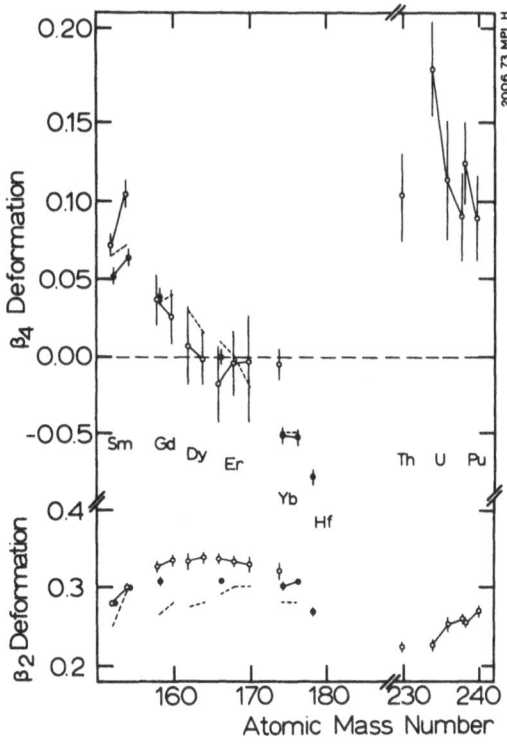

Fig. 17. Values for the mass (dots) and charge (circles) deformations in the
rare earth nuclei and the antinucleus. Dotted lines display the calculation of Götz et al.

from Coulomb excitation and the results of the calculation performed by U. Götz
et al.[27]. The deformation parameters are all scaled to a radius $R_o = 1.1\ A^{1/3}$ fm
thus assuming equal mass and charge radii. Differences outside the quoted errors
are observed particularly for the $\lambda = 2$ deformations. Another noticeable fact is
that there is no conclusive indication yet from Coulomb excitation experiments that
the charge hexadecupole deformation changes sign in the middle of the rare earth
region as was deduced from high energy α scattering.

The preceeding discussions have demonstrated that the interference effects in
elastic and inelastic scattering give us a tool for exploring nuclear surface
properties. The interpretation of the results in terms of a phenomenological potential which is used successfully in elastic scattering reproduced also the observed
effects in inelastic scattering. By this procedure the parameters that enter into
the description of the scattering process seem to be determined without ambiguities.
A comparison between deformation parameters characterizing the nuclear mass and
charge distributions yields an overall agreement in their signs and magnitudes.
Deviations in the order of 10% between these two sets of parameters are observed
in the case of quadrupole deformations. The reason for these deviations is not
understood but might indicate the limitation of the phenomenological approach.

REFERENCES

1. Alder, K., Bohr, A., Huns, T., Mottelson, B., Winther, A.: Revs. Mod. Phys.
   28, 432 (1956).

2. Kumar, K.: Phys. Rev. Letts. 28, 241 (1972).

3. de Boer, J., Winther, A.: in Coulomb Excitation by K. Alder and A. Winther,
   Academic Press (1966), p. 303.

4. Bertin, M.C., Taylor, S.L., Watson, B.A., Eisen, Y., Goldring, G.: Nucl. Phys.
   A167, 216 (1971).

5. Igo, G.: Phys. Rev. Lett. 1, 72 (1958).

6. Rawitscher, G.H.: Nucl. Phys. 85, 337 (1966).

7. Eisen, Y., Vager, Z.: Nucl. Phys. A187, 219 (1972); Rawitscher, G.H.: Nucl. Phys. 85, 337 (1966).

8. Eisen, Y., Eisenstein, R.A., Smilansky, U., Vager, Z.: Nucl. Phys. A195, 513 (1972).

9. Broglia, R.A., Landowne, S., Winther, A: Phys. Lett. 40B, 293 (1972).

10. Christensen, P.R., Chernov, I., Gross, E.E., Stokstad, R., Videbeak, F.: preprint (1973).

11. Malfliet, R.A.: in Proceedings of the Argonne Conference on Heavy Ions (1973).

12. Malfliet, R.A., Landowne, S., Rostokin, V.: ibid.

13. Broglia, R.A., Winther, A.: Phys. Reports 4, 153 (1972).

14. Ford, K.W., Wheeler, J.A.: Ann. of Phys. 7, 259 (1959).

15. Rawitscher, G.H., Spicuzza, R.A.: Phys. Lett. 37B, 221 (1971).

16. Edwards, V.R.W., Sinha, B.C.: ibid, page 225.

17. Satchler, G.R.: Phys. Lett 39B, 495 (1972) and references quoted therein.

18. Samuel, M., Smilansky, U.: Computer Physics Communications 2, 455 (1971).

19. Eisen, Y., Samuel, M., Smilansky, U., Watson, B.A.: to be published.

20. Brückner, W., Merdinger, J.C., Pelte, D., Smilansky, U., Traxel, J.: Phys. Rev. Lett. 30, 57 (1973); Brückner, W., Pelte, D., Smilansky, U., Traxel, K.: to be published.

21. Alder, K.: in Proceedings of the Osaka Conference on Nuclear Moments and Nuclear Structure, to be published.

22. Devons, S., Duerdoth, J.: in Advances in Nucl. Phys., Vol. 2, edited by Baranger, M. and Vogt, E. (Plenum Press, 1969).

23. Satchler, G.R.: Phys. Rev. Lett. 39B, 492 (1972).

24. Glendenning, N.K., Hendrie, D.L., Jarvis, D.N.: Phys. Lett. 26B, 131 (1968).

25. Aponick, A.A., Chesterfield, C.M., Bromley, D.A., Glendenning, N.K.: Nucl. Phys. A159, 367 (1970).

26. Hendrie, D.L., Glendenning, N.K., Harvey, S.G., Jarvis, D.N., Duhm, H.H., Saudinos, J., Mahoney, J.: Phys. Lett. 263, 127 (1968).

27. Götz, U., Pauli, H.C., Alder, K., Junker, K.: Nucl. Phys. A192, 1 (1972).

HYPERFINE INTERACTIONS IN HIGHLY IONIZED ATOMS

Gvirol Goldring

Department of Nuclear Physics, Weizmann Institute of Science, Rehovot, Israel

The subject of my talk is the magnetic interaction between nuclei in excited states and atoms that are ionized in passage through matter. This subject has been studied extensively in the last few years, both here and at other laboratories, and I would now like to give you a general survey of the state of the art to date.

A beam of ions acquires after passage through a thin layer of condensed matter (of the order of a $\mu g/cm^2$) a charge distribution which is independent of the thickness of the material traversed and of the initial charge. This is called the equilibrium charge distribution. It is on the whole rather well known, and this type of ionization is utilized extensively in the production and acceleration of heavy ion beams. The degree and pattern of excitation of these ions is, however, much less known, and in the study of hyperfine interactions the excitation is a central issue, as can be seen from the following expression:

$$H(0) = 0.168 \ \frac{z^3}{n^3} \ \text{MGauss}$$

which gives the effective hyperfine field generated at the nucleus by s-electrons in a hydrogen-like ion, as a function of the principal quantum number n, and shows the strong dependence of the field on excitation.

It has now been established that there are two extreme regimes of excitation encountered in such ions. In one, labelled hot ionization, the ions are produced in a complex mixture of a large variety of highly excited states. In the other, cold ionization, the ions are produced predominantly in the ground state. A rather interesting intermediate situation has also been dealt with recently and will be described later on.

I shall now attempt to give a plausibility argument as to why and how these various patterns of ionization occur. Later on I shall present some of the relevant experimental evidence.

The main features of the equilibrium charge distribution can be described
and explained qualitatively by means of the famous matching principle enunciated
by Niels Bohr: an electron in a given orbit in a moving ion is stripped off
effectively once the translational velocity of the ion reaches the "average orbital
velocity" of the electron: $v = \sqrt{\dfrac{2E_I}{m}}$ where E_I is the binding energy of the
stripped electron. For an ion moving with a given velocity one would therefore
assume all the electrons bound less tightly than the limiting value E_I to be
stripped off and all the others to remain. This is a rather crude first approximation. A quantitatively more satisfactory description has been given by Dmitriev[1]
who takes for his point of departure the probability of ionization of hydrogen as
a function of velocity (fig. 1). This function is ascribed universal validity
with velocities given in terms of the relevant characteristic velocity. On the
basis of this universal function and simple statistical arguments Dmitriev could
describe rather well the probability of ionization as a function of the velocity,
the atomic number and the charge state of the ion. Examples of such charge distributions are shown in figs. 2 and 3 for helium and oxygen respectively. We note
that, for example, the $(Z-1)^+$ charge state in both these cases is produced most
copiously when the velocity of the ion is matched to the ground state configuration
of that ion.

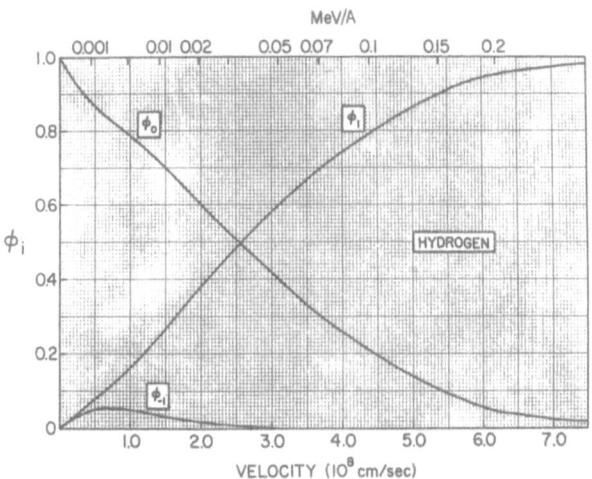

Fig. 1. Probability of ionization of H

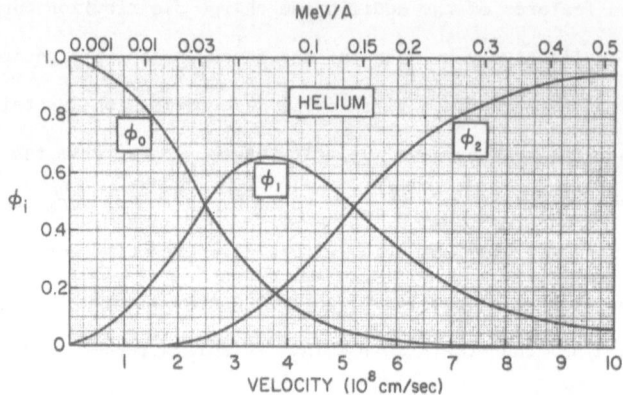

Fig. 2. Probability of ionization of He in various charge states

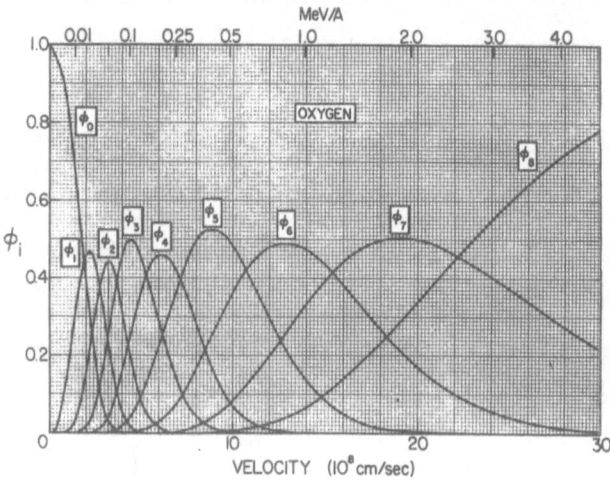

Fig. 3. Probability of ionization of O in various charge states

We now ask how sharp is this matching condition, how far off the optimum conditions can we go and yet produce approximately the same pattern? We can get a rough answer to these questions by observing either the universal function of Dmitriev or any one of the probability functions for the production of a given charge state as a function of velocity. The total width of the matching function appears to be about a factor of two in velocity. If we therefore start from ideal matching conditions we can reduce the velocity by about $\sqrt{2}$ and still get copious production of that particular charge state. Conversely, if we keep the velocity of the ion constant we expect to populate states of lower binding energy than the ground state, down to about $\frac{1}{2} E_I$. We therefore conclude from this argument that whenever we produce a certain charge state close to the maximum of its population, the ions will be produced in all the states having binding energies ranging from E_I to $\frac{1}{2} E_I$, where E_I refers to the ground state. We now find that atoms and ions fall into two distinct classes: most of them have a large number of states in the region indicated. These we expect to be produced in the hot excitation pattern. Hydrogen-like and helium-like atoms have a highly isolated ground state with no excited state in the matching region, the first excited state having a binding energy of about 1/4 that of the ground state. These we would expect to be produced in cold ionization.

For the experimental study of these problems we revert to the distinction between so-called static perturbations and time-dependent perturbations. Both refer to the evolution of the polarization of nuclei under the influence of some external or internal fields. If those fields are constant in time, or if the total interacting system is isolated, then the polarization will change as the sum of periodic functions of time, like any other dynamical variable in a conservative system.

For hyperfine interactions in particular these periodic functions are:

$$e^{\frac{i}{\hbar}(E_F - E_{F'})t} = e^{i\omega_{FF'}t} \qquad ; \qquad F \neq F'$$

where F and F' refer to the total angular momentum of the nucleus and the ion:

$$\vec{F} = \vec{I} + \vec{J}.$$

For s electrons there are only two values of F:

$$F = I \pm 1/2,$$

and consequently only a single frequency ω. For hydrogen-like atoms this is given by:

$$\omega = \frac{\mu_N g H(0)}{\hbar} (2I+1)$$

where, as before, the effective field H(0) is given by:

$$H(0) = 0.168 \frac{Z^3}{n^3} \text{ MGauss}$$

Experimentally one observes the gamma rays emitted in the de-excitation of the interacting nuclear state. The change in polarization is expressed by means of attenuation coefficients $G_k(t)$. For s-electrons these are given by:

$$G_k(t) = 1 - \frac{k(k+1)}{(2I+1)^2} (1 - \cos\omega t)$$

and if one integrates this over all lifetimes one obtains the integral attenuation coefficients:;

$$G_k = 1 - \frac{k(k+1)}{(2I+1)^2} \frac{(\omega\tau)^2}{1 + (\omega\tau)^2}$$

An entirely different situation obtains if the fields themselves change during the lifetime of the nucleus. In particular if the fields change in a random fashion with a characteristic correlation time τ_c, the attenuation coefficients are given by:

$$G_k(t) = e^{-\frac{k(k+1)}{3} \omega^2 \tau_c t}$$

$$G_k = \frac{1}{1 + \frac{k(k+1)}{3} \omega^2 \tau_c \tau}$$

Clearly, this will be the situation if the stripped ions are produced in a highly excited state, de-exciting in many steps during the lifetime of the nucleus. The exponential decay of the attenuation coefficient is characteristic of the spreading of the polarization into a large reservoir; in this particular case the radiation field.

We consider first instances of hot ionization. These were discovered in experiments with medium and heavy nuclei, in charge states up to 15^+. The hyperfine interaction was classified as time dependent from systematics such as the one shown in fig. 4, which relates to measurements of perturbation in 2^+ states in Te isotopes following recoil into vacuum[2]. Similar data were obtained in many other cases. In every case one obtains approximately constant g-factors for the various isotopes on the assumption of rapidly fluctuating fields, whereas static interactions would imply g-factors monotonically decreasing with mean life for any one

of these families, an extremely improbable state of events. In a number of cases
it was possible to evaluate the quantity τ_c, the average time between two sub-
sequent radiative transitions, from the data and it is consistently found to be
about 3 ps. The apparent decrease in the g-factor for ^{130}Te is also consistent
with this general view and is ascribed to the fact that the description of the
process as an interaction with fluctuating fields is expected to break down at
this point, namely at a mean life: $\tau \sim \tau_c \sim 3$ ps.

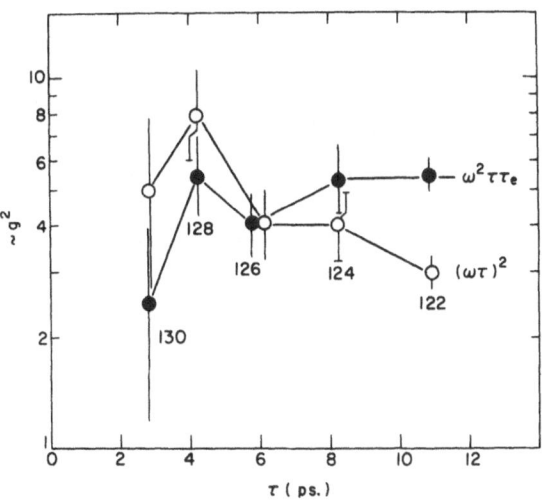

Fig. 4. A quantity proportional to g^2 derived from recoil into vacuum
measurements[2] involving 2^+ states of Te isotopes on two different
assumptions: 1. "$\omega^2\tau\tau_c$" - fluctuating fields and 2. "$(\omega\tau)$" -
static fields

Further evidence was obtained in a measurement in Freiburg[3] for the 2^+ states
of Y_b , which essentially provided a measure of the attenuation coefficient as a
function of time. This was achieved by a so-called plunger measurement. The
excited nuclei are allowed to recoil into vacuum and are then collected on a
stopping foil which freezes their polarization. By varying the distance of the
collector foil from the target, one can vary the time over which the interaction

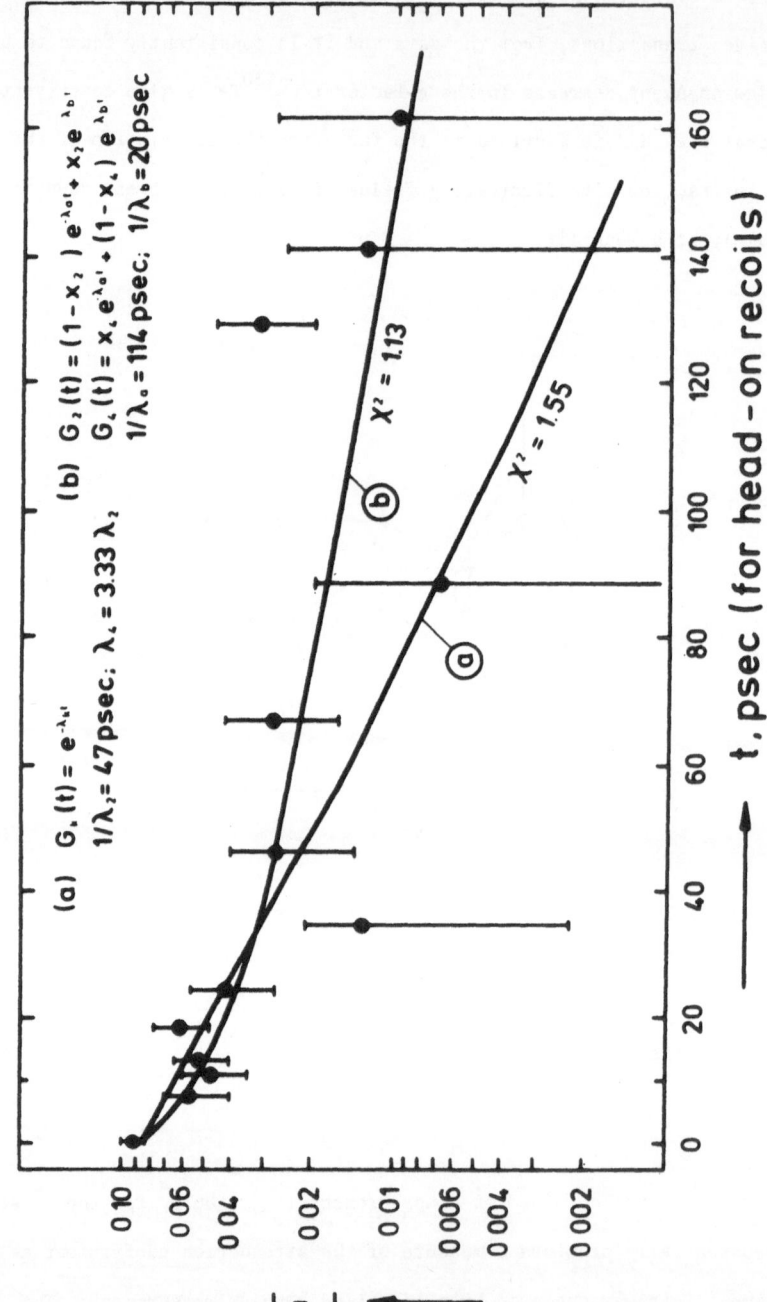

Fig. 5. Time dependence of attenuation in vacuum of Yb 2^+ states, in a plunger measurement[3]

is active and directly determine the time dependence of the attenuation. This is shown in fig. 5. It is generally consistent with an exponential decay (with some modifications which do not concern us here) and in particular with the value for τ_c quoted above.

Because of the high complexity of the atomic environment one cannot hope to calculate the hyperfine fields or the other relevant atomic parameters in detail. However, the mean values of those parameters over the entire atomic population can be determined experimentally and they can be utilized for comparative measurements of gyromagnetic ratios. Fig. 6 shows the systematics of gyromagnetic ratios of 2^+ states of deformed nuclei in the rare earth region determined in this way[4] and fig. 7 shows the effective mean hyperfine fields in the same region.

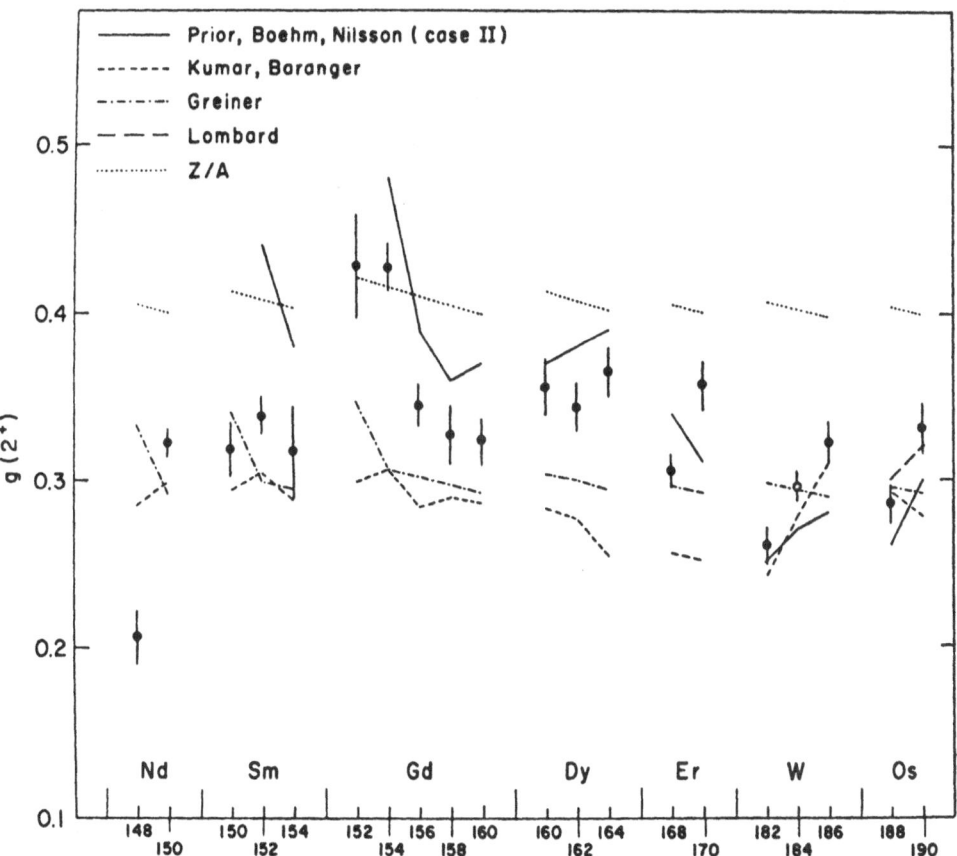

Fig. 6. Systematics of g-factors of rotational 2^+ states in the rare earth region, determined by recoil into gas measurements [4]

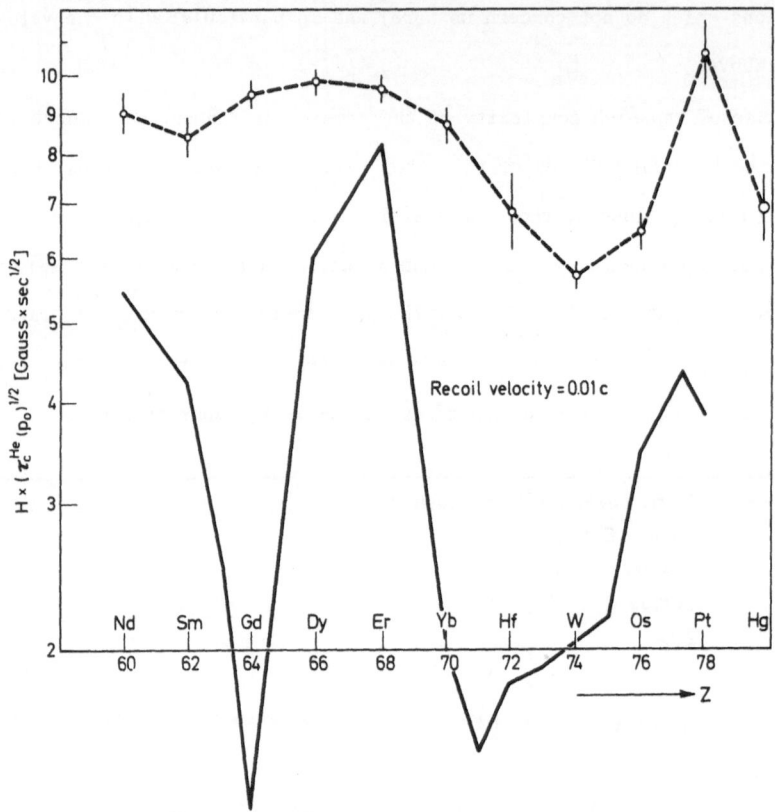

Fig. 7. A quantity proportional to the mean hyperfine field for heavy nuclei recoiling in helium with a velocity of $0.01c$[4,5,6]. As a comparison the absolute value of the internal field experienced by the same nuclei embedded in magnetized iron is also shown (in arbitrary units)

Instances of cold ionization have been found mainly in light atoms which were ionized to the $(Z-1)^+$ charge state. A remarkable example is provided by a recent experiment at the University of Oxford.[7] This relates to the 3^- state of ^{16}O at 6.13 MeV with a mean life τ of 25 ps, excited in the well-known $^{19}F(p,\alpha)^{16}O^*$ reaction through a resonance in the compound ^{20}Ne nucleus. In order to achieve the required high recoil velocity the roles of beam and target were reversed with a 49 MeV ^{19}F beam impinging on a hydrogen target (hydrogen absorbed on a titanium surface). The ^{16}O nuclei were allowed to recoil into vacuum and stopped on a movable foil just as in the Yb measurement described above. Alphas and gammas were detected at 0° in coincidence and provided a measure of the attenuation of the angular correlation as a function of time. The results are shown in fig. 8. The oscillatory behaviour of the perturbation is evident and bears witness to the static nature of the perturbation and to its origin in s-electrons. As the ions are predominantly in the 7^+ ionization state the perturbation is associated with 1s electrons.

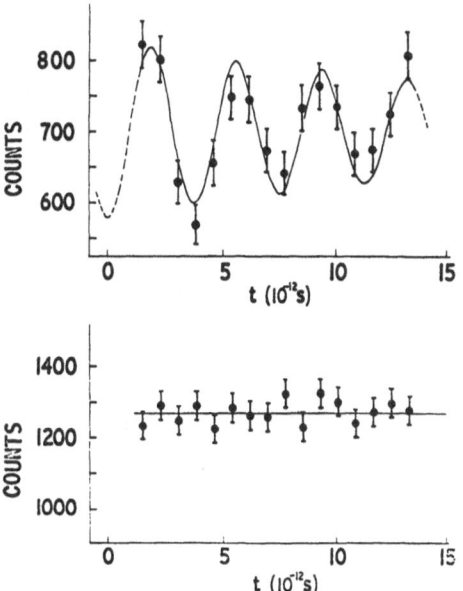

Fig. 8. γ-α coincidence yield vs. time of flight of oxygen ions recoiling with a velocity: v=0.059c, corresponding to: (a) the 6.13 MeV level, and (b) the 7.12 MeV short lived level [7]

With the very well known field of these hydrogen-like ions the gyromagnetic ratio of this level could be established very accurately. This measurement approaches in accuracy the high quality measurements of gyromagnetic ratios carried out with external magnetic fields and nuclear levels of about a thousand times longer mean lives.

The first proof for cold ionization was less direct and was again provided by
an integral measurement. This was a measurement in Heidelberg relating to the first
excited 2^+ state in ^{20}Ne [8] with a mean life of 1.2 ps. The level was excited in the
^{12}C(^{12}C,α)^{20}Ne* reaction and gamma-rays emitted from nuclei recoiling into vacuum
were measured in coincidence with backscattered alphas. The attenuation coefficient
G_4 was determined in this way as a function of recoil velocity and results are shown
in fig. 9. The remarkable feature revealed here is that the attenuation follows the
same curve as the abundance of the 9^+ charge state and this ties the perturbation to
the hydrogen-like ions. From the strength of the attenuation it was concluded that
an overwhelming fraction of the ions had to be in their ground state.

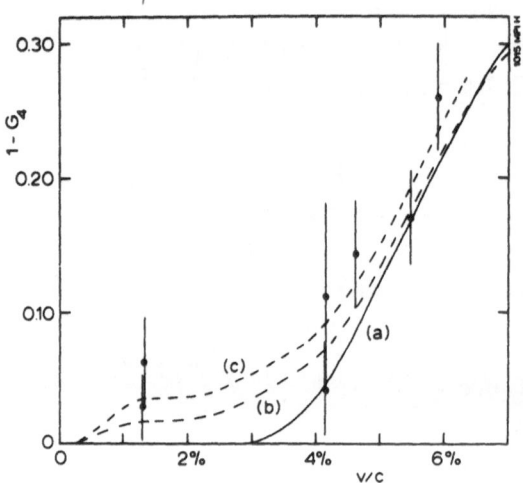

Fig. 9.   The coefficient G_4 for gammas emitted from the 2^+ state in ^{20}Ne following recoil into vacuum[8]. Curve A follows the probability function for
the 9^+ charge state

A further example was provided by a measurement in Oxford by a joint group
from the University of Oxford and the Weizmann Institute relating to the first
excited 2^+ state in ^{18}O of a mean life of 3.3 ps[9]. The level was excited by the
inelastic scattering of ^{18}O on a ^{12}C target in the arrangement shown in fig. 10.
The recoiling ^{18}O nuclei were analyzed in a magnetic spectrometer and separated
into the component charge states. Gamma distributions for 6^+, 7^+ and 8^+ ions,

the dominant charge states, are shown in fig. 11. The 8^+ charge state yields the standard unperturbed correlation. In comparison to this the 7^+ charge state clearly exhibits a strong perturbation and the 6^+ charge state a rather weak one, both consistent with predominant ground state population for these ions, the ground state of the 6^+ ion, like the ground state of neutral helium, having zero magnetic field. With the aid of independent information on the ground state population of the 7^+ ion it was possible to extract a value of the gyromagnetic ratio for the 2^+ state in ^{18}O from this measurement.

A summary of gyromagnetic ratios determined in cold ionization measurements in light nuclei is presented in the table below, together with some new results obtained recently at lower recoil velocities. These will be discussed presently.

TABLE I

nucleus	I^π	Excitation MeV	τ ps	$|g|$	method	active ionic state	reference
^{14}N	3^-	5.83	18	0.5 - 0.85	recoil into gas	1s	10
^{16}O	3^-	6.13	24	0.55±0.03	plunger	1s	7
"	"	"	"	0.54±0.05	"	2s	12
"	"	"	"	0.56±0.05	decoupling in external field	2s	12
^{18}O	2^+	1.98	3.3	0.2 -0.36	recoil into vacuum, charge selection	1s	9

A number of further measurements involving hydrogen-like ions are now in progress at the Weizmann Institute, in Bonn, in Cologne and in Oxford. They all relate to very light nuclei in the p shell or the beginning of the sd shell. This is clearly a promising tool for accurate and reliable measurements of gyromagnetic ratios in light nuclei. A number of experimental techniques have been developed for this purpose including recoil into gas and the decoupling of the nucleus from the atomic shell in an external longitudinal magnetic field. But with recoil velocities available at present these measurements are all restricted to a range of nuclei within, or just beyond, the p-shell. With the advent of a new generation of heavy ion accelerators this range is expected to be extended considerably but not, in the foreseeable future, beyond the sd shell. However, possibilities for carrying out measurements of a similar nature over a much larger range and also considerably simplified, now emerge from a series of measurements recently carried out in Rehovot and relating to atoms in the n=2 electronic shell[11].

These measurements all relate to the 3^- state in ^{16}O already referred to above and excited in the $^{19}F(p,\alpha)^{16}O^*$ reaction in the ordinary way with a proton beam of 1.38 MeV, observing the gamma rays in coincidence with backscattered alphas. The recoil velocity is in this case 0.011c, and the significant charge states are 1^+, 2^+ and 3^+. In the ground states these ions have all the electrons in the n=2 shell. According to our matching criterion, the n=2 state is still reasonably well isolated from n=3 and we therefore expect these ions to be produced predom-

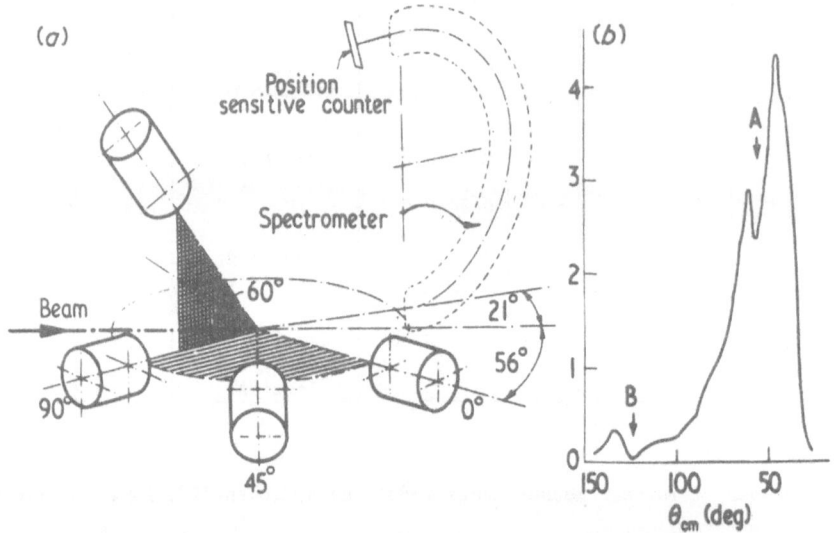

Fig. 10. (a) Schematic view of spectrometer and counter assembly in ^{18}O experiment[9]. (b) Differential cross section in arbitrary units of inelastically scattered ^{18}O, $E_{^{18}O}$ = 35 MeV. Arrows refer to spectrometer acceptance angular for: (A) $^{18}O^*$, and (B) the kinematically coincident ^{12}C

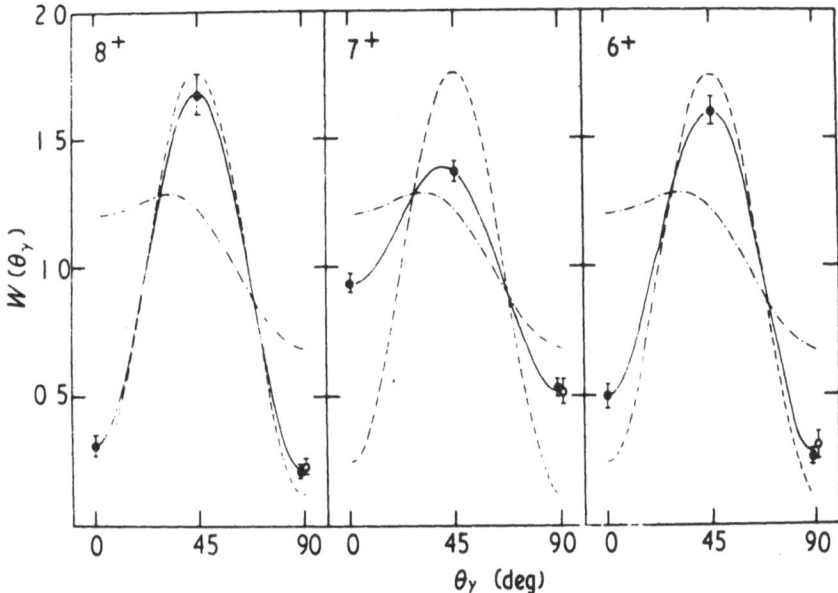

Fig. 11. Angular correlation of gammas in coincidence with ^{18}O ions in the 8^{+},

7^{+} and 6^{+} charge states. Also shown are calculated curves for a pure

2 → 0, I_z = 0 correlation and for maximum (hard core) perturbation,

as well as perturbed correlation least square fits to the data [9]

inantly in the ground state as far as the principal quantum number is concerned.

The n=2 state accommodates, however, both 2s and 2p electrons. Moreover, in those

complex ions there is a large number of individual levels all belonging to the n=2

state. In the light of our matching principle we would expect these levels to be

excited rather uniformly. We therefore expect in this case an intermediate regime

of excitation, which we might term "cool": predominant ground state occupation in

the n=2 state and near uniform distribution among the levels belonging to this

quantum number.

We can check these ideas with an experiment carried out in Rehovot two years

ago[11] in which the attenuation of the gamma correlation was measured as a function

of the recoil velocity. Results are shown in fig. 12 and compared to the curve

showing the abundance of unpaired 2s electrons calculated on an assumption of a

purely random distribution of 2s and 2p electrons. Underlying this computation is the observation that the hyperfine interaction associated with 2p electrons is expected to be quite small, and for the sake of simplicity it has been neglected in all the computations. The measurements strongly suggest a random distribution within the n=2 state

For the more sophisticated experiments to be described presently the hyperfine interaction frequencies were computed for each individual above level. As

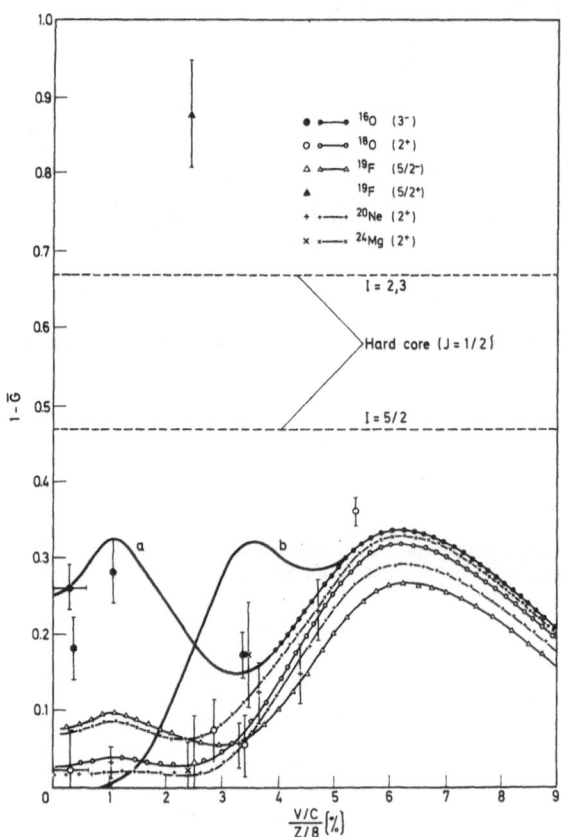

Fig. 12. Mean attenuation coefficients $\overset{\gamma}{G} = \dfrac{\Sigma k(k+1)G_k}{\Sigma k(k+1)}$ as a function of recoil velocity, for various nuclear levels. Note in particular the strong $^{16}O[3^-]$ attenuation at low velocities, referred to in the text. Computed curves for ^{16}O in low velocity region: (a) assuming uniform population of 2s, 2p as explained in the text, and (b) assuming strict ground state population also within the n=2 shell [11]

before, it was assumed that only the 2s field is active and that all n=2 levels are uniformly populated. The hyperfine frequency spectrum for this level distribution is shown in fig. 13.

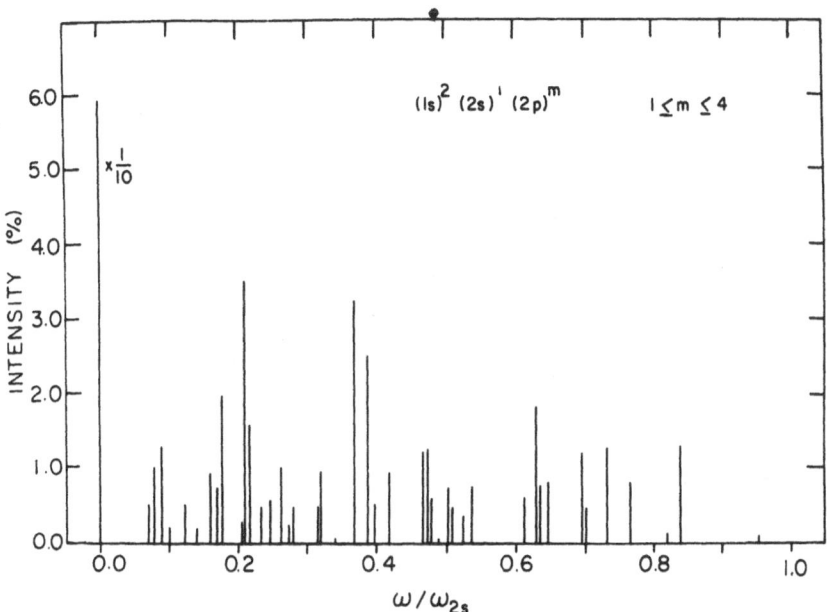

Fig. 13. Spectrum of frequencies for hyperfine interactions of ^{16}O nuclei with ions recoiling at a velocity of 0.01c, computed as explained in text. ω_o is the interaction frequency for the $(1s)^2$ (2s) configurations [12]

The first experiment I wish to present here is again a plunger type time differential measurement of the perturbation in vacuum with two gamma counters, one situated at 0° and one at 144° to the beam.[12] The gamma counters were Ge(Li) detectors making it possible by virtue of the Doppler effect to separate events of gamma emission in flight from emission within the stopping foil. The latter are the significant events, as they relate to nuclei experiencing the hyperfine interaction all the way from the target to the stopper foil. The measured attenuation as a function of time is shown in fig. 14 together with calculations based on an assumption of static hyperfine interactions with the frequency spectrum of fig. 13.

One gets in this way a value for this g factor of: $|g| = 0.54 \pm 0.05$ in very good agreement with the value determined at Oxford, and lending credence to the cool ionization pattern outlined above.

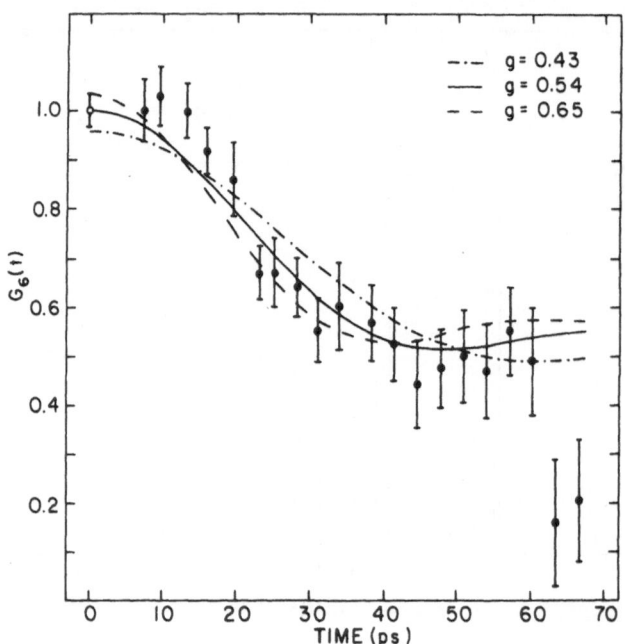

Fig. 14. Anisotropy as a function of time in plunger measurements for ^{16}O [3$^-$] recoiling at a velocity of 0.01c.[12]

The second experiment is a decoupling measurement in an external longitudinal field where the requisite high magnetic field was provided by a superconducting coil.[12] The results are shown in fig. 15. Calculations based on assumptions similar to those stated above yield a value for the g factor: $|g| = 0.56 \pm 0.05$ again in excellent agreement with the Oxford value.

It is evident that this regime of cool ionization, despite a certain lack of conceptual clarity as compared to cold ionization in hydrogen-like atoms, lends itself very well to precise formulation and interpretation. Its biggest attraction is the widely increased range of application. Fig. 16 shows a map of the relevant charge states in the space of ion velocity and the atomic number. The large

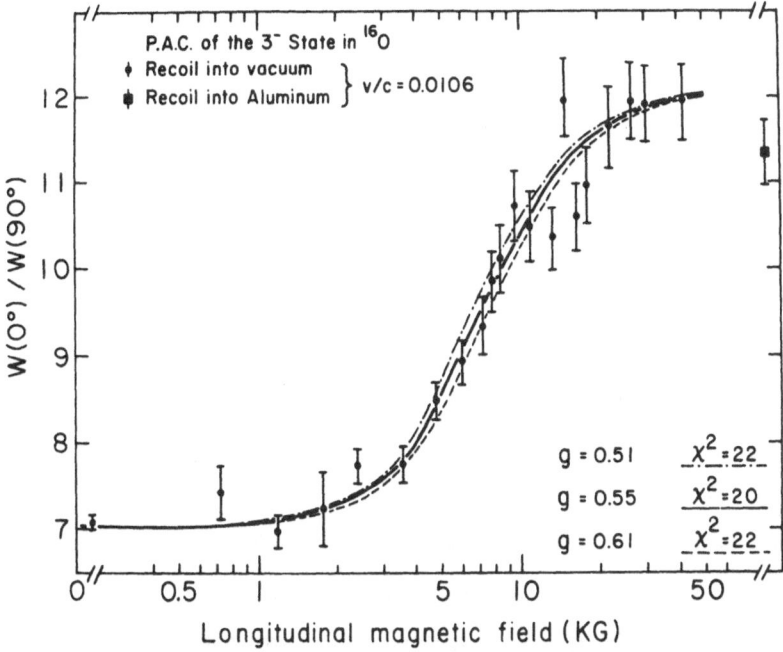

Fig. 15. Anisotoropy as a function of magnetic field in a decoupling measurement for ^{16}O [3^-] recoiling at a velocity of 0.01c.[12]

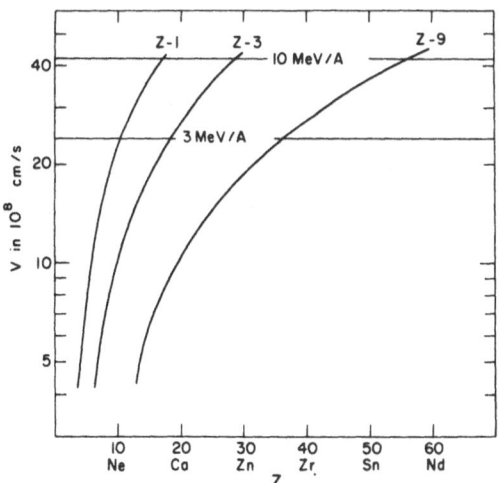

Fig. 16. Lines of maximum abundance of the charge states Z-1 and Z-3, Z-9 prescribing respectively the regions of large 1s and 2s contributions to the hyperfine interaction

increase in range due to 2s electrons stems from the fact that these electrons appear in a roughly constant fraction in all the charge states from $(Z-3)^+$ to $(Z-9)^+$, whereas 1s fields are linked uniquely to $(Z-1)^+$ hydrogen-like ions.

REFERENCES

1. Dmitriev, I.S.: Sov. Phys. - JETP 5, 473 (1957).

2. Ben-Zvi. I., et al.: Nucl. Phys. A121, 592 (1968).

3. Brenn, R., et al.: in "Hyperfine Interactions in Excited Nuclei", Gordon and Breach (1971).

4. Ben-Zvi, I., et al.: Nucl. Phys. A155, 401 (1970).

5. Speidel, K.H., et al.: Nucl. Phys. A185, 330 (1972).

6. Doubt, H.A., Fechner, J.B., Hagemeyer, K., Kumbartzki, G., Speidel, K.H., Kleinfeld, A.M.: private communication.

7. Randolph, W.L., Ayres de Campos, N., Beene, J.R., Burde, J., Grace, M.A., Start, D.F.H., Warner, R.E.: private communication.

8. Faessler, M., et al.: in "Hyperfine Interactions in Excited Nuclei", Gordon and Breach (1971).

9. Goldring, G., et al.: Phys. Rev. Lett. 28, 763 (1972).

10. Berant, Z., et al.: in "International Conference on Nuclear Moments and Nuclear Structure" Osaka (1972).

11. Berant, Z., et al.: Nucl. Phys. A178, 155 (1971).

12. Broude, C., Goldberg, M.B., Goldring G., Hass, M., Renan, M.J., Sharon, B., Shkedi, Z., Start, D.F.H.: to be published.

MEASUREMENTS OF ELECTRIC QUADRUPOLE MOMENTS BY THE PAC-TECHNIQUE

Erwin Bodenstedt

University of Bonn

The study of electric quadrupole moments of excited nuclei by perturbed angular correlations or angular distributions has become a very attractive field during the last few years. In this talk I should like to give a survey of the present situation.

For those who are not familiar with the technique I should like to make a few introductory remarks[1].

Let us assume that we have the nuclei of interest in an environment which produces a strong static electric field gradient at the nuclear site. Then the interaction between the field gradient and the electric quadrupole moment of the nuclei produces a torque which causes a precession of the nuclear spins. If we now produce an aligned system of nuclei by observing the first gamma radiation of a gamma-gamma-cascade in a defined direction we can observe the precession of this aligned system by the rotation of the angular distribution of the second gamma transition. This is done by a time differential measurement of the gamma-gamma-angular correlation. For magnetic hyperfine interactions this technique has been applied many times and is well-known as the spin-rotation method for the determination of nuclear magnetic moments.

There are two essential differences, however:

a) The hyperfine splitting by a static magnetic field, the so-called nuclear Zeeman-effect, produces equidistant levels with the consequence of one single interaction frequency, the Larmor precession frequency ω_L, whereas the electric interaction contains several interaction frequencies. If, especially, the electric field gradient has axial symmetry the interaction energy depends quadratically on the m-quantum number:

$$W_{e\ell} = \frac{e}{4} \cdot V_{zz} \cdot Q_I \cdot \frac{3m^2 - I(I+1)}{I \cdot (2I-1)}$$

b) The magnetic spin-rotation can easily be observed by applying external magnetic fields. It is impossible, however, to produce external electric field gradients of sufficient strength. It is necessary to use crystalline electric field gradients which fortunately have exactly the right order of magnitude.

The main problem in the derivation of absolute values for the electric quadrupole moment of an excited nuclear state, where the spin rotation in the crystalline electric field gradient has been observed, is therefore the calibration of the field gradient. In principle at least, this calibration is possible by nuclear quadrupole resonance measurements, nuclear acoustic resonance or by Mössbauer spectroscopy with nuclear ground states, the quadrupole moments of which are known by atomic beam experiments or muonic x-ray investigations.

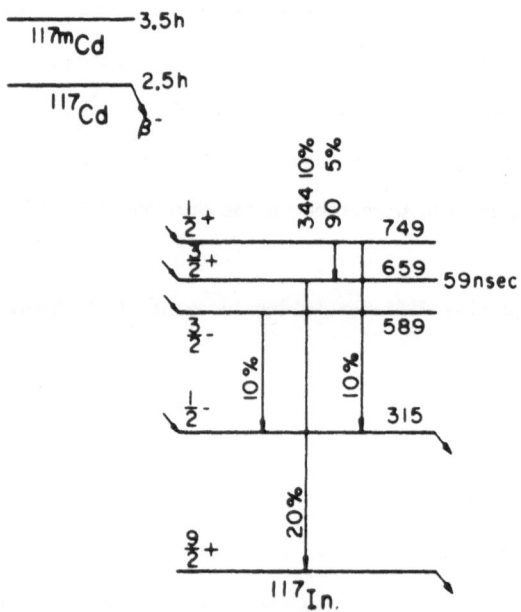

Fig. 1. Decay scheme of ^{117}Cd

In order to demonstrate the technique I should like to show two examples where the quadrupole moment of an excited state has been derived by differential perturbed angular correlation measurements.

The first case is the 659 keV state of ^{117}In ^{2}. This level has a halflife of 59 nsec and is populated in the decay of ^{117}Cd (see fig. 1). In this special case the spin is $3/2^+$ which has the consequence of an especially simple hyperfine pattern. Since there are only two electric hyperfine levels ($m = \pm 3/2$ and $m = \pm 1/2$) one has only one interaction frequency. The perturbed angular correlation has the form:

$$W(\theta) = 1 + A_2 \cdot G_2(t) \cdot P_2(\cos\theta)$$

with the perturbation factor:

$$G_2(t) = 0.2 + 0.8 \cdot \cos\omega_0 t.$$

The electric interaction frequency, ω_0, is defined by

$$h\omega_0 = \text{smallest level spacing in hyperfine level scheme}$$

The connection between this interaction frequency and the electric field gradient and the electric quadrupole moment is given by the relations:

$$h\omega_0 = \begin{cases} 6 \cdot \dfrac{e \; Q_I \cdot V_{zz}}{4I \cdot (2I - 1)} & \text{for half integer spin I} \\[3mm] 3 \cdot \dfrac{e \; Q_I \cdot V_{zz}}{4I \cdot (2I - 1)} & \text{for integer spin I} \end{cases}$$

Fig. 2 shows the observed spin rotation. In this diagram $A_2(t) = A_2 \cdot G_2(t)$ is plotted versus the time delay between emission of $\gamma 2$ and emission of $\gamma 1$. The 90 keV- 589 keV cascade has been used in this experiment. The crystalline environment was polycrystalline metallic indium. Metals are prefered because of the following reasons:

a)   There are no after-effects from perturbations of the electronic shell after the β-decay because of the availability of the electrons in the conduction band.

b)   In most cases the radioactive atoms can easily be put on substitutional positions of the host lattice because the metals form aloys with most other metals, the phase diagrams of which are usually very well investigated.

c)   Most metals crystallize in a hexagonal or tetragonal structure. These lattices produce axially symmetric electric field gradients of sufficient strength.

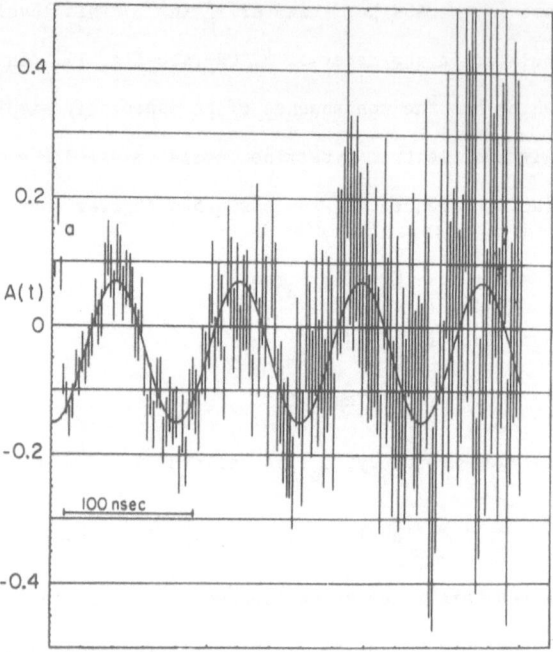

Fig. 2. Spin rotation measurement with the 90 keV- 589 keV γγ cascade of
[117]In in an environment of indium metal

In this case an absolute quadrupole moment could be derived since the field
gradient was calibrated by a nuclear quadrupole resonance experiment with indium 117
in the ground state, the quadrupole moment of which is well-known. The result for
the quadrupole moment of the 659 keV level is:

$$Q_I(659 \text{ keV}) = 0.58_6 \text{ b.}$$

As a second example I should like to discuss the measurement of the electric
quadrupole moment of the 482 keV state of [181]Ta [3], since this measurement shows a
few other typical phenomena.

Fig. 3 shows a partial decay scheme of [181]Hf and [181]W which populate excited
states in [181]Ta. The spin rotation measurement was performed in an environment of
metallic rhenium because the electric field gradient in this environment is known
from Mössbauer experiments with the 6 keV transition[4]. These Mössbauer experiments
need of course an extremely thin source because of the strong absorption of the

6 keV line, and the sources were therefore prepared by diffusion of the ^{181}W

activity into the surface of rhenium metal. In the Mössbauer absorption spectrum

the hyperfine splitting of both the $9/2^-$ state and the $7/2^+$ state are well resolved.

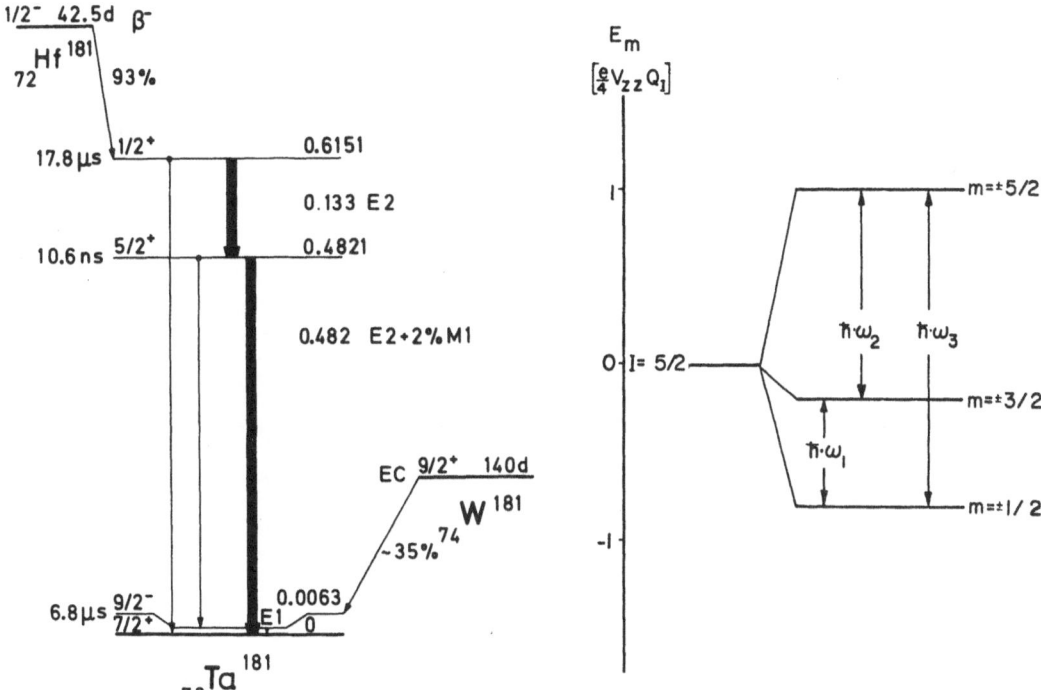

Fig. 3. Partial decay scheme of ^{181}Hf and ^{181}W, populating exciting states of ^{181}Ta

Fig. 4. Hyperfine splitting of an I=5/2 state in an axially symmetric electric field gradient

The quadrupole moment of the ground state of ^{181}Ta has been measured by different

techniques. The most reliable result is probably that obtained from muonic x-ray

spectra[5].

The angular correlation sources were prepared by melting metallic rhenium and

metallic hafnium, which was activated in a strong thermal neutron flux, in high

vacuum on a water cooled copper disk by an intense high energy electron beam.

Uniform aloys were prepared in this way. Several sources were produced with different hafnium concentrations in order to investigate a possible relation between interaction frequency and impurity concentration.

The spin of the intermediate state of the intense 133 keV, 482 keV γγ-cascade is 5/2. In the axially symmetric electric field gradient of the hexagonal rhenium lattice one obtains three hyperfine levels (m = ± 5/2, m = ± 3/2, and m = ± 1/2) and therefore three different interaction frequencies (see fig. 4). The theoretical shape of the attenuation factor $G_2(t)$ is shown in fig. 5 for different values of the axial asymmetry parameter η defined by

$$\eta = \frac{V_{xx} - V_{yy}}{V_{zz}}$$

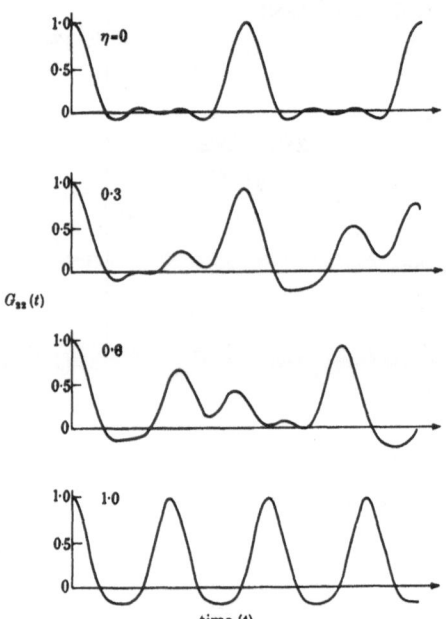

Fig. 5. Theoretical shape of the attenuation factor $G_2(t)$ for an I = 5/2 state for different values of the axial asymmetry parameter η

These curves are calculated for polycrystalline materials with randomly oriented field gradients. Figs. 6 and 7 show typical examples of measured spin rotation curves of hafnium 181 in rhenium for probes of somewhat different hafnium con-

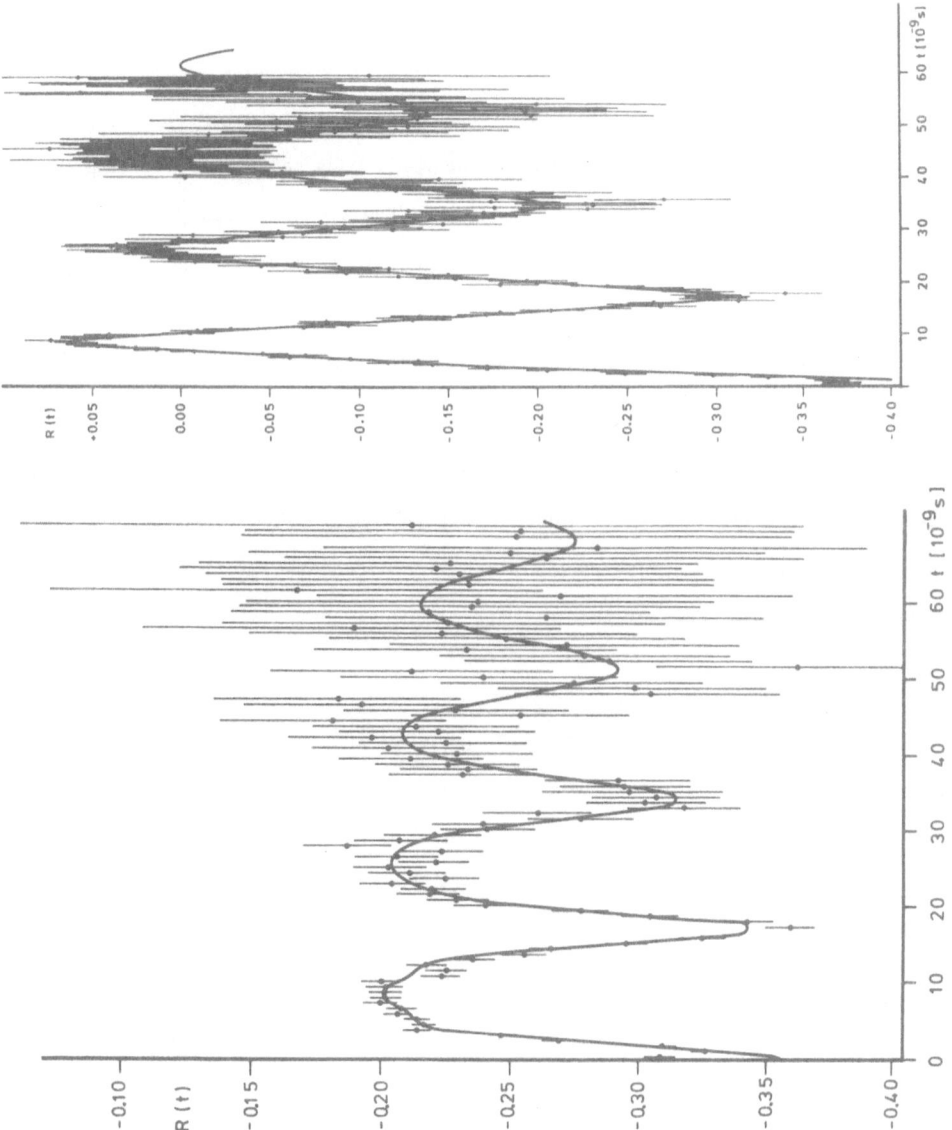

Figs. 6 & 7. Spin rotation pattern measured for the 5/2 state of ^{181}Ta in an
environment of rhenium metal. The asymmetry ratio

$$R(t) = 2\,\frac{N(180°)-N(90°)}{N(180°)+N(90°)} \text{ is plotted versus the time delay}$$

Fig. 6. The source contains a hafnium
impurity concentration of 0.08
weight percent

Fig. 7. The source contains im-
purities of 0.045 weight
percent hafnium and 0.9
weight percent tungsten

centration. The analysis of the data gave the following results:

a) The axial asymmetry parameter η is zero in all cases.

b) The amplitudes of the three interaction frequencies involved are different from source to source and deviate from the amplitudes calculated for random orientation of the field gradients. X-ray diffraction patterns proved that indeed our sources were more or less nicely developed single crystals and not polycrystalline probes.

c) The amplitude of the spin rotation curve is damped.

Quite often it is observed that spin-rotation patterns obtained for sources in crystalline environments show a damping. Usually this damping is caused by the fact that the crystals are not ideal. Lattice imperfections have the effect that the interaction frequency shows a distribution around the ideal value and then the damping is a consequence of the interference of slightly different interaction frequencies.

The least squares fit calculation gave for the interaction frequency ω_o a strong dependence of the hafnium concentration (see fig. 8). This behaviour is not unusual, similar effects are well-known from nuclear quadrupole resonance measurements. For this special investigation it was quite fortunate that an additional admixture of tungsten did not cause a measurable frequency shift. Therefore the field calibration from the Mössbauer experiments is reliable.

The result for the electric quadrupole moment of the 482 keV state was:

$$Q_I(482 \text{ keV}) = 2.51_{15}\text{b} \quad .$$

The results obtained up to now for quadrupole moments of excited nuclear states by the differential perturbed angular correlation technique are listed in table 1. There are some more isotopes where the quadrupole interaction frequency in metallic environments has been observed but the quadrupole moments could not be derived since a field gradient calibration was not yet possible.

Furthermore it should be mentioned that experiments are in progress in which the same isotope is studied systematically in many different metallic environments in order to get a better understanding of the theoretical background for the

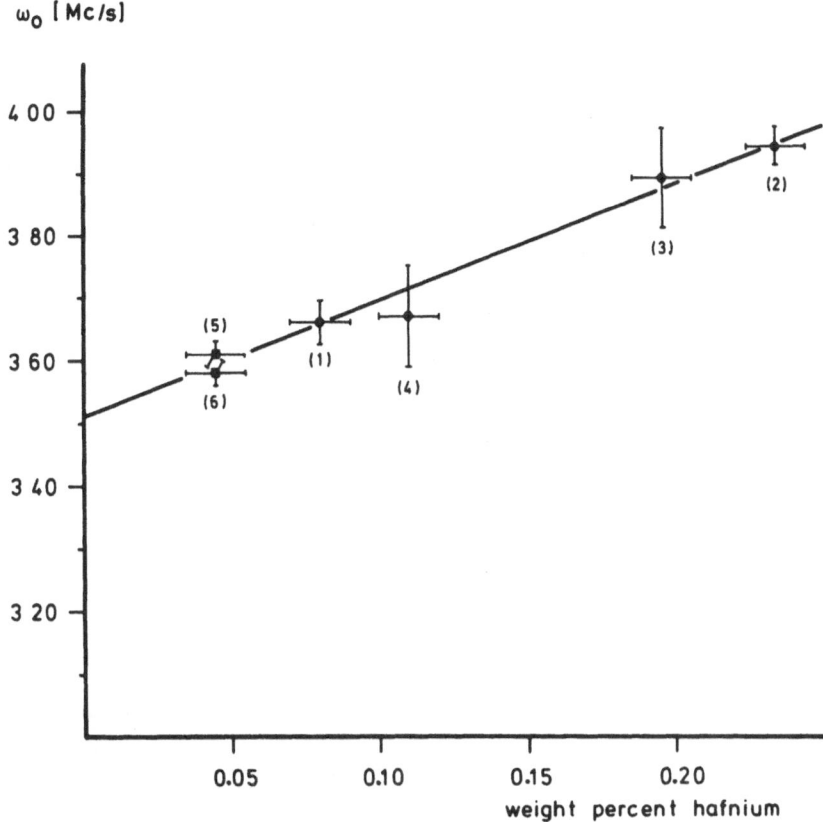

Fig. 8. Dependence between the interaction frequency ω_0 and the hafnium impurity concentration. The source of run (5) and (6) contains an additional impurity of 0.9 weight percent tungsten

effective field gradients at impurity atoms in metallic environments. A total number of nearly 40 different cases have been studied in the past.

The next topic of my talk is a short discussion of the problems involved in a calculation of the effective electric field gradient in metals.

There are essentially five different contributions:

a)   Point charge contribution of metal ions:

This contribution can easily be evaluated by a lattice sum calculation. Especially for the h.c.p. structure the calculation gives:

$$V_{zz}^{lat} = \frac{1}{4\pi\epsilon_0} Z \cdot e \cdot \frac{1}{a^3} \left\{ 0.0065 - 4.3584 \left(\frac{c}{a} - 1.633 \right) \right\} \tag{13}$$

TABLE I

Results of TDPAC - measurements of nuclear quadrupole moments

Isotope	level	spin	halflife	environment	Q_I	reference
^{44}Sc	68 keV		166 ns	scandium metal	0.19_3 b	6)
^{111}Cd	247 keV	5/2+	84 ns	indium metal	$+0.44_7$ b	7)
^{115}In	829 keV	3/2+	5.5 ns	indium metal	0.60_8 b	2)
^{117}In	659 keV	3/2+	59 ns	indium metal	0.58_6 b	2)
^{118}Sn	2.32 MeV	5^-	21.7 ns	$K(C_4H_2O_6Sb(OH_2))\cdot1/2H_2O$	0.095 ± 0.031 *)	8)
^{120}Sn	2.3 MeV	5^-	5.53 ns	"	0.021_8 b	9)
^{133}La	535.4 keV	3/2-	60 ns	lanthanum metal	0.36_8 *)	10)
^{140}Ce	2.083 MeV	4+	3.41 ns	$La_2(NO_3)_6 Mg_3(NO_3)_6\cdot24H_2O$	0.40_5 b	11)
^{181}Ta	482 keV	5/2+	10.8 ns	rhenium metal	2.51_{15} b	3)
^{187}Re	906 keV	9/2-	563 ns	rhenium metal	3.3_7 b	12)

*)Preliminary result

The only uncertainty is the charge state Z which has to be inserted into this formula. Usually the maximum charge state is inserted by which the uncertainty is shifted to the electronic contribution.

b)   Contribution of the quadrupole polarization of the metal ions:

The quadrupole polarizabilities of closed shell ions are tabulated[14] and these values can be used to calculate the induced quadrupole moments. Then a lattice sum calculation is possible. Usually, this term gives only a correction of a few percent.

c)   Sternheimer antishielding effect:

The effective electric field gradient at the nuclear site is very strongly influenced by the polarization of its own atomic shell. The Sternheimer correction factors γ_∞, defined by the equation:

$$V^{lat}_{zz\ eff} = V^{lat}_{zz}\ (1-\gamma_\infty)$$

are tabulated[14].

d)   Anti-shielding of conduction electron:

One has to take into account that the actual charge distribution of the conduction electrons deviates from a homogeneous charge distribution which would not produce any additional field gradient[15]. This deviation is certainly a reduction of the charge density at the site of the ions. This effect produces an enlargement of the effective charge of the metal ions. This effect has not yet been accurately calculated. The difficulty is that one needs rather precise radial wave functions of the conduction electrons. The order of magnitude is an enlargement of the effective electric field gradient by a factor between 1.5 and 3.

e)   Localized moment contribution:

The deviation of the charge distribution of the conduction electrons within their own atomic sphere from spherical symmetry produces a strong shielding effect. It was predicted by Watson et al[16], that in certain cases an overshielding might even happen which would have the effect that the sign of the electric field gradient changes. Large effects are expected especially if the conduction band

contains strong p- or d-electron contributions. No exact calculations exist up
to now.

Since the contribution a, b, and c can be calculated quite well and the contribution d can be estimated roughly the main uncertainty in the theoretical
prediction of the effective electric field gradients in metals comes from the contribution e. Rather good estimates exist therefore in cases where the contribution e can be neglected. If the contribution e is strong, however, no prediction
of the effective field gradient is possible at present.

It is quite interesting to compare the theoretical predictions with the
empirical information which has been obtained in the past by systematic studies of
the electric interaction frequencies in metals.

The most important result is that in most cases where an experimental value
for the effective field gradient could be calculated from the measured interaction
frequency and the known electric quadrupole moment it was found that this field
gradient is by a factor between 2 and 3 larger than the calculated lattice contribution which was corrected by the Sternheimer factor (this means the calculated
contributions e, b, and c). The missing part can therefore be attributed to the
above mentioned antishielding effect of the conduction electrons (contribution d).

This interpretation is confirmed in a few cases where also the sign of the
field gradient has been determined experimentally. The sign of the interaction
is directly measured in the Mössbauer experiments. It can also be determined in
γ-circular polarization correlations with aligned single crystals. The signs
observed so far agree with those calculated for the lattice contribution to the
field gradient.

Recently a case has been found, however, which gives evidence for the opposite sign. It is a Mössbauer study of the 6 keV transition in tantalum 181 in an
environment of metallic rhenium[4]. A special discussion of this case is quite
instructive:

The electronic configuration of neutral tantalum is the following:

$$K, \ L, \ M, \ N, \ (5s)^2 (5p)^6 (5d)^3, \ (6s)^2.$$

Apparently the unfilled 5d-electron shell is responsible for the strong localized moment effect (contribution e). Now we can compare this result with that obtained for hafniun in hafnium and tantalum in hafnium. A Mössbauer-experiment with the $(2^+, 0^+)$-transition of ^{178}Hf in metallic hafnium[17] gave definitely a positive sign for the electric field gradient in agreement with the sign of the lattice contribution. The absolute value

$$V_{zz}^{exp} = +9.0 \cdot 10^{17} \ V/cm^2$$

is by a factor 3 larger than the calculated lattice contribution to the field gradient:

$$V_{zz}^{lat} \ (1-\gamma_\infty) = +2.9 \cdot 10^{17} \ V/cm^2$$

Time differential perturbed angular correlation measurements with ^{181}Ta in metallic hafnium[48] gave for the electric field gradient at the site of the tantalum impurity in hafnium metal the value:

$$|V_{zz}^{exp}| = 5.5 \cdot 10^{17} \ V/cm^2$$

which is nearly a factor of 2 smaller than the field gradient at the hafnium site. Apparently the one additional 5d-electron of tantalum produces already the small shielding effect. If we now compare these results with that of ^{181}Ta in rhenium metal where the observed field gradient:

$$V_{zz}^{exp} = -6.1 \cdot 10^{17} \ V/cm^2$$

has to be compared with the lattice contribution:

$$V_{zz}^{lat}(1-\gamma_\infty) = +2.5 \cdot 10^{17} \ V/cm^2$$

we must consider the different electronic configuration of rhenium:

$$K, \ L, \ M, \ N, (5s)^2 (5p)^2 (5d)^5, (6s)^2$$

Rhenium has already five electrons in the 5d-shell and it is quite plausible that in this environment the 5d-contribution in the atomic sphere of the tantalum impurity is quite larger than in the hafnium environment. Therefore the overshielding effect of the 5d-contribution in this case is understandable.

A large effort has been involved in measurements of the temperature dependence of the interaction frequency. Since it is expected that this type of investigations could contribute essentially to the understanding of the mechanism which

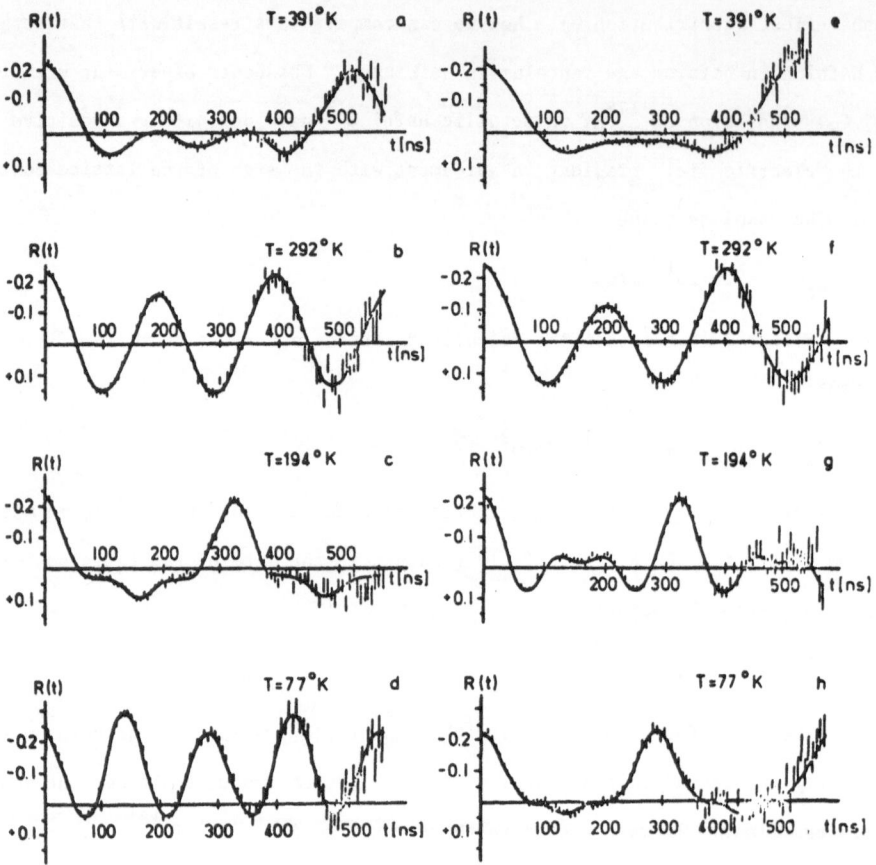

Fig. 9. Spin rotations of the I = 5/2 state of [111]Cd, observed in an environment
of indium metal at different temperatures. The two curves for each temperature were taken with the identical source by two different pairs of
detectors

produces the effective field gradients in metals, I should like to mention a
measurement of this type performed with [111]Cd in an environment of cadmium as well
as of indium[7]. Fig. 9 shows experimental spin-rotation curves observed for the
5/2[+] state of [111]Cd in indium at different temperatures. The curves exhibit
characteristic patterns for the spin 5/2 and for axially symmetric field gradients.
In this special case no damping of the spin-rotation is observed. On the other
hand we have again the phenomenon that the melting procedure in the source prepara-

tion did not produce polycrystalline probles with randomly oriented field gradients. The temperature dependence observed in this case is quite strong. The temperature dependence of the lattice contribution to the field gradient can be calculated, of course, since the temperature dependence of the lattice constants has been measured by the X-ray diffraction technique. Therefore, for the analysis of the data the temperature dependence of the contribution of the conduction electrons to the field gradient has been derived. We describe the conduction electron contribution by a factor F_{cond} defined by the equation:

$$V_{zz}^{exp} = V_{zz}^{lat} \; (1-\gamma_\infty) \cdot (1+F_{cond})$$

The results for the temperature dependence of F_{cond} for the environments indium and cadmium are plotted in table II. It is quite interesting to see that independent of the very different lattice structures of indium and cadmium the

TABLE II

Temperature dependence of F_{cond} for ^{111}Cd in indium and in cadmium

temperature in °K	77	194	292	400
$F_{cond}(^{111}$Cd in indium)	3.52	3.18	2.77	2.38
$F_{cond}(^{111}$Cd in cadmium)	2.62	2.14	2.12	1.87

temperature dependence of the conduction electron contribution is quite similar. It is found, however, that the indium environment gives a larger conduction-electron contribution than the cadmium environment, which might be due to the fact that cadmium contributes two electrons per atom to the conduction band whereas indium metal contributes three. The sign of the interaction has been determined in this case by a circular polarization measurement[19] and it agrees with that of the lattice contribution. The conduction-electron contribution to the field gradient results in this case predominantly from the antishielding mechanism (d) whereas the localized moment contribution plays a minor role.

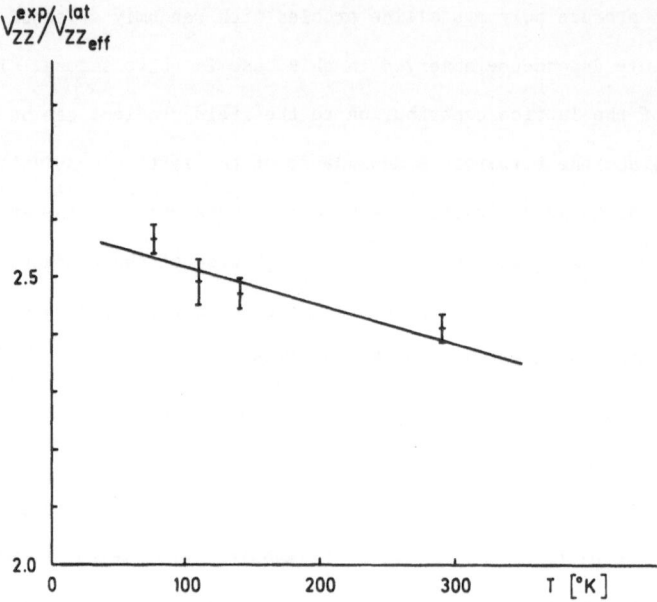

Fig. 10. Temperature dependence of the conduction electron contribution to the effective electric field gradient of ^{181}Ta in rhenium metal. The ratio $V_{zz}^{exp}/V_{zz}^{lat}$ $(1-\gamma_\infty)$ is plotted versus the temperature

The temperature dependence was also studied in the exceptional case ^{181}Ta in rhenium[3] and the result is shown in fig. 10. The ratio $V_{zz}^{exp} / V_{zz}^{lat}(1-\gamma_\infty)$ is plotted versus the temperature. It is found that the conduction electron contribution changes only slightly within the temperature region investigated.

The last topic of my talk concerns the nuclear information. Experimental values for the electric quadrupole moments of excited nuclear states can indeed contribute essentially to the understanding of the nuclear structure. I should like to show the kind of information involved in a few examples.

Let us consider in the first example the spectroscopic quadrupole moments observed for the 482 keV state, the 6 keV state and the ground state of ^{181}Ta. Tantalum 181 is a strongly deformed nucleus and the three states mentioned are assumed to be ground states of different rotational bands. We can now apply the Bohr Mottelson formula:

$$Q_I = \frac{2I^2-I}{(2I+3)(I+1)} \cdot Q_o$$

in order to derive the intrinsic quadrupole moments of the three states. Since the ratios of the interaction frequencies in these three states are quite precisely measured one can study in this way the change of the deformation connected with different single particle excitations. The result that the first excited state at 6 keV has an intrinsic quadrupole moment which is 3±1% smaller than that of the ground state and the 482 keV state shows an intrinsic quadrupole moment 6.4±1.4% smaller than that of the ground state. We think that this is a remarkable feature which should challenge nuclear structure theorists to try to understand the effect by microscopic calculations.

The second example is the 4⁺ state at 2.083 MeV of ^{140}Ce [11]. Excited states of ^{140}Ce are populated in the decay of ^{140}La (see fig. 11). The quadrupole moment was measured by observing the spin-rotation of the 487 keV- 1598 keV- angular correlation in an environment of lantanum-magnesium-double-nitrate. In spite of the rather high electric field gradient which is known from NMR experiments with the stable nucleus ^{139}La in the same environment only the beginning of a very slow

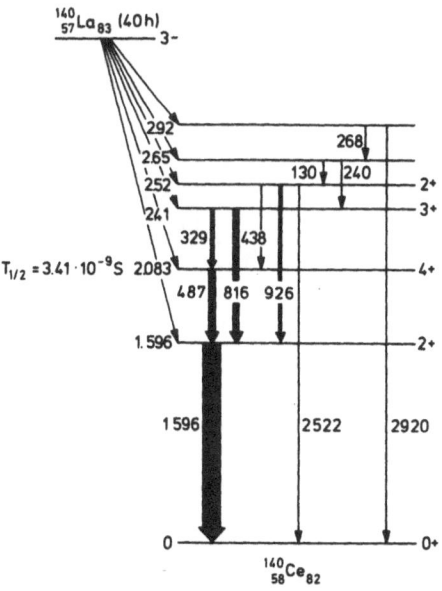

Fig. 11. Decay scheme of ^{140}La

spin rotation is observed (see fig. 12). The spectroscopic quadrupole moment which is derived from this measurement has the value:

$$Q_I(2.083 \text{ MeV}) = 0.40_5 \text{ b}.$$

The structure of the 4^+ state is known from the g-factor measurement[20] to be predominantly the configuration:

$$(g_{7/2}(\text{proton}), d_{5/2}(\text{proton}))_{4+}$$

The interesting aspect is now a comparison between this quadrupole moment and that of the ground state of ^{139}La the structure of which is essentially the single particle shell model state $g_{7/2}$(proton). Within the simple shell model lantanum

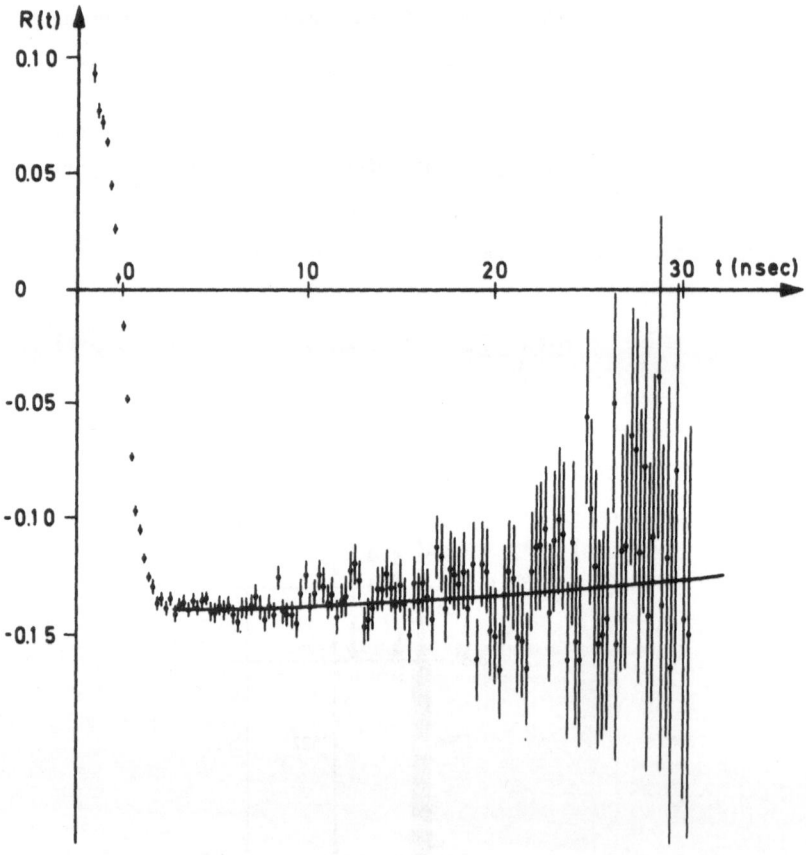

Fig. 12. Spin rotation of the 4+ state of ^{140}Ce observed in an environment of lantanum magnesium double nitrate

140 in the 4^+ state can be constructed from lantanum 139 in the ground state by adding one single proton in the $d_{5/2}$ orbit so that the spins couple to the value 4^+. Now one can calculate the spectroscopic quadrupole moments within the shell model for the two pure configurations mentioned. The results are:

$$^{139}\text{La}(I = \tfrac{7}{2}^+); \qquad Q_I^{theor} = -0.6667 \cdot <r^2> \qquad ;$$

$$^{140}\text{La}(I = 4^+); \qquad Q_I^{theor} = -0.1212 \cdot <r^2> \qquad .$$

The shell model predicts that the quadrupole moment of the 4^+ state of ^{140}Ce should be by a factor 5 smaller than that of the ground state of ^{139}La whereas it is larger actually. We draw the conclusion that in the actual structure in both cases many particles contribute to the observed spin and then the quadrupole moments should have the same order of magnitude.

The last example is the 3/2⁻ state of ^{133}La at an excitation energy of 535 keV. This state has a halflife of 60 nsec and was some time ago[21] interpreted as a so-called shape isomer. The main argument was the large quadrupole moment which was derived for this state from a spin rotation measurement in lantanum-magnesium-

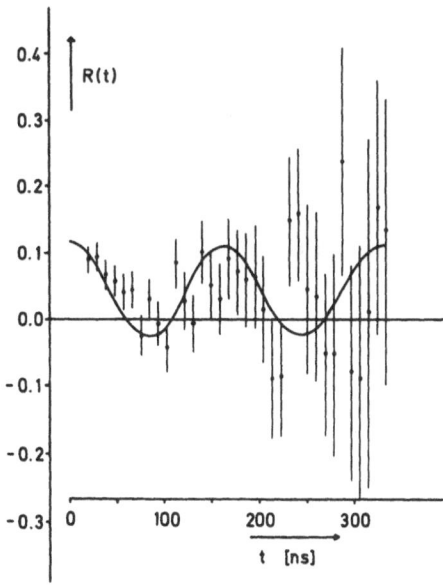

Fig. 13. Spin rotation of the 3/2⁻ state at 535 keV of ^{133}La, observed in an environment of rhenium metal (preliminary result)

double-nitrate. There were some doubts about the reliability of this result because the population of the excited states by electron capture decays from ^{133}Ce produces primarily high charge states which might produce after-effects in the insulating environment. In addition the evaluation contained a trivial error and the correct evaluation yields an even larger quadrupole moment which seems to be unrealistic.

The measurement was repeated two weeks ago in our group[10] by using lantanum metal as environment. In this case no after-effects are expected and we were able to observe a slow spin rotation (see fig. 13). This figure shows a preliminary evaluation of the first half of the data. Since the field gradient in lantanum metal is calibrated also by a nuclear quadrupole resonance experiment with ^{139}La an absolute value for the quadrupole moment could be derived. Our preliminary value is:

$$Q_I(535 \text{ keV}) = 0.36_8 \text{ b.}$$

We conclude from this result that the state of interest is not a shape isomer. But we think that this result should challenge additional investigations in order to see if in other cases the nature of shape isomers can be confirmed by direct quadrupole moment measurements.

REFERENCES

1. Frauenfelder H., Steffen, R.M.: in K. Siegbahn, "Alpha-, Beta- and Gamma-Ray Spectroscopy", N.H.P.C., Amsterdam (1965, p. 997 ff.

2. Haas, H., Shirley, D.A.: UCRL - 20426, p. 208 (1970).

3. Netz, G., Bodenstedt, E.: To be published in Nuclear Physics (1973).

4. Kaindl, G., Salomon, D., Wortmann, G: Phys. Rev. Letts. 28, 952 (1972).

5. DeWit, S.A., Backenstoss, G., Daum, C., Sens, J.C., Acker, H.A.: Nucl. Phys. 87, 657 (1967).

6. Haas, H.: Private communication.

7. Bodenstedt, E., Ortabasi, U., Ellis, W.H.: Phys. Rev. 6, B 2909 (1972).

8. Gerdau, E.: Private communication.

9. Wolf, H.J., Steiner, P., Gerdau, E., Müller, W.K., Lewandowski, E., Roggenbuck, A.: Z.f. Physik 232, 256 (1970).

10. Herzog, P., Klemme, B., Schäfer, G.: To be published (1973).

11. Klemme, B., Miemczyk, H.: Internat. Conference on Nuclear Mangetic Moments, Osaka (1972).

12. Haas, H., Shirley, D.A.: Private communication.

13. Das, T.P., Pomerantz, M.: Phys. Rev. 123, 2070 (1969).

14. Feiock, F.D., Johnson, W.R.: Phys. Rev. 187, 39 (1969).

15. Das, K.C., Ray, D.K.: Phys. Rev. 187, 777 (1969).

16. Watson, R.E., Gossard, A.C., Yafet, Y.: Phys. Rev. 140, A 375 (1965).

17. Gerdau, E., Steiner, P., Steenken, D.: in "Hyperfine Structure and Nuclear Radiations", N.H.P.C. Amsterdam (1968), p. 261.

18. Lieder, R.M., Buttler, N., Killig, K., Beck, E.: Z.f. Physik 237, 137 (1970).

19. Behrend, J.J., Budnick, D.: Z.f. Physik 168, 155 (1962)

20. Korner, H.J., Gerdau, E., Günther, C., Auerbach, K., Mielken, G., Strube, G., Bodenstedt, E.: Z.f. Physik 173, 203 (1963).

21. Gerschel, C., Perrin, N., Valentin, L.: Physics Letters 33B, 299 (1970).

STOPPING POWER EFFECTS IN NUCLEAR

LIFETIME MEASUREMENTS

Cyril Broude

Department of Nuclear Physics, Weizmann Institute of Science, Rehovot, Israel

A knowledge of gamma ray transition probabilities between nuclear states plays an important part in the direct testing of nuclear structure models. For a large range of levels and transition strengths, a direct measurement of the nuclear lifetime is the most readily accessible method for determining these quantities. As we heard from Prof. Goldring in this Symposium nuclear lifetimes are also involved in the derivation of the gyromagnetic ratios of excited states by the integral measurement of hyperfine interactions. There the measurement is of the nuclear rotation averaged over the nuclear lifetime, which must therefore be accurately known from independent measurements. Hence both directly and indirectly the precision of nuclear lifetime measurements is important in assessing nuclear structure models.

In this talk, I will review some recent work on the problems associated with the most widely used method for the measurement of nuclear lifetimes, namely the Doppler shift attenuation method (DSAM). Related to this, I will also talk about recent applications of the recoil distance method (RDM) which are directed towards making it applicable to shorter lifetimes. These developments have two aspects: to make it possible to use the recoil distance method to check the Doppler shift attenuation method by producing a range of lifetime where the two methods overlap, and to make possible the application of the lifetime independent hyperfine interaction experiment, the time differential method, to levels with shorter lifetimes.

It has not escaped my attention that the content of this talk as I have just described it is embarrassingly similar to that of a rapporteur talk[1] I gave at the Montreal International Conference in 1969; however, it is not intended to repeat that talk but it will be taken as the reference to the background to the subject.

In DSAM, the reference time scale is given by the characteristic stopping time of the excited ions recoiling through a slowing environment following a nuclear reaction. In the cases discussed here, the slowing environment is always a solid; these have stopping times around 0.5 ps and provide a range of lifetimes which can be measured from a few picoseconds to about 10^{-14} s. The lifetime dependent effect which is measured is the Doppler shifted energy weighted by the exponential decay of the excited recoil. For example, in the time integral version of the method, the measured mean Doppler shift of a gamma ray of rest energy E_o emitted by a level of mean life τ is simply

$$\bar{S} = \frac{E_o}{\tau} \int_0^\infty \frac{v_z(t)}{c} e^{-t/\tau} dt$$

where $v_z(t)$ is the time dependent velocity of the recoil projected on the direction of observation of the gamma rays. A measurement of this single quantity allows the mean life to be deduced if the recoil velocity as a function of velocity can be calculated from known stopping cross-sections.

DSAM is a method which predates the advent of the Ge(Li) detector but the widespread use of these detectors has resulted in a very large number of life time determinations, in many cases with several independent measurements of the same level by DSAM under varying experimental conditions, and by other methods. Obviously the reason is that very small energy shifts can be measured with the high energy resolution of these detectors. When the instrumental width is less than the Doppler shift, they also make possible the 'time differential' version of the method which measures not only the mean velocity of recoil but a gamma ray line shape due to the detailed time dependent velocity produced by the slowing down. Analysis of this kind of data is more complicated than mean shift analysis but is more informative as the detailed line shape contains information about the stopping process and so the ability (or otherwise) to reproduce the observed line shapes should give an indication of how good the slowing down calculations are and hence how reliable the derived mean life is.

As of a few years ago, the overall conclusion[1] from the redundant measurements both by DSAM under different conditions, and by other methods was that the degree of agreement obtained was surprisingly poor. This conclusion was reinforced as experimental techniques improved yielding measurements with smaller experimental

errors but still with poor overall agreement. At this point, there were one or
two attempts to make systematic measurements on the internal consistency of DSAM
as a function of stopping material and recoil velocity, but the sample size was
small. A review of those experiments can be found in reference 1.

It was then decided to carry out a fairly large survey of DSAM in which the
same mean life would be measured with the recoils slowing in a large number of
different solid elemental backing materials with all other experimental conditions
being kept the same. The experiment was done by Broude, Engelstein, Popp and
Tandon[2]. Experimentally it was very simple: a beam of alpha particles hitting a
target with a Ge(Li) detector fixed at 0° to the beam to measure the reaction
gamma radiation. The (α,p) reaction was chosen to give a reasonably small cone
of recoil in which all recoils could easily get out of the target into the backing
material. The mean recoil velocity was close to 1% of the velocity of light and
it was easy to populate a level with a mean life in the middle of the DSAM range.
The level chosen was at 3350 keV in ^{22}Ne, which decays by a 2070 keV gamma ray
and was known to have a mean life of about 0.3 ps. The reaction used was
^{19}F$(\alpha,p\gamma)^{22}$Ne; targets were made by evaporating 25 μg/cm^2 of ^{6}Li F onto a total of
39 different backing materials. The backings were either thick foils or evaporated
layers thick enough to stop the recoils. Some care was taken to prepare the surfaces of chemically active materials before evaporation of the target layer, and
to avoid oxidation before use.

Some examples of the data are shown in fig. 1. In each section is shown the
total capture peak, after background subtraction, of the 2070 keV line as observed
at 0° in a particular backing. The shapes vary dramatically as a function of the
stopping powers of the backing materials. The shapes are characterized by a
narrow stopped or unshifted peak whose width is largely determined by the Ge(Li)
detector resolution of 3.0 keV at this energy and a Doppler shifted peak whose
width is due largely to Doppler broadening as determined by the kinematic velocity
spread of the recoils due to the outgoing protons. The energy difference between
these two peaks corresponds to the recoil velocity of the center of mass and is
about 1% of the gamma ray energy, about 20 keV.

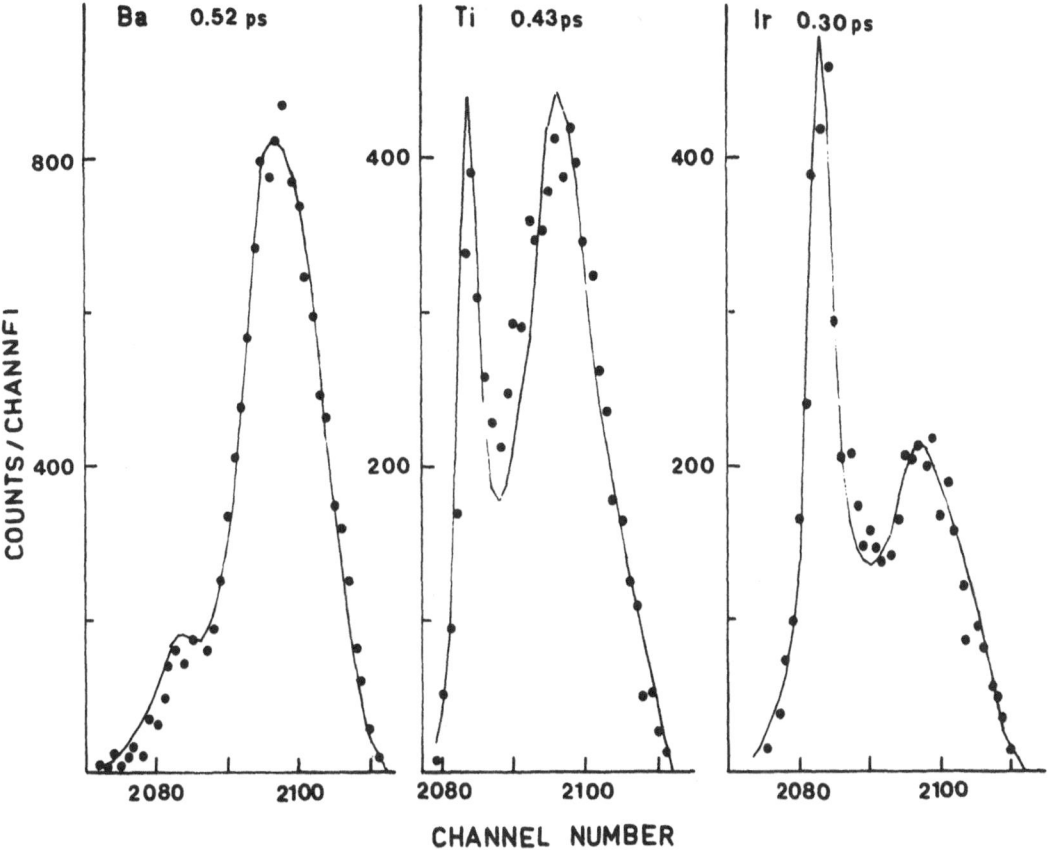

Fig. 1. Doppler shifted line shapes from the reaction $^{19}F(\alpha,p\gamma)^{22}Ne$ with three backing materials

The solid curves are the best least squares fits to the data by a computer programme which uses the theoretical Lindhard atomic and nuclear cross sections[3] to calculate the velocity as a function of time. The calculation integrates numerically over the variation in angle and velocity of the recoils caused by the kinematics and takes into account any specified recoil-gamma ray angular correlation. In cases such as this (α,p) reaction at about 6 MeV bombarding energy, the initial recoil spread is about a factor of two in velocity with a cone of recoil of about 30° half angle.

The examples shown here are not exactly typical of the quality of the fits; they are amongst the better ones. They have been chosen to illustrate that although the fits are good, they show very different derived life times as shown in the figure. In fig. 2 the derived mean lives from all 39 measurements are shown plotted against the atomic number of the stopping material. The first conclusion

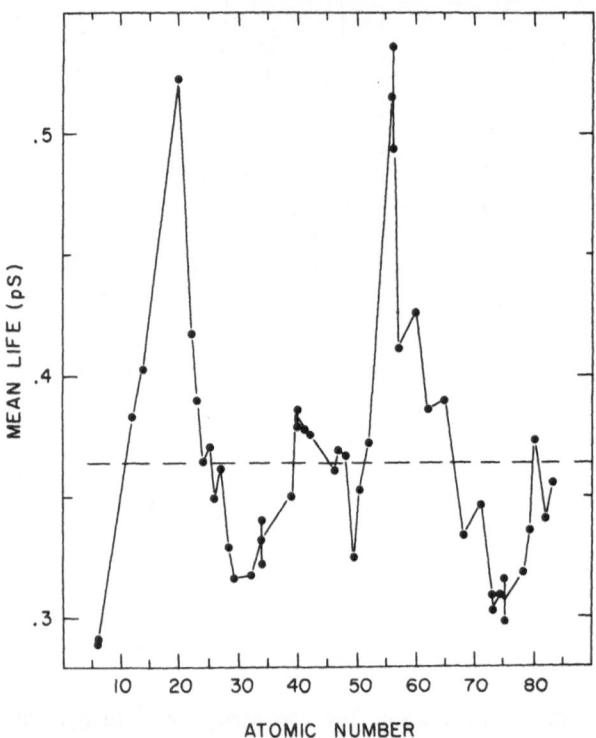

Fig. 2. Variation derived mean life of the 3340 keV level in [22]Ne with the atomic number of the stopping material

to be drawn is that the same method, consistently applied has produced a range of mean life much larger than the individual errors. Statistical errors have not been shown in this figure but the various repeated measurements indicate the accuracy of the data. Secondly, although many correlations were looked for with a variety of physical properties of the backing materials, e.g. density, stopping power, etc., only the plot shown, versus atomic number seems to produce a relatively smooth variation of the data. The size of the variation is considerable especially considering that the method is essentially an integral over a range of a factor of 2 in initial velocity which might be expected to smear out some of the fluctuations.

These data were viewed with the following reservations: although a consistent variation had been observed, there is no independent data to indicate the true value of the mean life. The dashed line in the figure is simply the mean of all the values derived in this experiment. There is no way from this experiment of knowing the absolute deviations from the correct mean life.

Secondly, it is possible that the observed effects are due to some peculiarity of our way of making the measurements, specifically that there are problems of ensuring that the recoils move out of the target into an environment of the pure backing material of density corresponding to the bulk material. A considerable

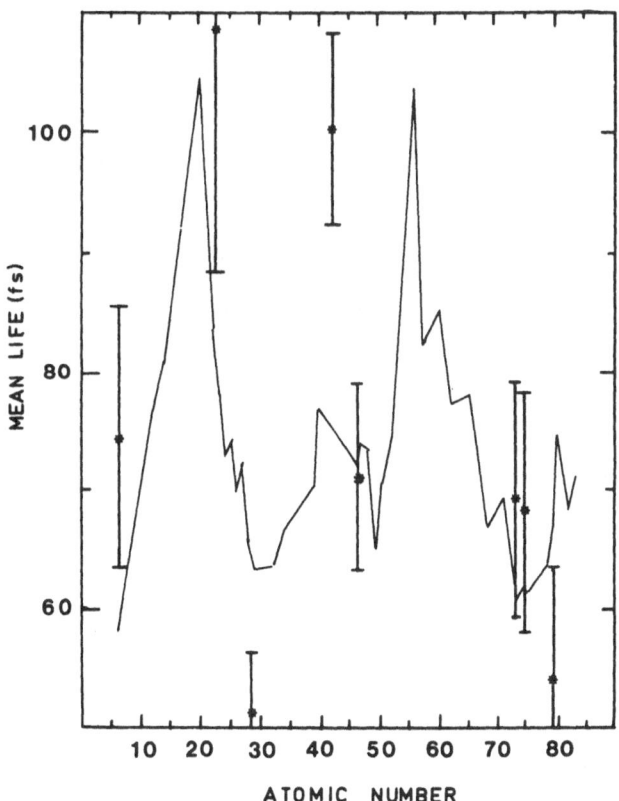

Fig. 3. Data of Bister et al. plotted as a function of atomic number of stopping
material. The curve of fig. 2 is also shown for comparison

number of the materials especially near the highest life time values· are liable to oxidation in varying degrees.

Thirdly, there is the possibility that our method of analysis has some peculiarities quite apart from the values of the stopping cross sections used which anyway are not quoted[3] with an accuracy of better than 20%.

Some confirmation of the effects was required in addition to evidence as to the generality of the effects with respect to varying atomic number of the recoiling ion, its mean life and initial recoil velocity. A private communication from Bister after publication of these data gave some independent confirmation of the effects. In fig. 3 is shown the data of Bister et al.[4], measurements in 7 backing materials from the reaction $^{13}C(p,\gamma)^{14}N$ with a recoil velocity of 0.4%, plotted versus atomic number. It should be realized firstly that the life time measured by Bister et al. is shorter than the earlier one by a factor of about 4 to 5, secondly that the initial velocity is 0.4%, a factor of 3 smaller and that the recoiling ion is ^{14}N instead of ^{22}Ne. Further, because the excitation was via a proton capture reaction, there was no kinematic spread in velocity. Finally, the analysis was completely independent of ours. Nevertheless, superimposing on the plot of these data the curve of fig. 2, simply scaled in mean life to distribute it roughly about the mean value for the ^{14}N data, a considerable agreement is seen between the positions of maxima and minima for the two sets of data. In fact, the ^{14}N data show a much larger variation of life time value but strongly correlated with the ^{22}Ne data. The errors shown are large but of course include the authors estimates of errors due to uncertainties in the stopping cross sections which for our purposes should not be included.

The next set of data is due to Broude and Danino[5] and is essentially a partial repeat of the ^{22}Ne experiment but for a different recoiling ion ^{31}P, and for a level with an independent mean life value. This is shown in fig. 4. Unfortunately, the data lack a measurement for a barium backing which will hopefully be added, but otherwise show the same kind of effect as the ^{22}Ne data shown earlier. The independent value[6] is also plotted. It has rather a large error for our purposes but tends to indicate that the correct value of mean life is somewhere in the

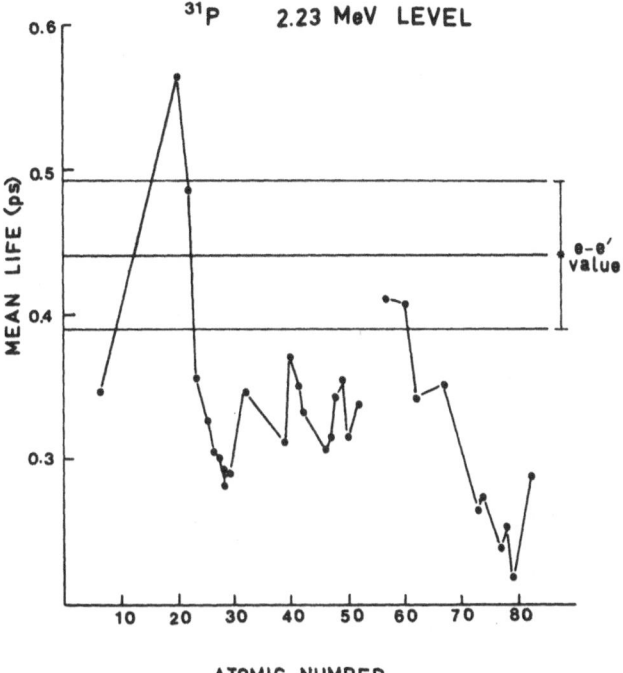

Fig. 4. Variation of derived mean life of the 2230 keV transition in ^{31}P with the
atomic number of the stopping material. An independent value from inelastic electron scattering is also shown

region of the longer values obtained by DSAM. It does indicate that it is not
very useful to obtain independent reference values of mean lives for these short
lived states as their accuracy is usually not adequate, and is in fact usually far
less than the nominal accuracy of the DSAM values.

It appeared worthwhile therefore to measure DSAM in several backings for a
life time which could also be measured by the recoil distance method. Such an
experiment has recently been carried out by Broude, Beck and Engelstein[7].

The advantage of the recoil distance method is that it is basically a model
independent method. In its basic form a beam of monoenergetic excited recoils is
caused to traverse a recoil flight path of known length. From a measurement
of the number, I_F, of ions which decay in flight and the number, I_S, which
decay after having traversed the gap, the mean life τ can be determined from

$$\frac{I_S}{I_S + I_F} = e^{-d/\tau v}$$

where v is the recoil velocity and d is the flight distance. Early recoil distance or plunger experiments involved relatively well collimated, monoenergetic recoils with velocities of several per cent of the velocities of light, and usually long life times. The aim now is to apply the method to values of v/c less than 1% and to life times short enough to be measured by DSAM even in those materials which have short stopping times. These requirements result in a number of deviations from the simple expression shown above for the basic plunger experiment. These include effects such as have already been mentioned with regard to DSAM i.e. a spread in velocity, a considerable cone of recoil angles and a finite stopping time at the end of the flight path. For this reason, the analysis of recoil distance experiments in general is more complex than suggested by the above equation.

Because these requirements are exactly those which entered into the analysis of the DSAM experiments, it is possible to use the same computational method as for DSAM, with an extra vacuum flight path between the target layer and the backing material which now represents the stopper of the recoil distance apparatus. In this way, the analysis of recoil distance at short life times can be analyzed including all effects computationally; this form of analysis has been applied by Broude et al.[7] as will now be described.

The level chosen for the experiment was at 1014 keV in ^{27}Al, excited via the reaction ^{24}Mg$(\alpha,p)^{27}$Al. At the same time, the gamma ray line at 1396 keV in ^{24}Mg was observed from inelastic alpha scattering. The latter was analyzed completely because it is a very severe test of the analysis since it has a very large recoil velocity range, i.e. effectively down to zero velocity, and a very large cone of recoil which means also that slowing in the target is very important in the analysis. In addition there are many independent measurements of this life time so it serves as a good check.

Figure 5 shows this line at two recoil distances with the computer fits. The far distance is effectively an infinite distance measurement because essential-

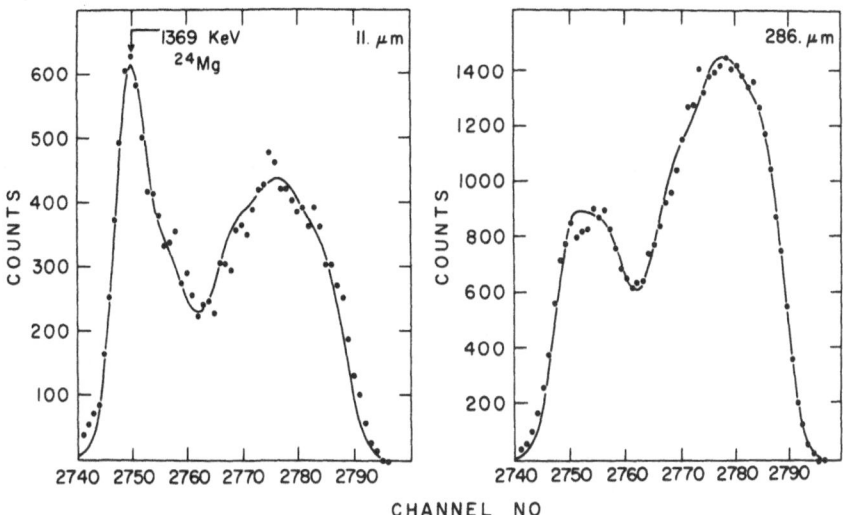

Fig. 5. Recoil distance line shapes with computer fits for the 1369-keV transition in ^{24}Mg at two separations

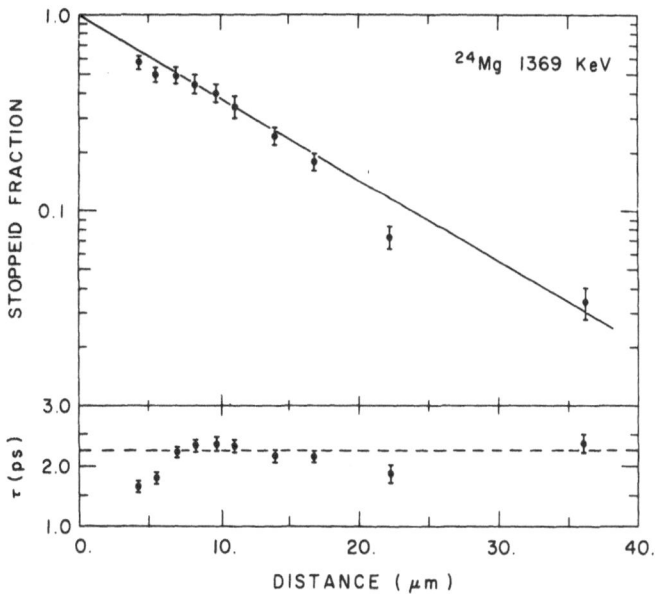

Fig. 6. Derived mean life (lower) and stopped fraction (upper) vs. recoil distance for the 1369 keV level in ^{24}Mg

ly no excited recoils can arrive at the stopper. The complex line shape shown is due to the angular distribution of the recoils and the slowing down and even stopping which occur in the target itself before the recoils even enter the recoil gap. This line shape has been used to determine the angular distribution coefficients in terms of the first 6 odd and even Legendre polynomials. These coefficients were then used in the analysis of all other distances, for example the line shown in the other half of fig. 5.

Figure 6 shows the results of analysis of all the data. In this method of analysis, a mean life value is derived at each recoil distance. When plotted vs. distance, the consistency of the data and their analysis is obtained from the consistency of the mean life values. An additional check is from the stopped fraction parameter derived from the fitting which should obey the exponential law of I_S/I_S+I_F as shown earlier. Deviations from this indicate non-linearity of the distance measurement. The data shown here indicate good consistency except for the two closest points. On the basis of this and other data which will be shown shortly for the 1014 keV ^{27}Al line, these two data points have been discarded.

TABLE I

Reported life time values for the 1369 keV level in ^{24}Mg

τ (ps)	Method	Reference
2.25 ± .09	Recoil Distance	Present
2.16 ± .16	Recoil Distance	9
2.04 ± .14	Coulomb Excitation	10
2.00 ± .10	Coulomb Excitation	11
1.95 ± .26	Resonance Fluorescence	a
1.92 ± .15	Resonance Fluorescence	12
1.90 ± .20	e,e'	a
1.69 ± .14	DSAM	a

a For references see ref. 14.

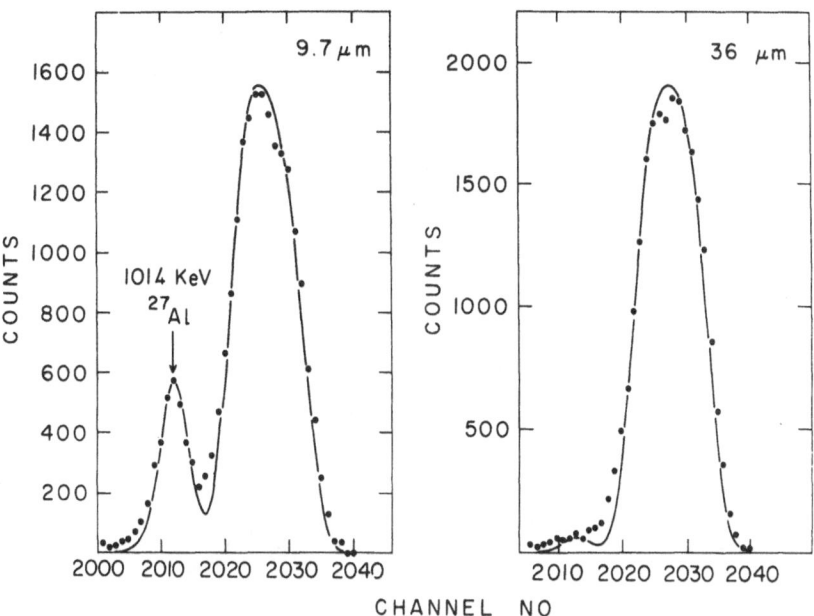

Fig. 7. Recoil distance line shapes with computer fits for the 1014 keV transition in ^{27}Al

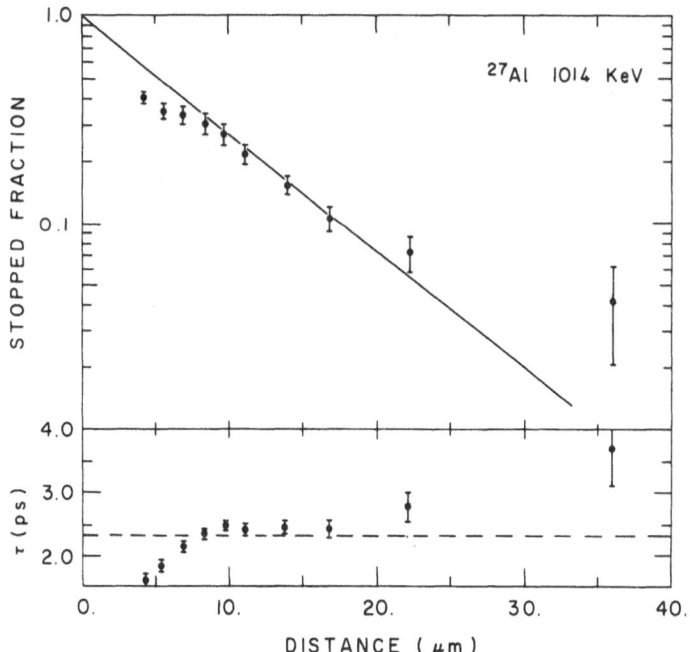

Fig. 8. Derived mean life (lower) and stopped fraction (upper) vs. recoil distance for the 1014 keV level in ^{27}Al

The result of this measurement together with all other measurements of this life time are shown in table I. It can be seen that the values are all reasonably consistent with the exception of the DSAM value.

Figure 7 shows line shapes at two distances for the 1014 keV line, one again being an infinite distance measurement. There the difficulties due to the kinematics do not exist because the recoils are all coned forward; the infinite distance line appears fully shifted. These data have been fitted with an isotropic centre of mass recoil distribution. Figure 8 shows the consistency of

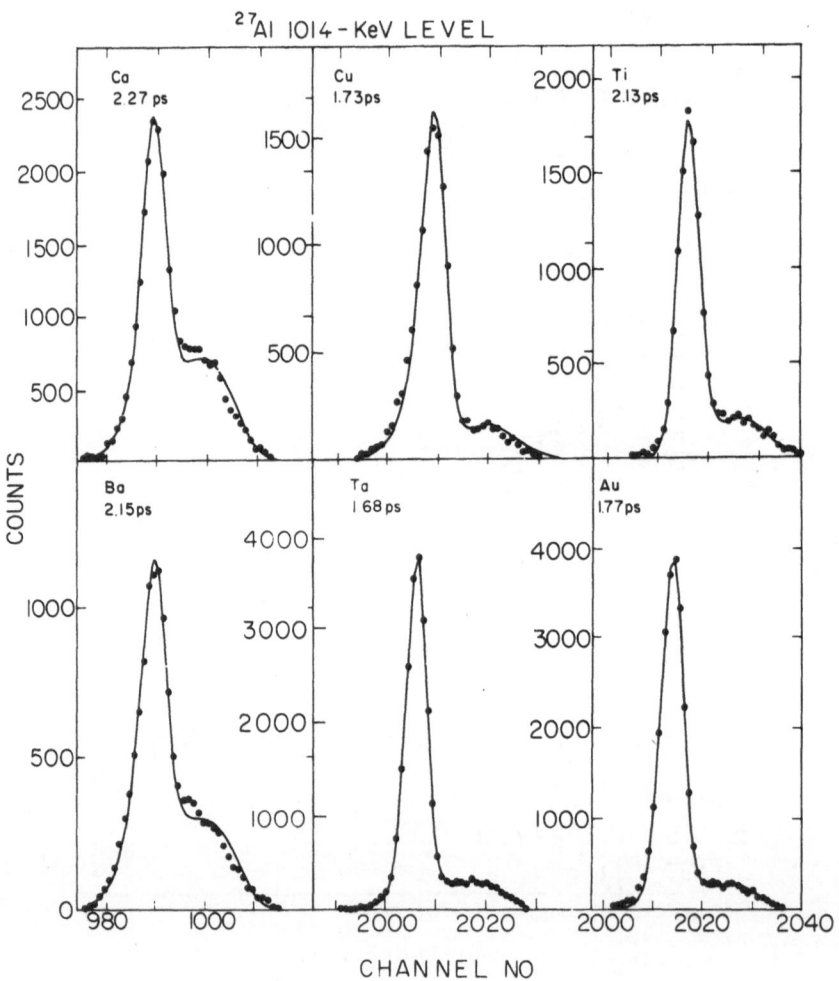

Fig. 9. DSAM line shapes and fits for the 1014 keV level in ^{27}Al

the mean life at various distances and shows the same effect as for the earlier
data: the closest two data points have been discarded.

Fig. 9 shows the data for the DSAM measurement of this same line in 6 backing
materials. The solid curves are the best fits which are all good. The backings
and derived life times are shown in each case. The backings were chosen to lie
on the extrema of the curve of fig. 2.

The present mean life values and all others for the 1014 keV ^{27}Al level are
shown in table II. The consistency of the plunger measurements with other methods
is acceptable. The DSAM shows a large spread. Fig. 10 shows the data plotted
versus atomic number with again a scaled version of the full mean life dependence
curve of fig. 2, plotted also. This indicates that once again there is a strong
correspondence between the life time dependence of this measurement and the
earlier one although the variation here seems to be smaller. Plotted also is

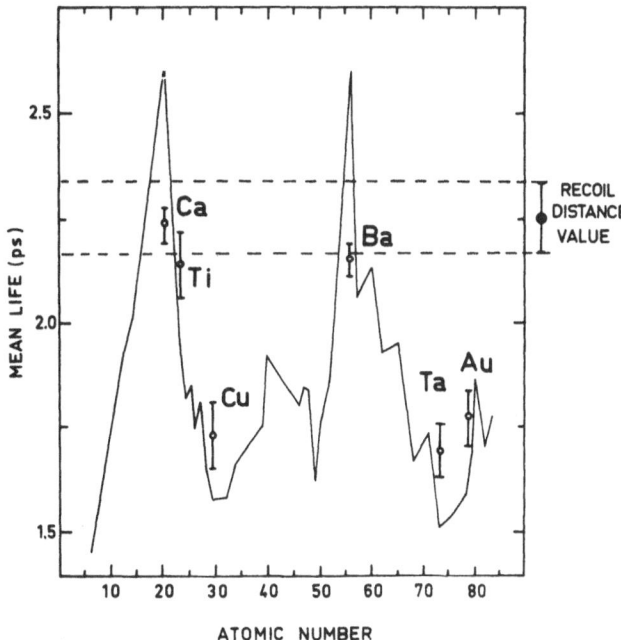

Fig.10. Variation of mean life of the 1014 keV level in ^{27}Al as a function of
atomic number of the stopping material. The recoil distance value is
also shown and the curve of fig. 2 is plotted for comparison

the recoil distance value which agrees only with the longest value obtained from
DSAM. This conclusion has been suggested for example by the data of fig. 4.

Reported life time values for the 1014-keV level in ^{27}Al

τ(ps)	Method	Reference
2.37 ± .12	Recoil Distance	Present
2.20 ± .30	Resonance fluorescence	a
2.13 ± .19	Resonance fluorescence	13
1.50 ± .40	DSAM (Al)	a
2.27 ± .04	DSAM (Ca)	Present
2.13 ± .07	DSAM (Ti)	"
1.73 ± .08	DSAM (Cu)	"
2.15 ± .04	DSAM (Ba)	"
1.68 ± .06	DSAM (Ta)	"
1.77 ± .07	DSAM (Au)	"
2.2 ± $^{.4}_{.3}$	DSAM (High Velocity)	8

a   Reference given in ref. 13.

As a result of this Symposium, I have become aware of the unpublished work of
Hauser, Neuwirth and collaborators at Cologne. It is not possible for me to
comment fully on their work as there has not been sufficient time for me to become
familiar with it. The results which can be mentioned are shown in fig. 11.
In this figure, the data points shown as open circles with error bars are the
measurements by this group of the electronic cross sections for ^{7}Li ions recoiling in a range of materials. These cross sections for a ^{7}Li ion energy of 100 keV,
are plotted versus the atomic number of the stopping material. These stopping
cross sections have been derived by a Doppler shift method described by Neuwirth,
Hauser and Kühn[15]. It will be seen that these measurements indicate peaks in the
electronic cross sections at atomic numbers 20 and 56 as do all of the DSAM data
shown previously. The solid line represents Lindhard theory[3] . The conclusions

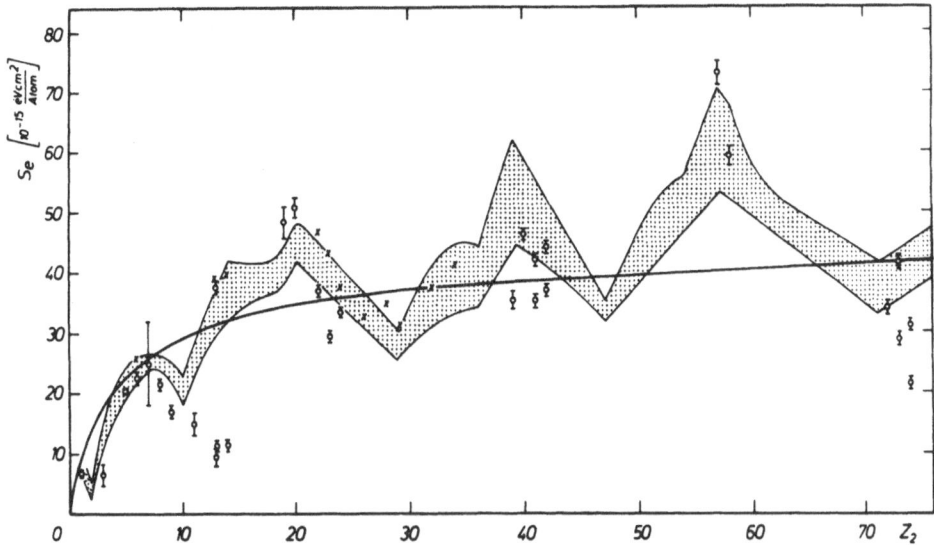

Fig.11. Plot of electronic stopping power for 100 keV ^{7}Li ions plotted versus
atomic number of the stopping material as measured by Hauser et al. and
shown by the open circles with error bars

of these workers is that the cross sections are proportional to velocity as
required by the Lindhard theory, but the absolute values vary with atomic number
as shown in fig. 11.

It can be said that although qualitatively the variations of the cross sections
of fig. 11 and those of the DSAM life time data agree with each other, the observation that the longest life times are the correct ones, i.e., in backings of atomic
number near 20 and 56, is not consistent with the cross sections for these
materials deviating strongly from Lindhard theory. These data are of course not
observed for the same slowing ion and so the inconsistency is not conclusive.
However, it was a previously observed fact[2] that the line shapes in the DSAM measurements cannot be fitted well with large changes in the normalization of the
electronic cross sections alone.

To summarize the present situation, qualitatively similar oscillatory behaviour
in the mean life as derived from DSAM has been observed for 5 different light
recoils in independent measurements. These nuclei are ^{7}Li, ^{14}N, ^{22}Ne, ^{27}Al and

^{31}P. Quantitatively it is difficult to compare the observed amplitudes of the effect for the 5 experiments but the DSAM effects appear to be largest in the measurements on ^{14}N, about the same for ^{22}Ne and ^{31}P and smallest for ^{27}Al. However, this is presumably due to the differences in the conditions of the experiments and not due directly to the atomic number. The main correlating parameter appears to be the nuclear life time in each experiment which can be classified as very short (.08 ps) for the largest effect (it also involved the lowest recoil velocity), intermediate (0.4 ps) for ^{22}Ne and ^{31}P and long (2.4 ps) for ^{27}Al. There are two possible effects which can arise from these conditions: that the various life times sample the relative electronic and atomic cross sections differently and that the mean distance travelled before radiation is a function of mean life so that for shorter mean lives surface effects on the backings, such as oxidation layers, are more important than for long life time recoils which can get into the bulk material before decaying very much. The final conclusion which is tentative is that the longer DSAM values seem to correspond to the correct mean life. This is surprising in that it is the slow stopping materials which are most reactive and one would expect surface and even bulk oxidation of the materials which would lead to erroneous results in these cases.

Future efforts in part will be directed to making recoil distance measurements applicable to shorter lifetimes. We have now in analysis, data taken at recoil distances down to 3 microns giving observed recoil distance effects in levels which have quoted DSAM lifetimes as short as 1 ps. It is hoped that such measurements will become routine enlarging the region of overlap between the DSAM and recoil distance. As described earlier, at these mean lives the recoil distance method analysis is complicated and begins to merge with DSAM because of the finite stopping time of the plunger stopper material. However, at least to 1 ps it is believed that it is valuable to attempt recoil distance checks of lifetimes measured to date only by DSAM.

REFERENCES

1. Broude, C.: Int. Conf. on Properties of Nuclear States, Montreal (1969) (University of Montreal Press) Invited Rapporteur Talk.

2. Broude, C., Engelstein, P., Popp, M., Tandon, P.N.: Phys. Lett. 29B, 185 (1972), and to be published.

3. Lindhard, J., Scharff, M., Schiott, H.E.: Mat. Fys. Medd. Dan. Vid. Selsk 33 (1963).

4. Bister, M., Antilla, A., Pilparinen, M., Viitasalo, M.: Phys. Rev. C3, 1972 (1971).

5. Broude, C., Danino, Y.: to be published.

6. Kossanyi-Demay, P., Lombard, R.M., Bishop, G.R.: Nucl. Phys. 62, 615 (1965).

7. Broude, C., Beck, F.A., Engelstein, P.: to be published.

8. McDonald, A.B., Alexander, I.K., Haüsser, O., Ewan, G.T.: Can. J. Phys. 49, 2886 (1971).

9. Alexander, T.K., Broude, C., Haüsser, O., Pelte, D.: Int. Conf. on Properties of Nuclear States, Montreal (1969) University of Montreal Press, Contribution 4.50.

10. Hausser, O., Hooton, B.W., Pelte, D., Alexander, T.K., Evans, H.C.: Phys. Rev. Lett. 22, 359 (1969).

11. Vitoux, D., Haight, R.C., Saladin, J.X.: Phys. Rev. C3, 718 (1971).

12. Swann, C.P., Phys. Rev. C4, 1489 (1971).

13. Imada, H., McIntyre, J.A.: Nucl. Phys. A184, 574 (1972).

14. Endt, P.M., Van der Leun, C.: Nucl. Phys. A105 (1967).

15. Neuwirth, W., Hausser, U., Kühn, E.: Z. Physik 220, 241 (1969).

STATUS OF THE DARMSTADT PROJECT

Peter Armbruster

GSI Darmstadt

1. INTRODUCTION

Dreams and plans for building a heavy ion accelerator in Germany are rather old. Christoph Schmelzer, having returned from CERN in 1960, began to think about his Universal Linear Accelerator for acceleration of all ions up to energies above the Coulomb barrier. It took a long time to get a project started. A study group was founded, the UNILAC Group, in 1963. This study group had been working for six years as guests of the MPI Heidelberg by the time the final decision was made in December 1969 to build a new laboratory in Darmstadt, GSI, housing a heavy ion machine. Favorable but dangerous winds, blown by the world-wide community of super heavy nuclei speculators, started our small vessel on its way. In 1970 the final decision was made as to which accelerator should be built. The conservative concept of UNILAC was selected, and the plans to build a helix accelerator were abandoned.

The laboratory at Darmstadt had to start from scratch. In 1971 the actual work in Darmstadt began. The tasks which had to be solved when I joined the institute in January 1971 were:

1) to build a laboratory

2) to construct a heavy ion machine

3) to prepare a research program

In this report I would like to talk about these three subjects.

Before I begin let me make some remarks concerning the organization of GSI (Gesellschaft fur Schwerionenforschung). Four-hundred people, including 120 scientists, will be the permanent staff of GSI. Up to now about half of the

personnel is hired and working with us. The total investments will be 142 million
DM, not including salaries and daily running costs: DM 66 million for the
laboratory

DM 47 million for the
accelerator

DM 29 million for the investments for experiments
and computers.

The purpose of GSI is to construct a heavy ion machine and to do fundamental
research using heavy ions. The research will be done in close collaboration in
fully integrated mixed groups of GSI and outside users. Outside users are German
universities and research institutions such as the MPI and the federally-owned
research centers. Fields of research are:

a. nuclear physics

b. nuclear chemistry

c. atomic physics and solid-state physics

d. ion source and accelerator improvement

e. technological applications of heavy ion research.

2. THE LABORATORY[1]

The institute is situated 7 km north of Darmstadt in a forest area. It is
easily accessible by car from Frankfurt airport (20 km) and the surrounding
universities in the Rhein Main area.

The topology of the accelerator and the experimental areas as the heart of
the laboratory and the decision to avoid a splitting of the laboratory in single
institutes dictate the general layout of our research facility. Parallel to the
accelerator there are two buildings housing offices and labs, connected by two
other buildings which form bridges to the main buildings. All parts of the whole
installation can be reached without leaving the house. The total area of the
institute amounts to 23,000 m^2, comprising:

5800 m^2 accelerator + supplies

5900 m^2 experimental areas

4500 m^2 offices, library, computers, cafeteria

5900 m^2 labs and workshops

1000 m^2 central power supply

Fig. 1 gives a model of the laboratory. Fig. 2 is a view taken from the chimney of the central power station (see fig. 1) in direction of the accelerator and the experimental halls. Injection, stripper and main experimental hall connected by the accelerator tunnel are seen in the foreground. The prefabricated main buildings are set up behind them.

3. THE ACCELERATOR[2,3]

A universal accelerator for all elements and masses, bringing the particles to a constant specific energy W/A, needs a total accelerating voltage U, which strongly depends on the mass of the ions:

$$U = \frac{W/A}{q/A}$$

The specific charges available from ion sources run from 0.5 for light elements to 0.05 for uranium 12^+. The voltage needed to accelerate uranium to a certain velocity would be ten times larger than for alpha particles. A one-stage accelerator for uranium needs a voltage of 160 MV to achieve 8 MeV/nucleon.

The average ionic charge of a fast heavy ion is velocity-dependent. Whenever it passes matter, it will lose or capture electrons in charge changing collisions. The further away from its average value the actual value of the charge is, the higher the cross sections for charge-changing collisions will be. During acceleration, charge-changing collisions in the rest gas should be avoided. In most accelerator systems, they lead to a loss of part of the beam. The vacuum in heavy ion machines should guarantee not to lose more than 1%. The pressure in heavy ion machines should be lower than 10^{-7} torr.

The way out of the dilemma of high accelerating voltages and extreme vacuum is to combine systems and to adjust the ionic charge to the average value corresponding to its actual velocity. This helps save voltage and reduce the vacuum requirements.

Fig. 1. Model of the GSI research facility

Fig. 2. View from the chimney, see fig. 1, in direction of the accelerator and
experimental halls

240

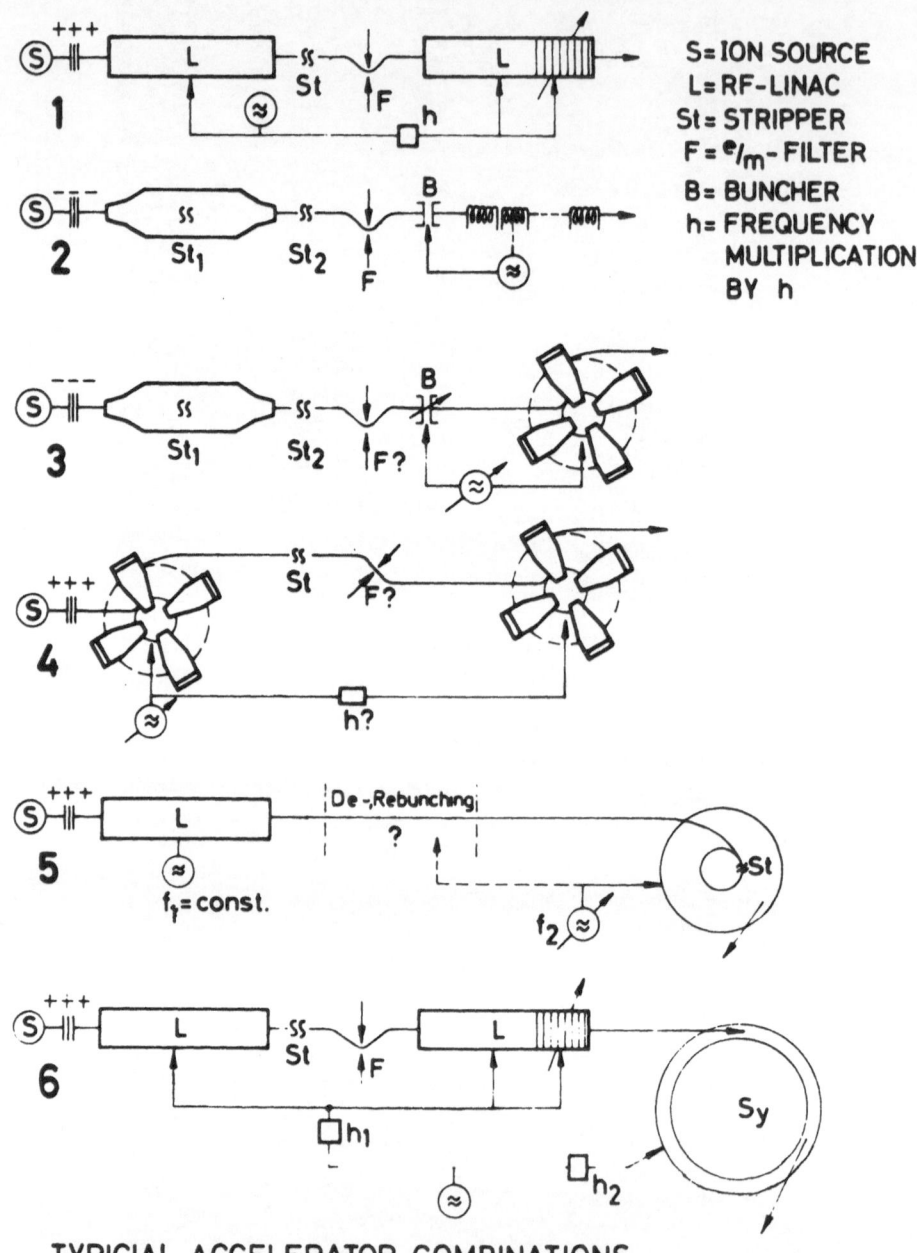

S = ION SOURCE
L = RF-LINAC
St = STRIPPER
F = e/m- FILTER
B = BUNCHER
h = FREQUENCY
     MULTIPLICATION
     BY h

TYPICIAL ACCELERATOR COMBINATIONS

Fig. 3. Typical heavy ion accelerator combinations. S = Ion source; L = RF Linac; St = Stripper; F = e/m-Filter; B = Buncher; h = Frequency multiplication[3]

Of the many possible ways to combine accelerator systems, we chose a two-step linear accelerator (fig. 3). The built-in velocity profile of its drift tubes gives a constant specific energy for all masses A. W/A = const. Compared to other accelerators, where the constringents, holding field for cyclotrons $W/A \sim (q/A)^2$ or maximum attainable voltage (W/A = q/A), cause a strong mass dependence of the final specific energy, the linear accelerator with its constant dependence meets best the physical fact that the Coulomb barrier for masses A > 100 remains about constant throughout all masses.

The layout of the UNILAC is given in fig. 4. The detailed specifications are given in data sheets published in our UNILAC Project Reports[4].

4. DESCRIPTION OF THE SYSTEM

Ion sources	duoplasmatrons A $\lesssim$ 130
	penning sources A $\gtrsim$ 80
DC injector	320 KV DC
Beam transport	mass resolution A/ΔA 250, 10 cm·mrad
Wideröe	12 keV/nucleon injection. 4 sections, 27 m long.
	27 MHz, as β only 0.5% End energy 1.4 MeV/nucleon.
	1.2 MW.
Stripper	CO_2 jet or foil stripper. Helix section to compensate for energy loss. Charge separator for selection of one charge state for post accelerator and another for "stripper" experimental hall.
Post accelerator	108 MHz, three sections a) first Alvarez b) second Alvarez ⎫29m, 3.4 MW
	c) single resonators.
	Single resonators allow energy variation between (5.9 ± 2.6) MeV/nucleon.
	Total power installed 2 MW. Duty cycle δ = 25%, 8 MW/pulse possible.

Computer control of the system is provided.

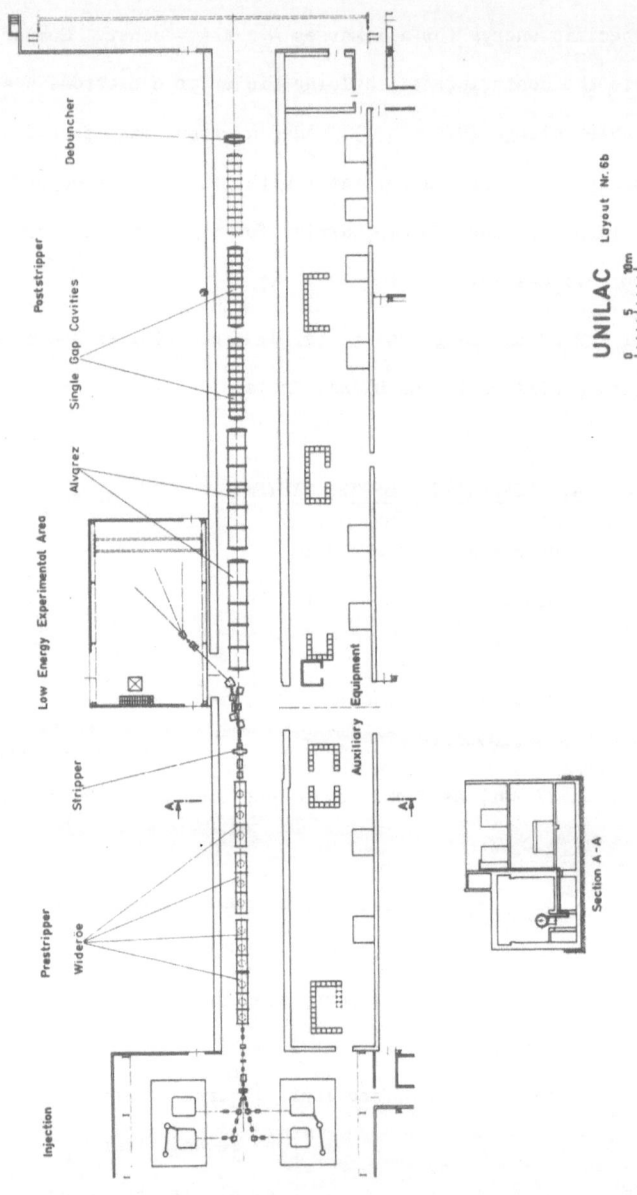

Fig. 4. Layout of the UNILAC[3]

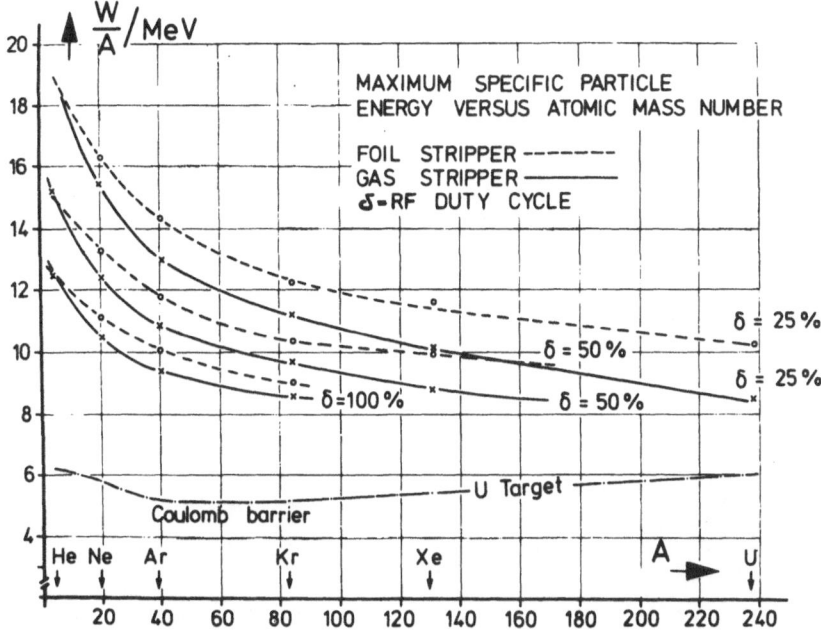

Fig. 5. Maximum specific energies W/A versus mass number A for the UNILAC. Full

curves: gas stripper, dashed curves: foil stripper. δ = macroscopic duty

cycle[3]

Beam specifications of first beam available in early 1975[5]:

1) Energy as shown in fig. 5.

2) Time structure: micro 36 nsec, macro 5 ms beam, 15 msec no beam.

3) Energy resolution: 1%

4) Time resolution: 1.2 ns without buncher

5) Intensity: Ne 1.5 10^{13}/sec; Xe 10^{11}/sec, increased up to 10^{13}/sec

within the first year of operation.

6) Beam emittance: (2-3) cm·mrad.

The following slides will inform you of the stage of construction of the

various parts of the accelerator. The various parts have been tested in the past

years as prototypes, partly as small-scale models. They are now being manufactured

by different companies in Europe, with the responsibility for the total system

remaining at our institute. Only part of the slides shown in the talk are reproduced in the written manuscript. Fig. 6 shows the electroplating of an Alvarez prototype. The electroplating is performed in special installations built directly on the site.

Fig. 6. Electroplating of Alvarez structure

The construction schedule is tight. The first beam - with reduced specifications - is expected to be available by the end of the year 1974. The accelerator will be brought to full power during the year 1975. In 1975, the first experiments already can be made. Experimental equipment should be ready to be tested by early 1975.

5. THE EXPERIMENTS

There are three experimental areas:

1) a low-energy area $E \stackrel{<}{=} 12$ keV/nucleon

2) a medium-energy area $E \stackrel{<}{=} 1.4$ MeV/nucleon

3) the main experimental area $(3 < E < 10)$ MeV/nucleon

In the low-energy area the test injector, a 300-K-DC machine, can be used to make low-energy heavy ion experiments. This facility gives strong beams, 10^{15}/sec, for all ions. Main applications:

a) ion source tests and experiments

b) atomic physics at low energies

c) irradiation with heavy ions, material testing, sputtering, defects, ion implantation.

The beam separation at medium energies provides a beam of fixed energy, 1.4 MeV/nucleon, which is coupled concerning the ion selection to the main beam. This parasitic beam has an energy too small to induce nuclear reactions. It can be handled without any radiation protection. The only harm it does is its inherent beam power of less than 1 KW. There will be three beam lines in the "stripper" hall - that is the name of this area - for atomic, solid state and applied physics.

The main experimental hall is (40×60) m^2 large. A 30-ton crane covers the whole area. Besides on the south an experimental area behind a 1 m concrete shielded wall of 600 m^2 is provided. The control room for the experiments is found there. Low-level chemistry and physics laboratories will be situated as near as possible to the beam behind this wall. Fig. 7 shows the preliminary layout of the experimental area (March 1973) with a possible first generation of experimental equipment. This layout will be built within the following years step by step.

Beam splits allow simultaneous experimentation at different positions, provided users can be found who are willing to do experiments with the same ions of the same energy at the same time.

There are three main beam lines. The central beam, containing the full beam of several charge states, will be used mainly for isotope production. The experi-

246

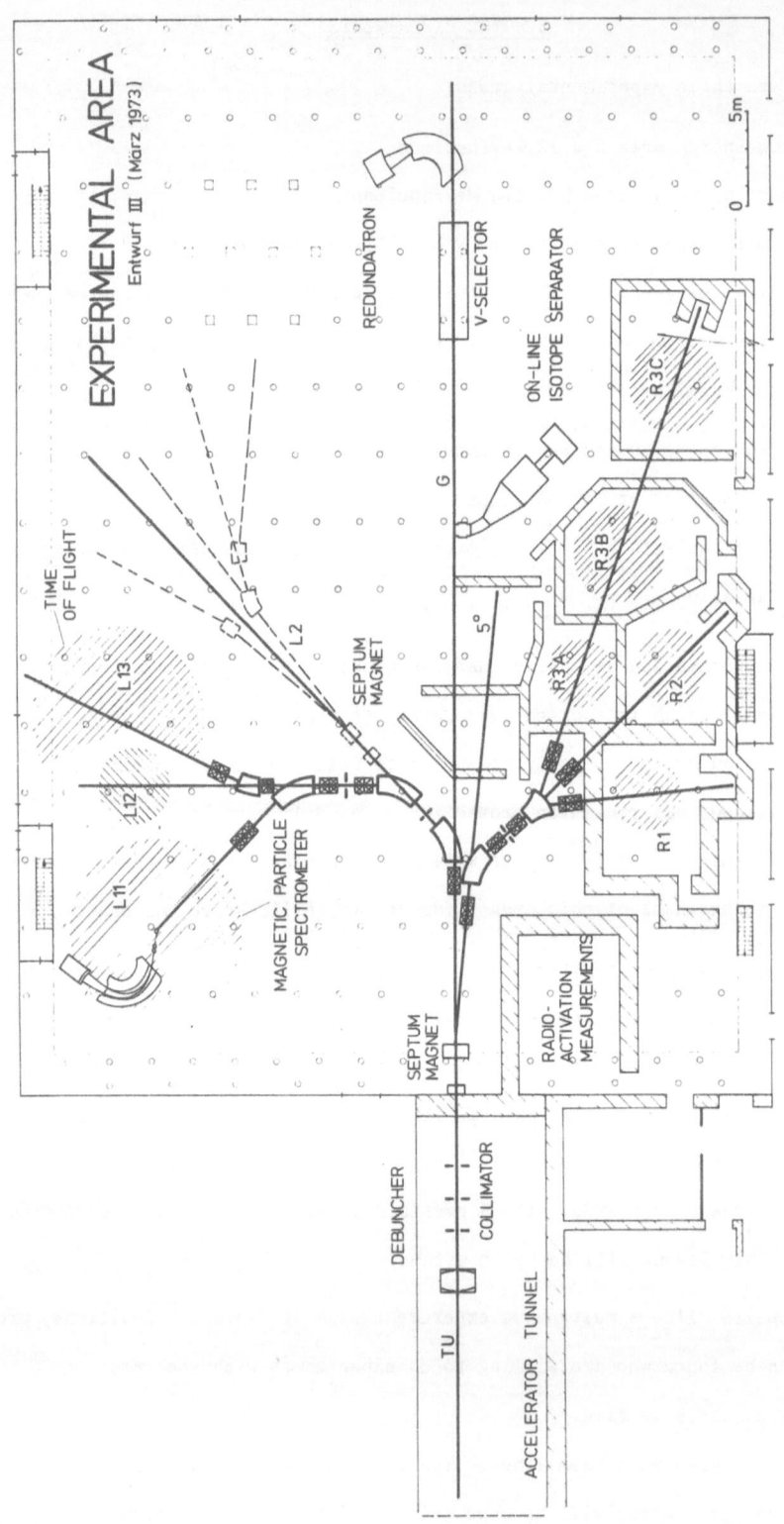

Fig. 7. Preliminary layout of main experimental area (March 1973)

ments south of the central beam will be supplied by one charge state of the beam. Its intensity may be lower by a factor of 2-3 compared to the central beam; its energy resolution is as delivered by the machine. A switching magnet brings the beam into three caves. In beam-gamma spectroscopy, a large scattering chamber, an atomic physics position, the possibility of installing a special scattering chamber for fission experiments is planned. On the northern side of the central beam, improved energy resolution ($5 \cdot 10^{-4}$) is provided. A magnetic spectrometer will be installed, a test beam for accelerator operation, a TOF spectrometer and some other high-resolution beams will be found there. A second septum magnet making use of the beam dispersion necessary to improve the energy resolution allows use of those parts of the beam which would otherwise be lost.

In the last part of my talk I would like to cover three special topics of our experimental program which have the individual fingerprints of the speaker:

a) an isotope production machine for low cross sections

b) a coincidence equipment for the investigation of 2-body disintegrations, mainly fission

c) atomic collision experiments leading to the excitation of inner shells, a topic discussed in these Proceedings in the paper presented by K. Dietrich.

6.   ISOTOPE PRODUCTION[6]

Isotopes which would be produced at light particle machines should not be made with a heavy ion accelerator. Assuming the same production cross sections (a first optimistic view), the same available beam intensities (a second optimistic view), the small effective target thickness, which is of the order of a few mg/cm^2, still reduces the production rates by a factor of 100-1000 compared to light particle machines. The very neutron-rich, proton-rich and heavy isotopes are the domain of a heavy ion accelerator. Isotopes with half lives shorter than minutes, with production cross sections of a few millibarns, or superheavy elements with production cross sections of several nb, must be isolated. Separation must be fast and efficient.

Recoil spectrometers, spectrometers separating the unslowed reaction products, have separation times of 10^{-6} sec. The thin targets (1 mg/cm^2) will no longer be a disadvantage to other systems, as even ISOL systems cannot have effective target thicknesses larger than a few mg/cm^2. The kinematics of heavy ion reactions makes the recoils to be emitted in forward direction. Thus 20% of the fusion reaction products and even more than 1% of the fission products are emitted into the acceptance angle of a recoil spectrometer.

The reaction products emitted into the forward direction must be separated from the beam of projectiles. High selectivity is required. 10 nb correspond to a ratio of 10^{-12} reactions products to projectiles. Most effective separation will be achieved not by a single separator using the techniques of tomorrow, but by a series of separators using the techniques of today. Each of these separators should separate the isotopes according to a different physical quantity. The selectivity is obtained by the redundance of the system in determining mass and charge. The apparatus which I propose to call "Redundatron" is shown in fig. 8. It combines measurements of v, dE/dx, B$\bar{\rho}$ and E. Four independent mass and three

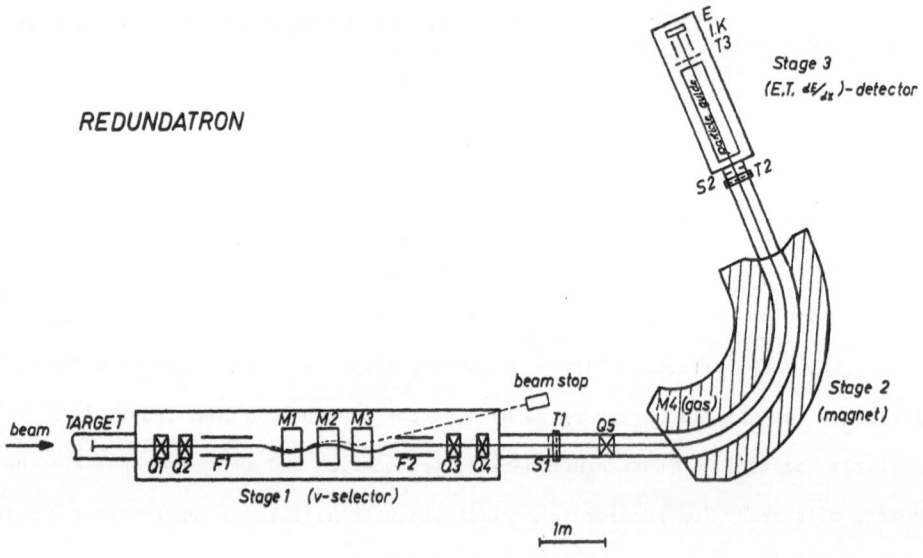

Fig. 8. "Redundatron", a 3-stage recoil spectrometer for investigating heavy-ion reaction products

charge determinations follow from different combinations of the four quantities measured. The first stage of the system is a v selector. It separates the primary beam from the reaction products and defines the velocity for the next separation stage. The second stage is a magnetic deflection with gas filling. The gas-filled magnetic separator separates according to $B\bar{\rho} = mv/\bar{q}$. The average value $\bar{q}$ depends, as does the specific ionization, only on v and Z. If v is known, $B\bar{\rho}$ is a quantity which can be calibrated in a similar way as the specific ionization. The two-stage set-up allows separation with high efficiency (20%) within 10^{-6} sec. The third stage is a purely electronic device, a time-of-flight, energy, dE/dx telescope. Ionization chambers allow measurement of dE/dx with (1-2)% accuracy for heavy recoils. At the slit positions between the stages, in-beam nuclear spectroscopy may be applied either in order to be of some help in the further identification or to investigate nuclear properties of the isotopes.

7. SPECIFICATIONS

η	= 20% for fusion products
η	= 1 % for fission products
selectivity	$\sim 10^{-15}$, number of particles in exit window/to number of projectiles at entrance slit
velocity resolution	~ 100
mass resolution	< 180
charge resolution	< 80
acceptance	= 30 cm·mrad first stage
	100 cm·mrad second stage
	200 cm·mrad third stage

8. FISSION STUDIES BY KINEMATIC COINCIDENCES

The most successful method for studying the fission reaction is the so-called double-energy method. This method relies on the fission kinematics and obtains the mass values of the fission products from velocity or energy measurements. Three laws of conservation give four equations: mass conservation, energy conservation, and momentum conservation in two coordinates. Three quantities must be measured to determine the set of parameters characterizing the reaction completely.

A method applying the same principle to heavy ion reactions is the time-difference method. It can be shown that the measurement of two angles of the reaction products referred to the beam direction (θ_1 and θ_2) and the measurement of the difference of their arrival times in two detectors ($\Delta t = t_1 - t_2$) is sufficient to determine the mass unambiguously.

$$R_1 = R_2 = R$$

$$D_1 = \frac{R}{tg\theta_1} \quad , \quad D_2 = \frac{R}{tg\theta_2}$$

$$\frac{m_1}{m_c} = \frac{1}{2} \left(1 + \frac{\Delta t \cdot v_c}{D_1 + D_2}\right) \quad \text{with} \quad v_c = \frac{m_{Pr}}{m_c} v_{Pr} \quad , \quad \text{and} \quad m_c = m_t + m_{Pr}$$

Specifications:

$A/\Delta A$	> 230
ΔQ	< 4.2 MeV
Δp	< 50 MeV/c
$D_1 + D_2$	= 4 m
$R_1 = R_2$	= 60 cm
$d_1 = d_2 = d_t$	= 3 mm
$\delta \Delta t$	= 0.2 ns

A measured spectrum for $^{32}S \rightarrow {}^{40}Ca$ is shown. It was obtained by Rehm, Müller and Henning in TUM[8], fig. 9.

10. DETECTION OF SUPERHEAVY QUASIATOMS[9]

The formation and observation of a transient superheavy atomic system becomes possible if the following conditions hold:

1) $Z = Z_1 + Z_2 > 100$

2) $b_{min} \lesssim a_{shell}$, that is the wave functions overlap. It follows:
   $$v/v_0 > 2 \cdot 10^{-2} \cdot Z/n$$

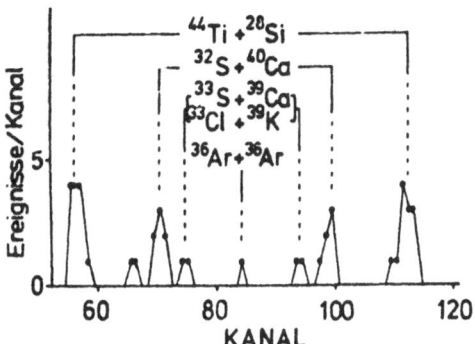

Fig. 9. Time-of-flight difference method applied to measure the mass spectrum of the reaction products from ^{40}Ca bombardment by ^{32}S 8

3) collision time > orbital time of electrons, that is the collision is nearly adiabatic. It follows: $v/v_0 < Z/n$.

Conditions 2 and 3) give: 4 MeV < E_{Pr} < 10 GeV for a system of Z = 145.

4) Production of vacancies which may decay by emission of characteristic x-rays.

5) Collision time, that is the lifetime of the quasiatom becomes comparable to the electric dipole transition time.

$$\tau_{coll} = 2.10^{-15} \frac{n^3}{Z^2} \sec$$

For I on U and n=3 the maximum collision time is $\tau_{coll} = 5 \cdot 10^{-18}$ sec at an optimum collision energy E_{opt}=8 MeV.

$$\tau_{mn} \propto \frac{1}{E_{mn}^3 D_{mn}^2} = 37 \cdot 10^{-10} \frac{n^2}{Z^4} \sec$$

The lifetime of a M-shell vacancy in Z = 145 is $\tau_{mn} = 10^{-17}$ sec.

$$\frac{\tau_{coll}}{\tau_{mn}} = 5.2 \times 10^{-6} n Z^2$$

This ratio is 0.25 for a M-shell vacancy at the optimum collision energy. It is 10^{-7} for a 1 MeV El-transition in a nucleus.

Broad lines are observed in the bombardment of U, Th, Au with (11-57) MeV iodine ions. The energy of the peaks coincides with the 4f-3d transitions of elements Z=132, 143 and 145 at 8.0, 9.5 and 10 keV. From the width of the line ΔE=2 keV, a collision time 10^{-18} sec follows. The cross section for production of these lines is about 100 b. The spectrum iodine on gold at 11 MeV is shown in fig. 10. The future application of the observed phenomenon will be a study of electronic structure of superheavy elements without producing them.

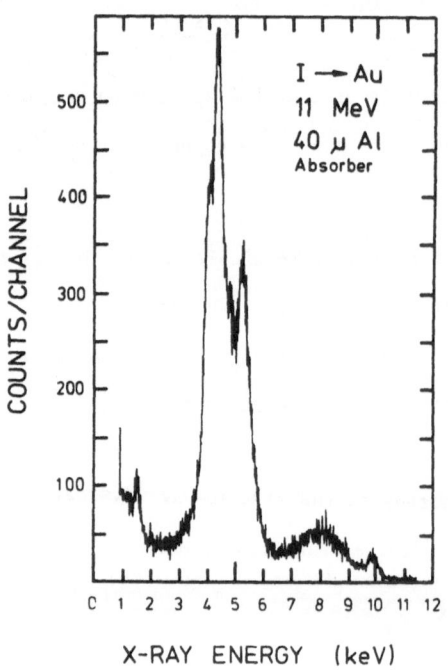

Fig. 10. X-ray spectrum from 11 MeV iodine bombardment of gold. The broad line at 8 keV is interpreted as the (4f-3d) transition in a transiently formed superheavy quasiatom of element 132

REFERENCES

1. UNILAC-Projekt-Bericht Nr. 7, GSI-Bericht 73-1.

2. Blasche, K., Böhne, D., Schmelzer, Ch., Stadler, N.: UNILAC, a variable energy heavy ion linear accelerator. In Nuclear Reactions Induced by Heavy Ions, Eds. R. Bock and W.R. Hering, North-Holland Publishing Company, Amsterdam, 1970, p. 518.

3. Schmelzer, Ch. Journal de Physique, Tome 33, Colloque C-5, supplement au no 8-9, 1972, p. C5-195.

4. UNILAC-Projekt-Berichte 5-7, Datenblätter.

5. Böhne, D.: Status Report on the UNILAC Project. Proceedings of the 1972 proton linear accelerator conference, Los Alamos, Oct. 10-13, 1972, p. 25 ff.

6. GSI-Bericht 73-3.

7. Armbruster, P.: Ein Laufzeitdifferenz-Massenspektrometer zur Untersuchung der Kernspaltung mit schweren Ionen. GSI-Bericht 73-2.

8. Henning, W., Müller, R., Richter, M., Rother, H.-P., Rehm, D.E., Schaller, H., Spieler, H.: Messung der elastischen Streuung identischer schwerer Ionen mit den Reaktionen $^{32}S + ^{32}S$ und $^{40}Ca + ^{40}Ca$. In: Jahresbericht, 1972, Beschleunigerlaboratorium der Universität und der Technischen Universität München, p. 25.

9. Mokler, P.H., Stein, H.J., Armbruster, P.: X Rays from Superheavy Quasiatoms Transiently Formed during Heavy-Ion-Atom Collisions. Phys. Rev. Lett. 19, 13 (1972), p. 827.

THE NEW ACCELERATOR FACILITY IN REHOVOT

Gvirol Goldring

Department of Nuclear Physics, Weizmann Institute of Science, Rehovot, Israel

Our accelerator project started as a rather vague idea back in 1970.
Since then it has changed and evolved in time, continually conforming to a changing
pattern of technological capabilities and scientific demands. Even as it stands
now the project is more an expression of directions and aims than rigid targets of
achievement. These aims, the hopes and the still open questions I shall now attempt
to portray.

The new facility is being planned in response to two major objectives: first
there is the pressing need to advance and improve on the present facilities afford-
ed by the EN tandem accelerator of the Heineman Laboratory whose performance is not
up to the requirements of a considerable fraction of current projects or projects
now in the planning stage. In particular, there is the desire to widen the scope
of some projects that have been successfully developed at this laboratory and let
them unfold to their natural and inherent limits. This requires higher energy and
a larger choice of accelerated particles. Secondly, there is a long range aim to
provide the laboratory with the capability for a major and far reaching development
of its accelerator facilities at a later date, in particular in the heavy ion
field.

The entire project is now conceived as a two stage operation. The first
stage is a 14 UD Pelletron accelerator, purchased from the NEC company at Middleton,
Wisconsin, and the second stage - a cryogenic (superconducting) linear accelerator
to be operated as a post accelerator to the Pelletron. This is now in the stage of
research and development.

14 UD PELLETRON

The Pelletron is basically a tandem Van-de-Graaff of 14 MV at the terminal, in which the charge transport is handled by an inductively charged chain of metallic "pellets" - cylinders of 2.5 cm diameter and 3.5 cm length. This is a far smoother and more stable charging system than the conventional belt. The accelerating tube is of a novel design. Its main new features are: a cylindrically symmetric electrode structure with a linear voltage characteristic, normally associated with inclined field tubes, and a high vacuum capability. The tube normally operates in the range of 10^{-8} - 10^{-7} torr. This capability, extremely important for heavy ion acceleration, is achieved by a bakeable construction of ceramic-titanium, stainless steel and aluminum O-rings, and by a very open electrode design providing the tube with a high pumping speed at every point.

In the beam handling and interior beam optics there are several options open in the NEC design. In our own accelerator we have consistently emphasized high beam quality (sometimes at the expense of other quality parameters, including cost), and this for two reasons: high beam quality is highly desirable, first for the projects that follow the current style of work in the laboratory and secondly - for the matching to a future cryogenic accelerator

The beam handling components of such an accelerator (in addition to the accelerating tubes) fall into three distinct classes: the ion source, the stripper and the specifically optical components both outside and inside the accelerator proper, the latter including as a possibility a charge selector in the high voltage terminal.

There are now a number of ion sources available for the production of negative heavy ion beams, some commercially and some in various stages of development and design. Probably the most promising of these are the various variants of sputter-ing ion sources which appear to be effective over a wide range of ions. Some are apparently capable of efficiently ionizing matter in very small quantities and are therefore suitable for the handling of separated isotopes.

The stripper which transforms the injected negative ions into positive ions can be either low pressure gas differentially pumped (in the Pelletron by special

pumps located in the terminal) or a foil, usually carbon. The foil stripper has
the advantage of generating ions of higher average charge particularly for very
heavy ions. On the other hand - the foil stripper has two severe limitations:
there is a limit to the intensity of the beams (or intensity per unit area) that
can be passed through a foil without destroying it and there is a lower limit to
the thickness of foils that can be practically employed. At present the limiting
thickness is about 5 μg/cm^2. This is considerably higher than the optimum thickness for heavy ions and gives rise to excessive scattering and beam broadening
in the foil. With an intense negative beam available in the low energy section
of the accelerator it is therefore possible that a higher intensity beam (in
particular: a beam limited to small phase space dimensions) can be generated by
a gas stripper for some given high charge states, although the relative abundance
of that particular charge state may be considerably higher in a foil stripper.

Recently, considerable improvement has been achieved in the current transmission capacity of foils by heating them to a narrowly controlled temperature
(500°C - 570°C) during operation.

Another possible improvement in stripping is through the employment of gases
of high molecular weight, in particular fluorocarbons. It is not quite clear at
the moment what the increase in average charge will amount to at the maximum voltage of ouraaccelerator. Also there are some as yet unsolved problems in the pumping of these gases in the terminal. Our Pelletron will have facilities of both
gas and foil strippers.

In the pelletron accelerators one is considering for the first time the introduction of sophisticated optical elements into the confines of the accelerator
proper. Possible locations for such elements are in the regions of the acceleration
tubes and in the terminal which is quite spacious: 360 cm long and 150 cm in diameter.

One element which has been incorporated in the design of our accelerator is
a triplet quadrupole lens in the middle of the low energy acceleration tube. This
serves two objectives: it combats the lens effect of the transition region at the
entrance to the tube; and by focussing the beam to a small point at the stripper

with large angular divergence, it minimizes the deterioration in beam quality
caused by scattering in the stripper.

The large divergence of the beam issuing from the stripper necessitates the
introduction of an additional lens - in the terminal - to converge the beam into
the high energy acceleration tube. This lens has to be able to provide a variable
image position because this position - the proper object for the high energy tube -
varies with charge. For charge 14^+ the object is close to the tube entrance
moving progressively back for lower charges. For charge 1^+ the object is practically at infinity.

A further possible element of interest is a charge selector in the terminal.
This would, ideally, select a single charge state for acceleration in the high
energy tube out of the broad distribution issuing from the stripper. The importance of such an element stems from the fact that some of the blocked ions, if
accelerated would be improperly matched to the high energy tube and would strike
the electrode structure, liberating electrons and, presumably, causing the notorious
beam loading which is the practical limitation at present to heavy ion beam intensities in tandem Van de Graaffs. It is reasonable to believe that a charge selector
of even modest resolution would appreciably increase the range of beam currents
energy and the atomic species of the accelerated ions.

The earliest designs of charge selectors involve a system of electrostatic
deflection plates (cf. fig. 1) allowing the beam either to be processed by the
selector or to be passed through unmodified. On closer inspection this design

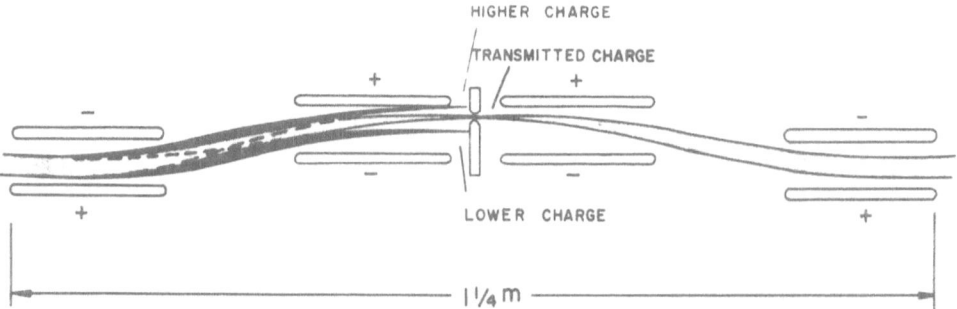

Fig. 1. Conventional design of charge selector

turns out to be rather unsatisfactory because the operation of the selector requires a waist in the beam well within the structure of the selector and therefore far from the entrance to the high energy tube. This would make such a selector unsuitable for the high charge states for which it is most needed.

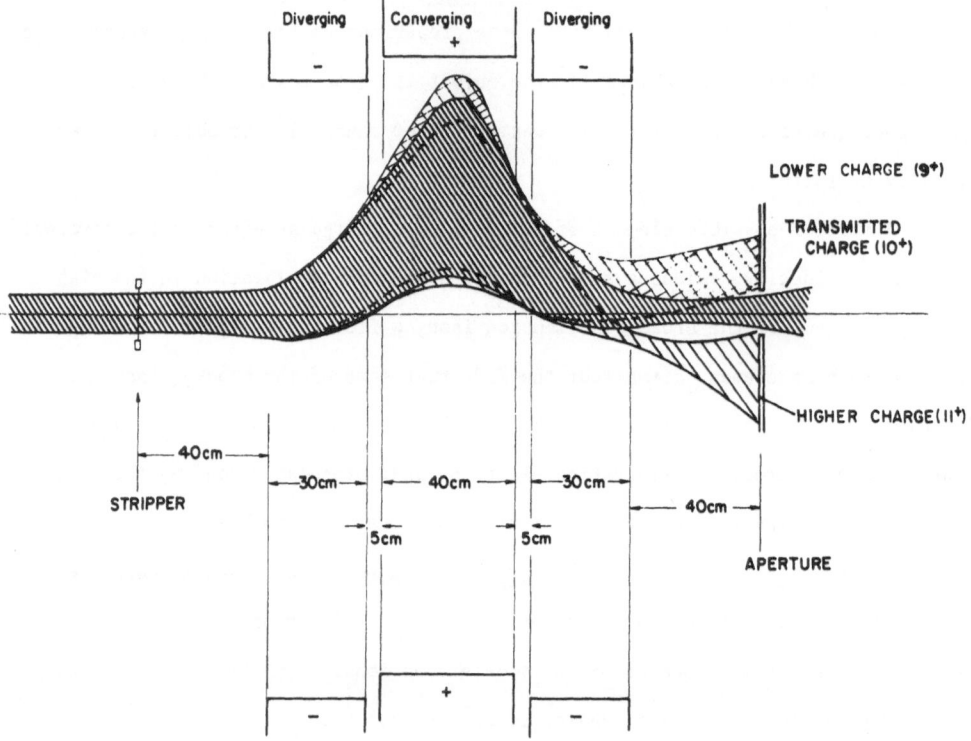

Fig. 2. Proposed new design of charge selector. The scale of the dimension perpendicular to the beam is 20 times the longitudinal scale

An alternative design, suggested by D. Larson, consists of a triplet of off-axis quadrupole lenses. These emit the required beam on-axis and with a waist beyond the lens structure. At the location of this waist, the other beams are defocussed and deflected off-axis. A clean beam can therefore be selected by placing a suitable diaphragm in the waist position. The waist is also conveniently located for optical manipulation of the beam, matching it to the high energy tube. The proposed novel charge selector is shown in fig. 2. This charge selector has

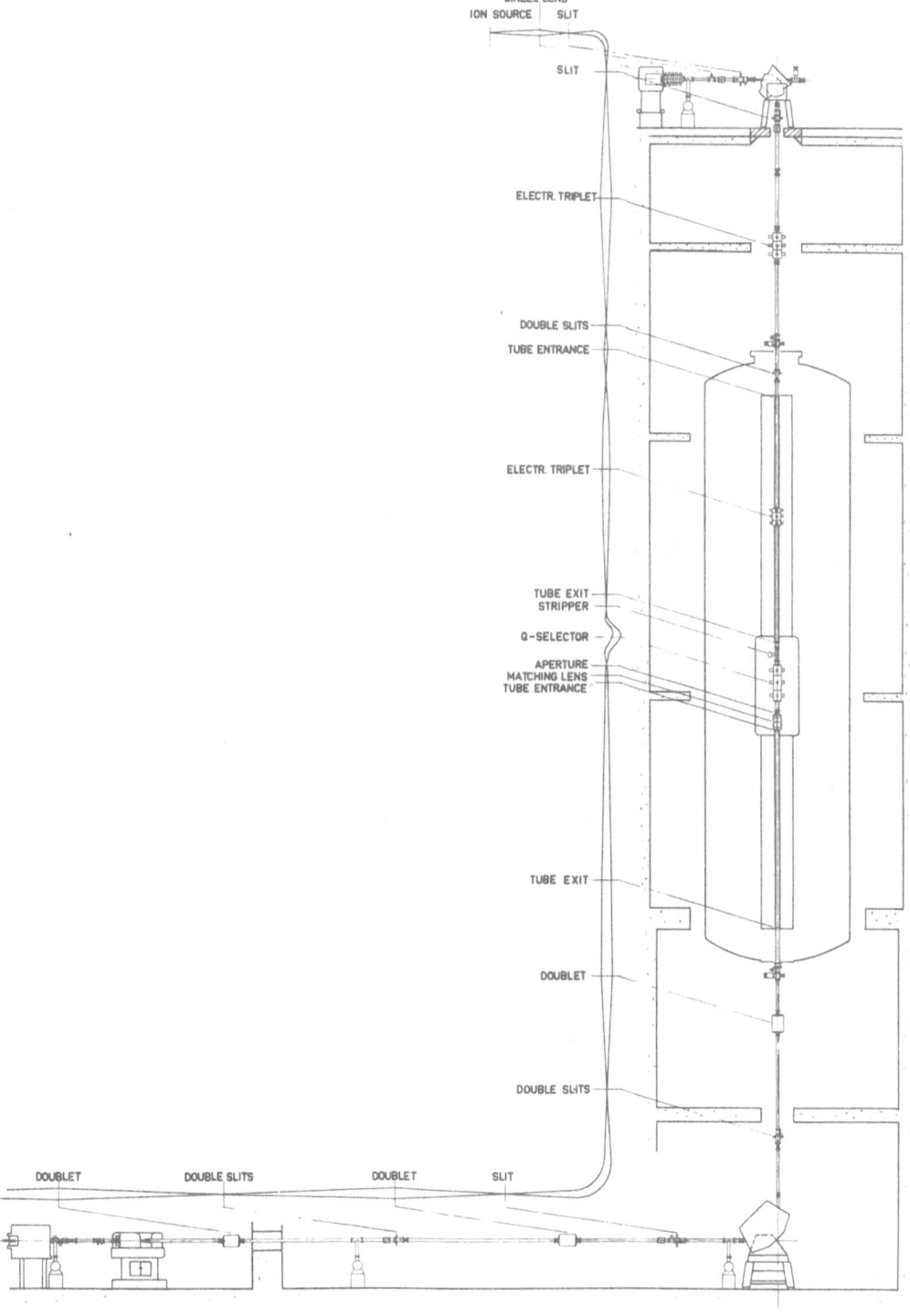

Fig. 3. Schematics of the Pelletron lay out and the beam optics

a theoretical resolving power of ten or better. It is designed to handle charges of $\geq 5^+$. Lower charged ions (in practice: light ions up to and including lithium) would be passed through the selector unprocessed. The optical design of the entire installation is shown in fig. 3. It provides for a high resolution analysis of the beam issuing from the ion source, a waist at the optically sensitive region of the entrance into the low energy tube, the low energy and terminal optical systems and conventional beam handling beyond the accelerator.

The full development of the terminal optical system will be a high priority project of our laboratory in the immediate future. The project will include experimentation with a test model to be set up on an external beam of the EN tandem or our 3 MeV Van de Graaff accelerator, and also the design and construction of a diagnostic system for monitoring the beam in the terminal. This work will be carried out in close collaboration with NEC.

As for the other beam handling components: ion sources and improved strippers, no development work is envisaged in those areas in Rehovot at the present time,

Fig. 4. Perspective view of Pelletron installation

mainly due to man power limitations. These components will be acquired commercially. Because of the rapid progress in those areas it is our aim to postpone the
final decision and acquisition as long as possible, with the exception of the components which are part of the basic accelerator.

Two target halls will be available for work with the Pelletron which can be
engaged by a rotating 90° analyzing magnet: the existing target hall of the EN
tandem and a new one to be constructed to the north of the Pelletron. This one
will have more massive shielding, it will allow the continued and concurrent use
of the EN tandem (using the old target hall) and it will be placed in such a
location as to make it possible to introduce a linear accelerator at a later stage,
employing the same target hall. The control rooms for the Pelletron operation and
the new target hall will all remain concentrated in the area of the present control
room, suitably enlarged. A pulse-relay station will be provided for the new target
hall.

Plans and perspective views of the Pelletron installation are shown in figures
4 and 5.

With the prevailing uncertainties regarding the performance of the ion source,
stripper and charge selector, it is evidently impossible to forecast with any
precision the performance of the Pelletron with heavy ion beams.

For light ions the following performance can be reasonably expected: micro-
ampere beams of 28 MeV protons and deuterons and 42 MeV helium and tens of nano-
amperes of 110 MeV oxygen and 170 MeV sulphur and chlorine.

It is also clear that whatever the initial performance of the Pelletron -
there will assuredly be room and opportunity for improvement, and the accelerator
can be expected to reach its full inherent capability only after a concerted
research effort of several years. The Pelletron itself will therefore fulfil in
large measure both requirements of the accelerator project: an immediate major
improvement in the capabilities of the laboratory and a long range ability of
significant advancement.

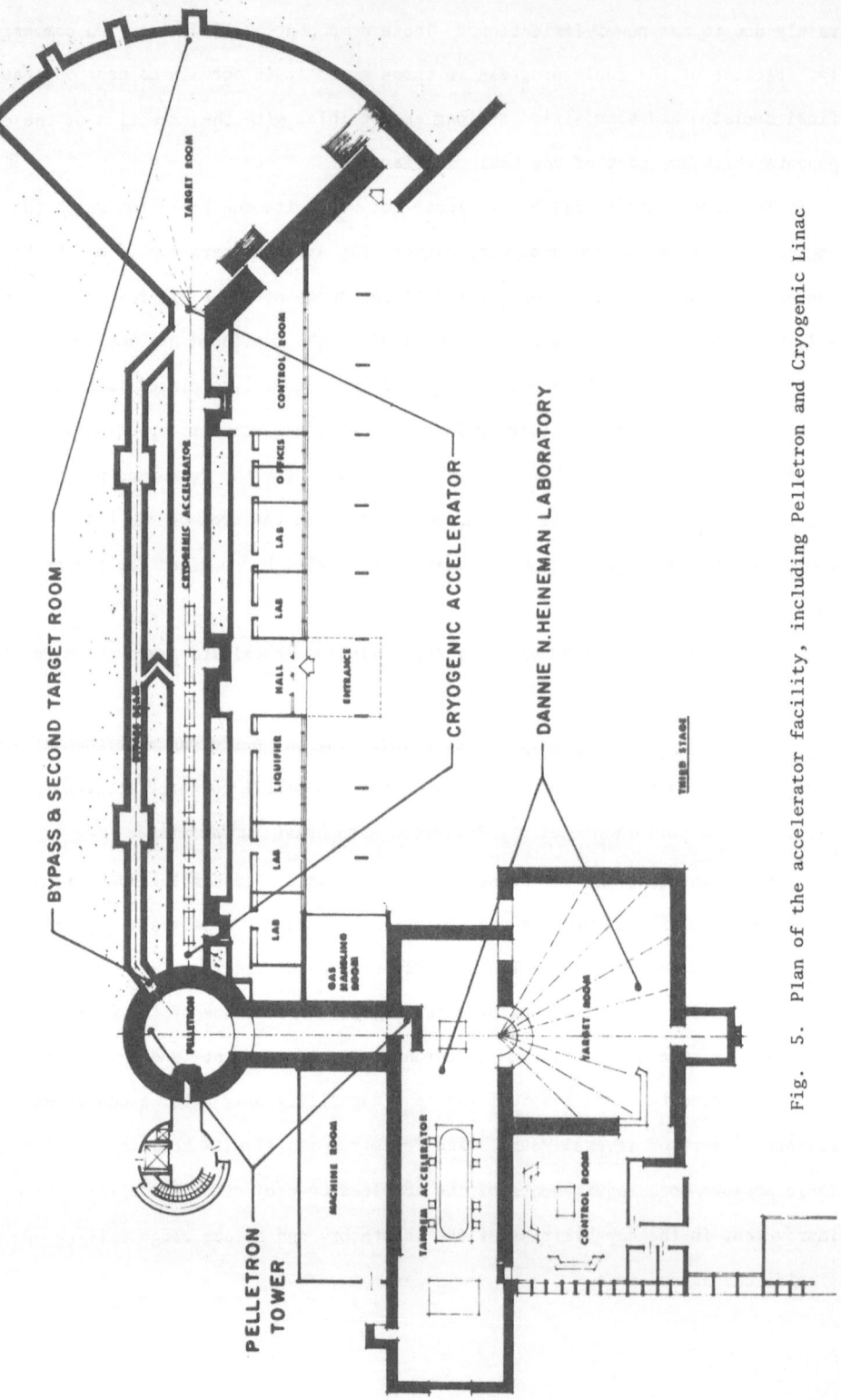

Fig. 5. Plan of the accelerator facility, including Pelletron and Cryogenic Linac

CRYOGENIC LINAC

The cryogenic superconducing linear accelerator is conceived as a future extension of the Pelletron. A superconducing accelerator is attractive in this context for two reasons: the low energy dissipation associated with superconducting structures allows a smooth, continuous wave operation. In addition it allows and indeed favours an accelerating structure made up of many individual units, each of which can be independently controlled in phase and amplitude. This makes the accelerator very versatile in the control of the energy and the optical properties of the beam. The high Q value of superconducting resonant structures imparts an inherent high degree of stability to such an accelerator. The sum total of these features makes the superconducting linac an ideal associate for a Pelletron (or Van de Graaff) type accelerator because the peformance of the two is quite well matched.

Looked at from another side - there is now considerable research activity in superconducting accelerators of different designs and purpose at Stanford, Karlsruhe and Argonne National Laboratory. Of all the contemplated designs, the heavy ion accelerator following injection by a high energy, high quality Pelletron is by far the simplest as it side steps some of the most difficult problems in other types of accelerators (intense beams in electron accelerators, low velocity injected beams in other heavy ion accelerators).

It is also clear that any improvement in the injected beam intensity or quality will have a beneficial effect on the planning and construction of the linac. The optical development work on the Pelletron beam that is planned in this laboratory will therefore be of significance also in the combined operation of both accelerators.

Our own involvement in this field has up to the present been entirely linked to the research and development activity at Stanford University. Dr. Ilan Ben-Zvi of our staff has actively participated in this work in the last two years and in particular in the design and processing of both the test cavity for the heavy ion linac and what is now hoped to be the actual eventual accelerating cavity, both made out of niobium. The latter cavity, of which three specimen are now being

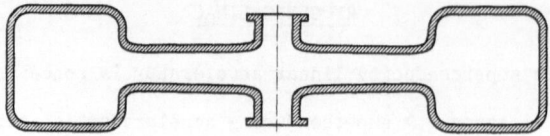

Fig. 6. Design of superconducting accelerating cavity

constructed is shown in fig. 6. It is designed to work at a frequency of 440 MHz
and to provide a peak accelerating field of 10 MV/m. The design and expectations
are based on the proved performance of an electromagnetically similar test cavity.

The value quoted for the accelerating field is quite conservative. Even so -
the present design target is to construct an accelerator of 20 MV acceleration,
based on this low value. This would call for 160 cavities. They would probably
be housed in 8 large cryostates of 20 cavities each. The total length of this
structure is expected to be about 25 m, inclusive quadrupole lenses and pumping
stations but exclusive of the chopper-buncher and other beam processing equipment
at injection. The accelerator will therefore occupy about half of the space
available in the accelerator hall (cf. fig. 4).

Consistent with the emphasis on beam quality, no further strippers are planned
at present beyond the terminal, either in the high energy tube or immediately
preceding the linac.

The accelerating cavity will accept a beam of a minimum velocity of 0.07c.
What this signifies in terms of mass number or energy of the accelerated particles
is impossible to tell so long as the heavy ion performance of the Pelletron is not
better known. Indeed, the overall performance of even a perfectly functioning
linac can be assessed only after the performance of the Pelletron and the associated
optical equipment has been studied and determined in detail.

Present work at Stanford is directed towards the construction and testing
of the accelerating cavities. This is expected to be concluded in three or four
months from now. The next step, also actively planned now, is the operation of two
cavities as a model accelerator with a proton beam. This test is planned in con-
junction with a full size cryostat. Some form of chopper will also have to be
constructed at this stage. It is hoped that this stage of the work will be conclud-

ed late in 1974. Beyond this there are at the moment no definite plans either at Stanford or in Rehovot.

Our own efforts in this field are now exclusively concerned with the training of key personnel. Dr. Ben-Zvi has been our pioneer in this field and others are expected to follow, both at Stanford and in Karlsruhe.

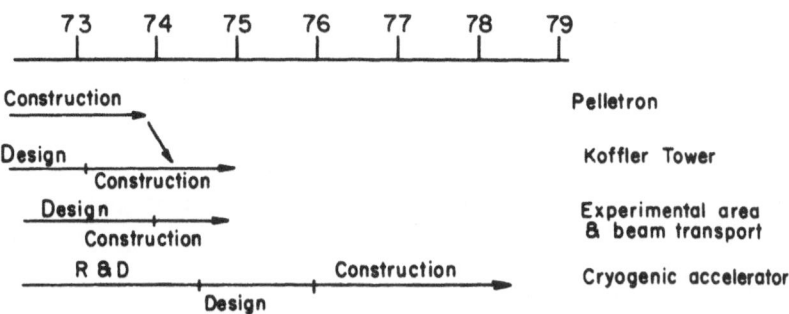

Fig. 7. Estimated time schedule of the accelerator facility

COST ESTIMATES

Item	Million Dollars (1972)
Pelletron	
Equipment	2.6
Building	1.7
Additional Target Hall	
Equipment	0.4
Building	0.4
Cryogenie Linac	
Equipment	2.5
Building	0.5

TABLE I. Estimated cost of the various parts of the project

In conclusion and as a general perspective, the last figure and table show the estimates of time schedules and costs for the entire operation. Any reference to the cryogenic accelerator should be taken as an expression of faith and hope. No more significant estimate is at present available, or indeed possible.

In its organization and structure the new facility will follow the pattern established in the Heineman Laboratory: the facility will be operated jointly by most major nuclear physics laboratories in the country: The Weizmann Institute, the Hebrew University in Jerusalem, the University of the Negev in Beer Sheva and also at a later date, hopefully, the Technion in Haifa.

SOME NEW DEVELOPMENTS IN DIRECT REACTIONS INDUCED BY

HEAVY IONS[*]

W. von Oertzen

Max-Planck-Institut für Kernphysik

and

Lawrence Berkeley Laboratory

Reactions induced by heavy ions have been extensively studied in recent time (see for example the recent topical conferences). Specifically, direct reactions induced by heavy ions like elastic and inelastic scattering and transfer reactions rely on rather complicated experimental techniques for particle identification and methods of theoretical analysis. There has been in recent time considerable progress in both experimental techniques and in the understanding of the reaction mechanisms. At present stage nuclear structure studies with direct reactions induced by heavy ions should indeed yield the information people believed should be obtainable. The situation is, however, by no means clear in all respects and there is considerable work to be done on the new improved accelerators to establish the heavy ion reactions as the specific spectroscopic tool they are expected to be.

I will at first shortly discuss a few new experimental techniques, then show some rather specific examples of heavy ion reactions which show their unique possibilities, and finally discuss some new concepts which were developed for the understanding of the transfer reactions.

1. PROGRESS IN EXPERIMENTAL TECHNIQUES

The study of heavy ion reactions relies on an adequate identification of the reaction products and on high energy resolution. It seems to be rather obvious by

---

[*] Work performed under the auspices of the U.S. Atomic Energy Commission.

now that the system which will have both, a complete identification of the reaction product, as well as the large solid angle and intrinsic energy resolution to allow measurements with thin targets - is the magnetic spectrometer with a focal plane detector. Such a system is used at the 88-inch cyclotron in Berkeley[1]. In principle, four quantities have to be measured to identify completely a particle and to determine its momentum (energy) spectrum.

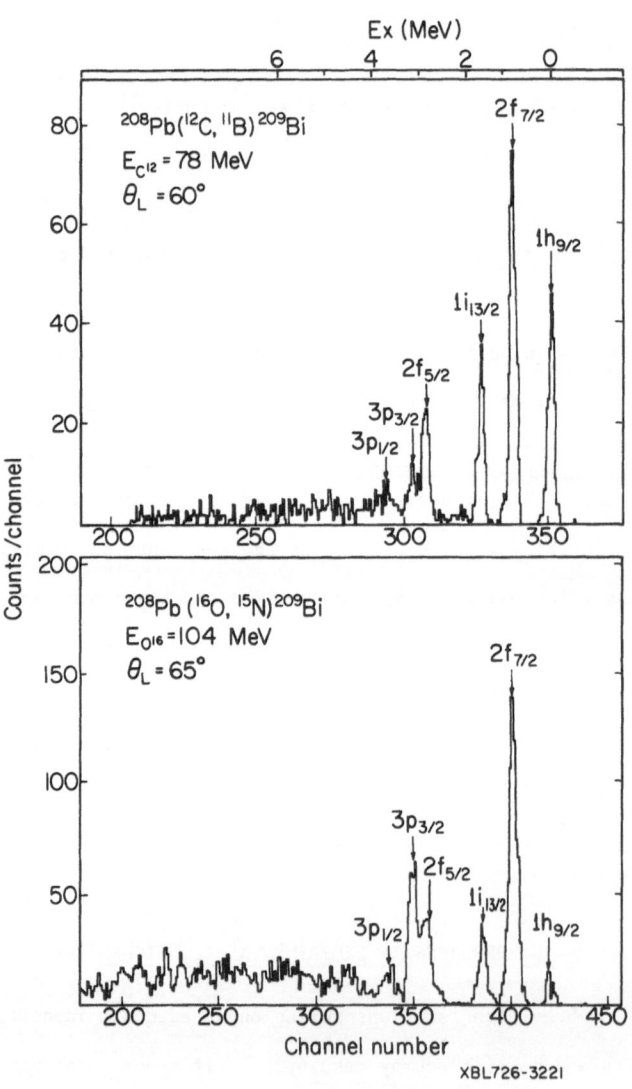

Fig. 1. Spectra of single nucleon transfer induced by ^{16}O and ^{12}C obtained using a magnetic spectrometer (Ref. 19)

These quantities are: Z-nuclear charge, q-charge state of ion while being analyzed in the magnet (at sufficiently high energies q = Z with 100%), m-mass and E-energy or p-momentum. In the system used in Berkeley[1] a resistive-wire proportional counter is placed in the focal plane. The measurement of $\Delta E/\Delta X$, position (or radius ρ) and time-of-flight t, (using a scintillator foil as start detector and a scintillator behind the proportional counter) yields three parameters.

$$(1) \quad \left(\frac{\Delta E}{B\rho}\right) \cdot \sim M^2 \left(\frac{Z}{q}\right)^2 \; ; \quad (2) \quad \frac{\Delta E}{t^2} \sim Z^2 \; ; \quad \text{or} \quad \Delta E \sim \frac{Z^2 M}{E} \; ; \quad (3) \quad B\rho \cdot t = \frac{M}{q}$$

Three parameters are usually sufficient at high energies and light projectile masses. Figure 2 gives an illustration of the two-dimensional matrix of the parameters $\Delta E/\Delta X$ and t.

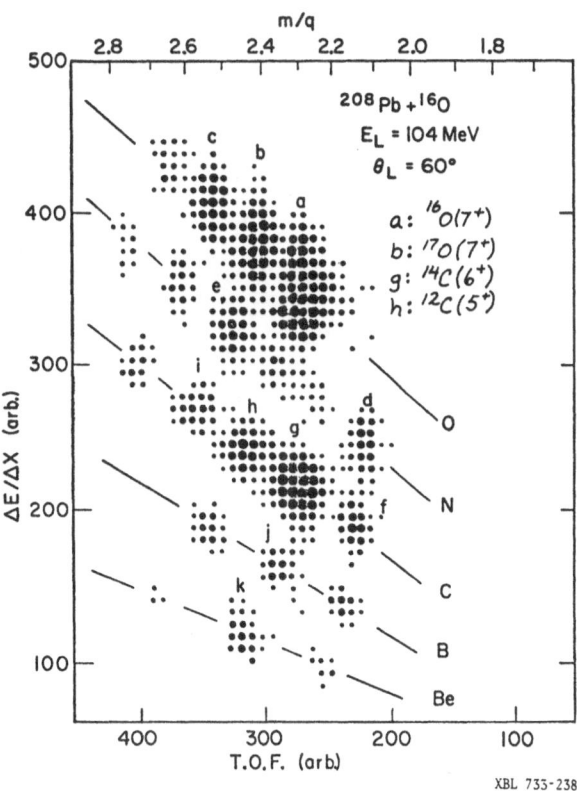

Fig. 2. Two-dimensional matrix of parameters $\Delta E/\Delta X$ and TOF for reactions products from bombardment of ^{208}Pb with 104 MeV ^{16}O. a: ^{16}O(7$^+$), b: ^{17}O(7$^+$), c: ^{18}O(7$^+$), d: ^{15}N(7$^+$), f: ^{13}C(6$^+$), k: ^{10}Be(4$^+$)

With a solid angle of 1-2 msr sufficiently thin targets can be used in experiments to obtain resolutions of 100-150 keV at 100 MeV particle energy. Figure 1 gives as an illustration single nucleon transfer reactions induced by ^{16}O and ^{12}C on ^{208}Pb. Still using a large solid angle spectra of single nucleon transfer reactions with good resolution are rather expensive (in terms of accelerator time) compared to conventional transfer reactions. Accelerator time can be saved in using measurements of γ-rays.

A recently[2] applied method which employs coincidences between γ-rays and reaction products gives (with thick targets) an excitation function from the shape of the γ-ray line. The γ-rays are affected by the changing Doppler-shift due to the changing velocity as function of the depth within the target substance. In the same way angular distributions have been obtained from the shape of free γ-ray

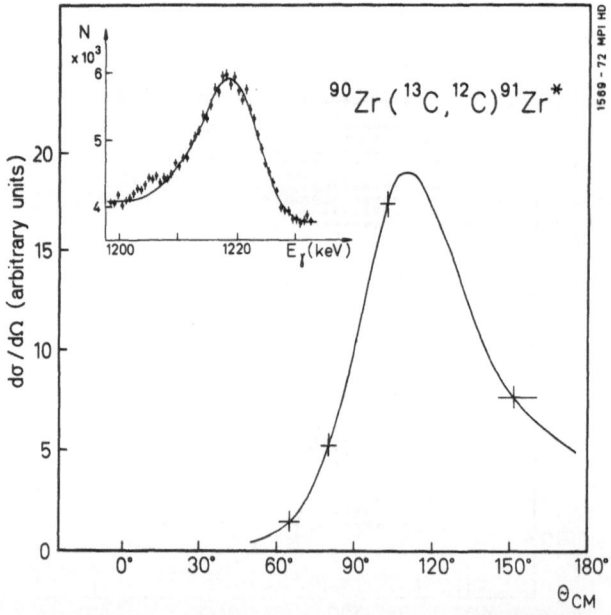

Fig. 3. Angular distribution of the reaction ^{90}Zr(^{13}C,^{12}C)^{91}Zr* (1.205 MeV) at 35 MeV incident energy derived from the shape of the free γ-rays (shown as insert in the figure). The crosses on the full curve illustrate typical errors (Ref. 2))

in (^{13}C,^{12}C) neutron transfer reactions. Figure 3 illustrates in an example how a complete angular distribution is obtained from one measurement of γ-rays. In this case the Doppler-shift observed depends on the reaction angle. This method is, of course, only applicable to excited states, however, is extremely efficient in terms of accelerator time. (The angular correlation of the γ-rays - if not isotropic - has to be known).

Finally the higher energy of the heavy ions from the new accelerators will bring considerable advantages for experimental and theoretical reasons. Target problems which play a considerable role in heavy ion reactions become less restrictive due to smaller energy losses. For many transfer reactions the kinematical restrictions (as discussed later) become smaller and a large abundance of reaction products is observed (fig. 2).

2. SOME SPECIFIC ASPECTS OF HEAVY ION REACTIONS

A.   Nuclear and Coulomb Effects in Elastic and Inelastic Scattering

The elastic and inelastic scattering of heavy ions on target nuclei with large Z, as for example of ^{16}O on ^{58}Ni or ^{208}Pb (at 60 MeV or 104 MeV, respectively) exhibit features which are determined by a strong Coulomb interaction competing with the nuclear forces. The signs of the two forces are opposite and this fact leads to peculiar properties of the elastic and inelastic cross sections. Figure 4 illustrates[3] the shape of angular distributions in elastic scattering; the real potential induces fine structure in the grazing region, where the differential cross section deviates from Rutherford scattering. Even more drastically the effects are seen in inelastic scattering which depends on the derivative of the nuclear and Coulomb potentials. The nuclear effect is now strongly localized (the derivative of a Woods-Saxon potential is peaked at the nuclear surface) and a cancellation occurs at a given radius due to the opposite signs of the two terms. A pronounced dip is thus observed in the angular distributions at the angle where the scattering orbit goes through a distance in the interaction region where the cancellation occurs (fig. 5). Both effects are extremely sensitive to details of the total potential, i.e., to both the real and imaginary part[4].

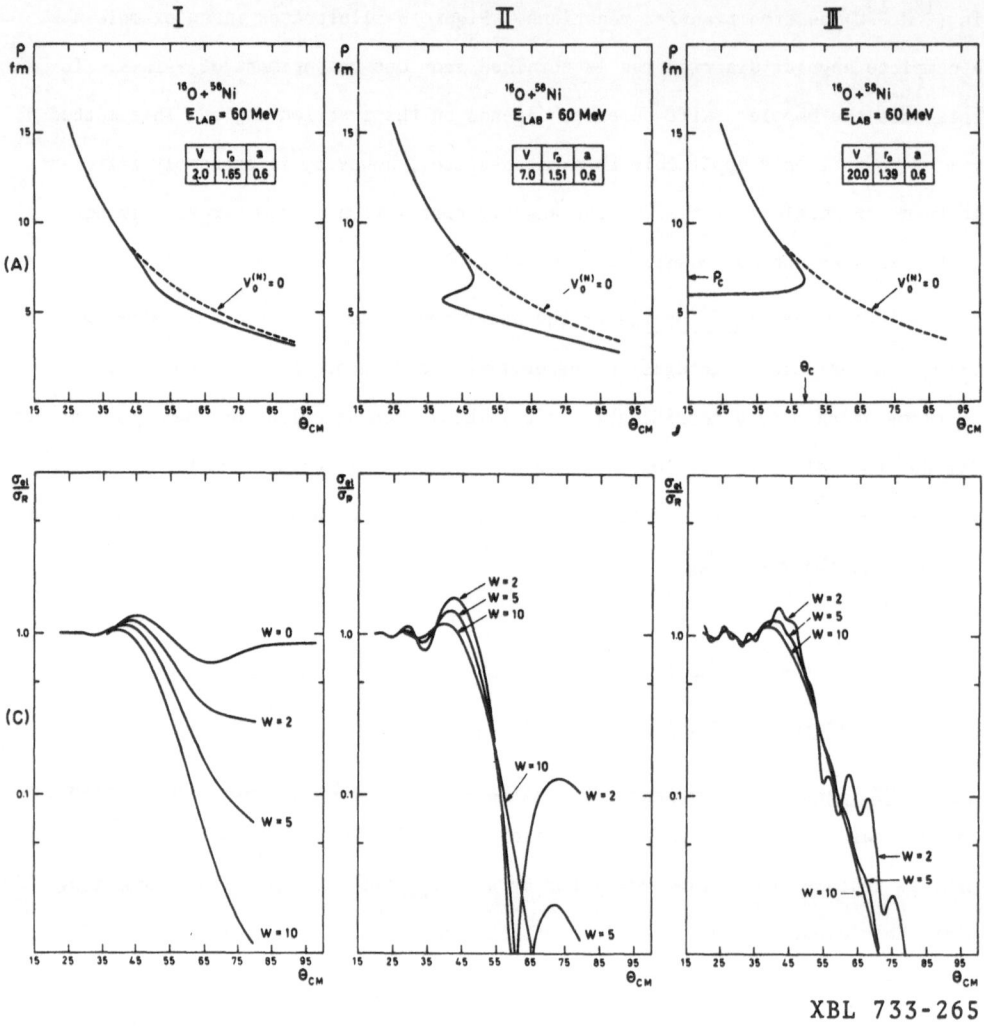

XBL 733-265

Fig. 4. Deflection function (impact parameter ρ as function of scattering angle θ)
and angular distributions of elastic scattering for different strength of
nuclear potential (Ref. 3)

B. Elastic Transfer

Interference effects in themselves are usually very sensitive to details of
the reaction process. The second example also involves interferences between two
competing processes. In elastic transfer, a transfer reaction of the type A(B,A)B
with B = (A+c), interferes coherently with the elastic scattering A(B,B)A. In

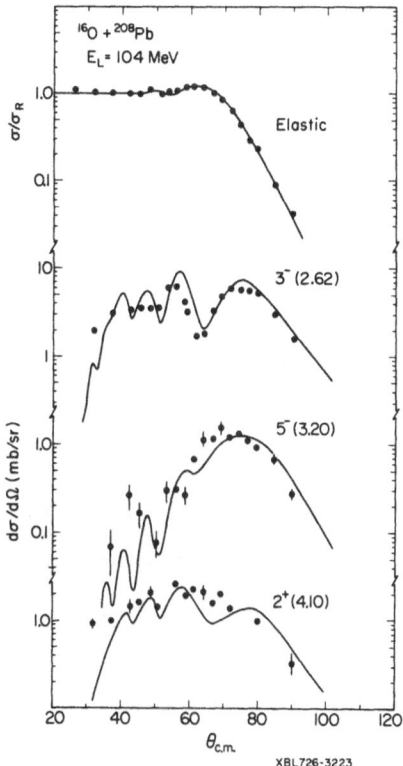

Fig. 5. Angular distributions for elastic and inelastic scattering for ^{16}O on ^{208}Pb. The minimum in the inelastic scattering to the 3$^-$ state occurs at the angle where σ/σ_R starts to deviate from unity. As shown in fig 4 at this angle the real potential becomes the same as the Coulomb potential (Ref. 4)

the center of mass system the scattering angles are connected by the relation $\theta_B = \pi - \theta_A$. The two reactions are indistinguishable and their interference gives rise to structures in the angular distributions which are similar to those observed in Mott-scattering[5,6]. Using the semi-classical description of the transfer process (see also section 3) $\sigma_{tr}(\pi - \theta) = P_{tr}\, \sigma_{el}(\pi - \theta)$, with P_{tr} transfer probability, we obtain for the total differential cross section

$$\sigma_{total}(\theta) = \left|\sqrt{\sigma_{el}(\theta)} \;+\; (-)^{A+\ell} \sqrt{P_{tr}\, \sigma_{el}(\pi - \theta)}\right|^2 .$$

The sign $(-)^{A+\ell}$ takes into account the proper antisymmetrization of the cores

(A - number of fermions in the core) and the symmetry property of the bound state of the transferred particle c (ℓ - angular momentum in the bound state). For $P_{tr} \equiv 1$ and independent of angle the expression for Mott-scattering is obtained. The interference structure depends on the Sommerfeld parameter $\eta = Z_1 Z_2 e^2 / hv$ just like in real Mott-scattering and, of course, on the ratio of the two amplitudes interfering (in Mott-scattering the forward and backward scattering amplitudes are equal yielding symmetry by 90° cm). In systems where elastic transfer can occur the region of interference is at those angles where it has comparable amplitude with the elastic scattering. Figure 6 shows as an illustration the scattering of ^{19}F on ^{18}O and ^{16}O. In these systems transfer of a proton and triton can occur. Two aspects of this example are worthwhile to mention. 1) The extraction of the spectroscopic information - the spectroscopic factors for the decomposition of ^{19}F into triton or proton plus ^{16}O or ^{18}O core, does not depend, as usually, on an absolute cross section but on the shape of an interference pattern. 2) In the present experiment[6] the ground state of ^{19}F was not resolved from closely lying states of 150 keV excitation. However, as can be seen from fig. 6, the small transfer cross section is amplified due to the coherence with elastic scattering. The unresolved states adding incoherently (typical strength is shown by dotted line in fig. 6) just fill slightly the minima in the angular distribution and do not affect the information in the data.

These aspects in this type of experiment could be rather important for heavier ions, because the energy resolution often will not be sufficient to separate the final states if projectiles with masses greater than 40 are involved.

C.   Multi-Nucleon Transfer

A third important aspect of heavy ion induced direct reactions is the possibility to transfer many nucleons or large amounts of nuclear matter. Quite a few experiments have been reported where exotic (neutron rich) nuclei are produced in a high energy induced transfer reaction[7]. The transfer of many nucleons possibly has to be considered as a many step process. In a semi-classical description

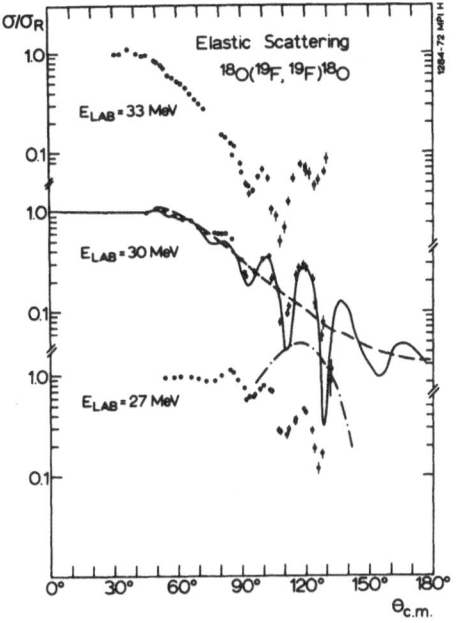 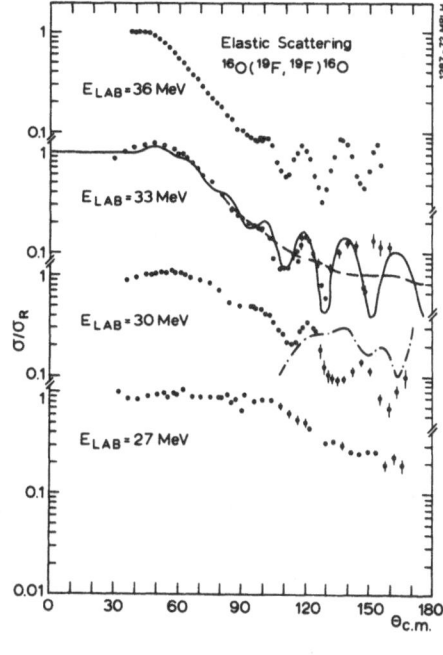

XBL 731-21

Fig. 6. Elastic scattering of ^{19}F on ^{18}O and ^{16}O at energies of ca 5 MeV above
the Coulomb barrier. The full curve represents calculations including
the transfer of a proton ($2s_{1/2}$) and a triton ($4s$), respectively (Ref. 6).
The dotted curves correspond to non-interfering cross sections

the reaction will consist of a product of many single probabilities (transfer of
x individual nucleons)

$$\sigma_{tr} = P_1 \cdot P_2 \cdot P_3 \cdots P_x \cdot EF \; \sigma_{e\ell}.$$

with $P_i \approx 10^{-1} - 10^{-2}$.

The cross sections for multi-nucleon transfer reactions therefore are expected to
be rather small, unless certain correlation effects occur (see also section 3a)
and lead to an enhancement expressed in terms of enhancement factors EF. Different
correlated groups could be transferred one after the other and particular effects
of two-step process will show up in these cases.

Typical two-step processes could be involved in reactions like (^{16}O, ^{13}C) which
could be viewed as a transfer of two protons and a consecutive transfer of a

neutron. Generally, the transfer of many nucleons will consist of a coherent sum of many processes in which different substructures in the transferred nucleons can contribute (for example ($^{16}O, ^{14}C - ^{13}C$) or ($^{16}O, ^{14}N - ^{13}C$) as possible sequences for the ($^{16}O, ^{13}C$) reaction).

Recent calculations[8] for example have shown that the interference between inelastic scattering before transfer and transfer before inelastic scattering processes can lead to a flattening of the shape of angular distributions (fig. 7) (if the interference is destructive). In this coupled channel calculation the inelastic scattering is considered to have a very large probability, the transfer process is, however, only treated in first order (CCBA).

The transfer of large amounts of nucleons has also to be viewed in terms of macroscopic properties of nuclei as discussed by Swiatecki[9]. Important parameters

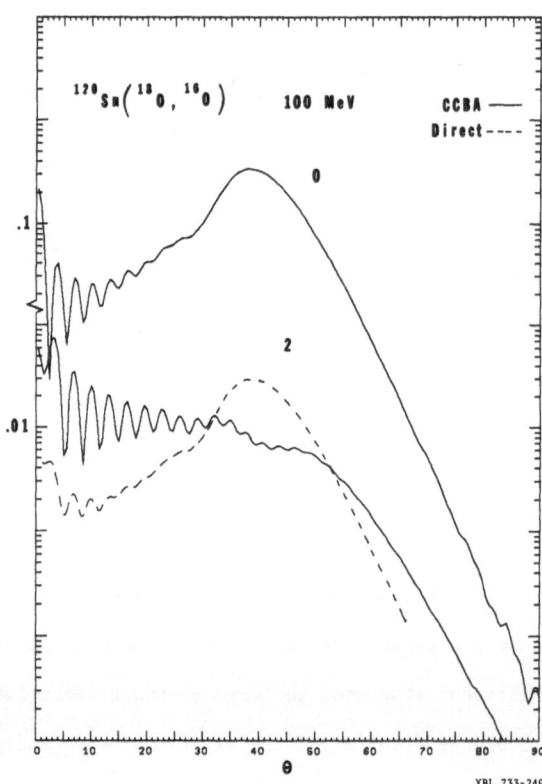

Fig. 7. Calculations using a coupled channels Born approximation for 2n-transfer involving inelastic processes in the initial and final channel (Ref. 8)

in this respect are the liquid drop parameter Z^2/A, and asymmetry.   Figure·8 shows
as an example the potential energy as function of asymmetry.   For two colliding
masses, m_1, m_2, there seems to be a critical asymmetry m_1/m_2 (depending on Z^2/A)
below which the lighter mass nucleus is eaten up by the larger target nucleus -
forming a heavier compound or residual nucleus - or for larger values of m_1/m_2
the two nuclei redistribute their masses in such a way as to achieve a symmetric
two-body configuration.   These aspects will be important for the formation of
compound nuclei and the transfer of large amounts of nucleons.

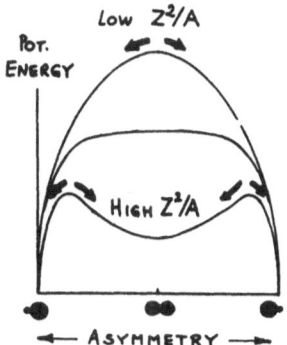

Fig. 8.   Stationary potential energy of two nuclei as function of the asymmetry

(ratio of their masses) for different values of the parameter Z^2/A

(Ref. 9)

3. PROGRESS IN CONCEPTS FOR THE INTERPRETATION OF TRANSFER REACTIONS

A.   Semi-classical Models and Window Effects

It has been realized more than 20 years ago by Breit and co-workers[10] that
heavy ion reactions can be described by semi-classical models provided the
Sommerfeld parameter η , $\eta = Z_1 Z_2 e^2/hv$ is large relative to unity.   If $\eta \gg 1$, the
minimum distance in the relative motion becomes large compared to the Broglie
wavelength λ, using relation

$$R = \frac{\lambda}{k} \left(1 + \frac{1}{\sin \theta/2} \right) = \eta \lambda \left(1 + \frac{1}{\sin \theta/2} \right) \qquad (1)$$

Where θ is the scattering angle and R is the minimum distance in the classical
orbit determined by the Coulomb field. The semi-classical models assume that the
orbits can be described by classical equations and the transfer process by quantal
methods, i.e., it has only small influence on the scattering path. It
has been realized only recently that the classical orbit description leads to
severe restrictions of changes of the important quantities, k, η, θ, if a sizable
cross section has to result. It has been found[14] that sub-Coulomb transfer reactions have only large cross sections if the minimum distances R^i_{min} and R^f_{min} are
equal. Thus, we obtain, $R^i_{min} = R^f_{min}$, as a condition which relate changes in η and
k in a reaction. An optimum Q-value is obtained (assuming $\theta_i = \theta_f$)

$$Q_{opt1} = \frac{Z_3 Z_4 - Z_1 Z_2}{Z_1 Z_2} \; E^i_{CM} \tag{2}$$

which depends on the amount of charge being transferred in the reaction. This
relation still holds approximately at energies above the Coulomb barrier. Modifications are mainly introduced due to possible large (or small) amounts of angular
momentum transfer and due to absorption processes. Thus, transfer of charge
between a light projectile and a heavy nucleus (1p, 2p α-transfer) is always
observed with negative Q-values. All other reactions with non-optimum Q-values
(like pick-up of charge from a heavy nucleus) are strongly depressed. Actually
the cross section can be shown to depend on a few simple factors which can be
discussed independently in a semi-classical model[11]. For a given scattering orbit
we have

$$\frac{d\sigma}{d\Omega}(\theta) = \sqrt{\sigma_i/\sigma_{Ri} \cdot \sigma_f/\sigma_{Rf}} \; \bar{\sigma}(\theta) \cdot P_t(\theta) \cdot F(\Delta D) \tag{3}$$

The cross section depends on a scattering probability. This scattering probability
is appropriately described by an average Rutherford cross section $\bar{\sigma}(\theta)$ multiplied
with the rates of absorption in the incident and final channel $\sqrt{\sigma_i/\sigma_R \cdot \sigma_f/\sigma_R}$.
The factor $F(\Delta D)$ gives the Q-value dependence as function of $\Delta D = |R^i_{min} - R^f_{min}|$
which is a measure of the overlap of the ingoing and outgoing scattering states.
$P_t(R)$ is the transfer probability which is mainly the form factor squared.

The three factors can be easily calculated numerically using the semi-classical
model. It has been shown[12] that the elastic cross section can be rather well
described by an exponential function of R_{min} (see fig. 9).

$$\frac{\sigma}{\sigma_R} = \begin{cases} 1 & , \quad R > R_o \quad ; \\ \\ 1 - e^{-(R-R_o)/\Delta} & , \quad R \leq R_o \quad ; \quad \Delta \simeq 0.5 \text{ fm} \end{cases}$$

Similarly, it has been shown that $F(\Delta D)$ has a gaussian shape for large angles.

$$F(\Delta D) \sim e^{-\Delta D/\alpha R \lambda^2} \tag{5}$$

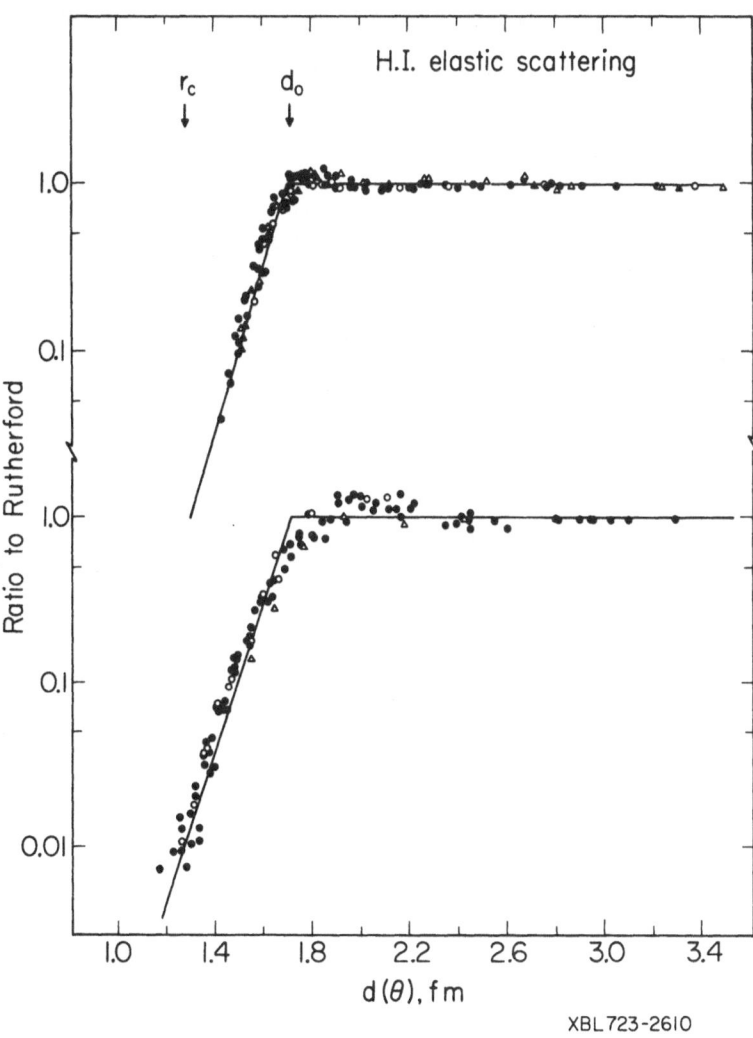

XBL 723-2610

Fig. 9. Ratios of elastic scattering cross sections to Rutherford scattering at different energies and target nuclei transformed to a common scale, $R_{min} = d_o(A_1^{1/3} + A_2^{1/3})$ (Ref. 12)

With α - the bound state decay constant determined by the binding energy E_B and reduced mass M_C of the transferred particle, $\alpha = (2M_C\ E_B \hbar^{-2})^{1/2}$. The transfer probability is typically also an exponential as function of the minimum distance.

$$P_t(R) \sim e^{-2\alpha R/(\alpha R)^2} \tag{6}$$

These simple formulae can help in many cases to study the chances of a certain experiment. The strong Q-value dependence of the cross section which is virtually contained in all three factors can give extremely small cross sections in the regions of interest. Typical shapes of these windows are shown[13] in Fig. 10

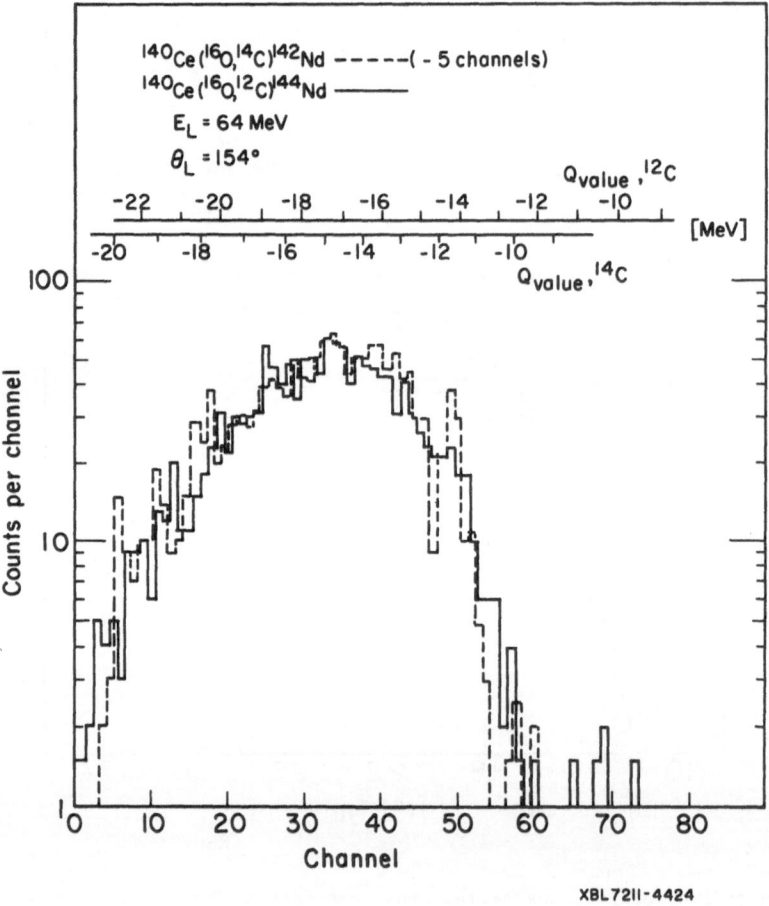

XBL 7211-4424

Fig. 10.   Spectra of ^{14}C and ^{12}C nuclei from ^{16}O induced reactions on ^{140}Ce illustrating the window effect in the spectra (Ref. 13)

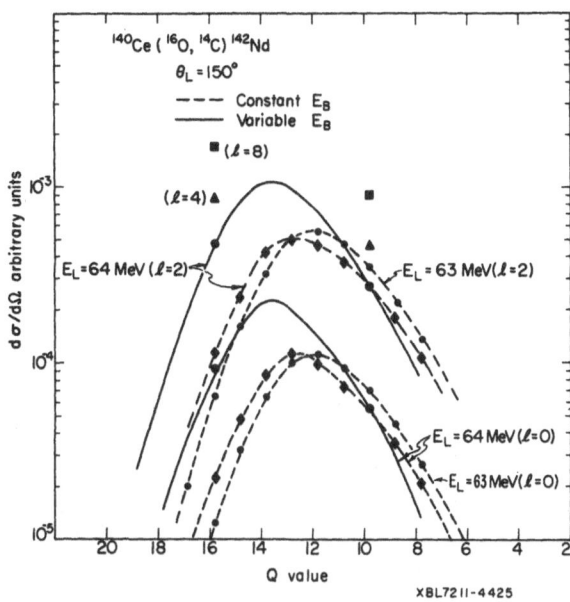

Fig. 11. Calculations (using DWBA) of Q-value windows for the $(^{16}O, ^{14}C)$ reaction
on ^{140}Ce at various incident energies and for different angular momentum
transfer (Ref. 13)

(experimental) and Fig. 11 (theoretical). Depending on the angle of observation,
different factors in expression (3) determine the shape of the Q-value window.

Any reaction product which is emitted from the surface of the nucleus (as a
result of a direct or compound reaction) will have to follow a certain trajectory
which is determined by the parameters: Charge product Z_3Z_4, radius R_o where the
particles originate (also absorption radius), and angle of observation θ. If the
particles start with tangential velocity, their energy in the final channel has to
be determined by Eq. (1). The optimum Q-value in this case would be

$$Q_{opt2} = E^f - E^i_{cm} = Z_3Z_4/2R_o \left[1 + \frac{1}{\sin \theta/2}\right] - E^i_{cm} . \qquad (7)$$

The final energy consists typically of a potential energy part, which is the
Coulomb potential at R_o, and a kinetic energy which is determined by the centrifugal barrier (and θ). For an angle θ larger θ_o the reaction yield as function of E^f
or Q scans the absorption probability and the transfer probability as function of
minimum distance in a similar way as does the variation of the scattering angle.

In Fig. 12 a schematic representation is given which illustrates the close relationship between the occurrence of a Q-value window and an angle window (for $\theta > \theta_o$). Both values θ_o and $Q_o = Q_{opt}$ can thus be related to an absorptive radius R_o. At angles $\theta < \theta_o$, where the nuclei never touch the matching of orbits of the initial and final channel as discussed for sub-Coulomb transfer reactions[14] becomes the most important factor in determining the reaction yield (Q_{optl}).

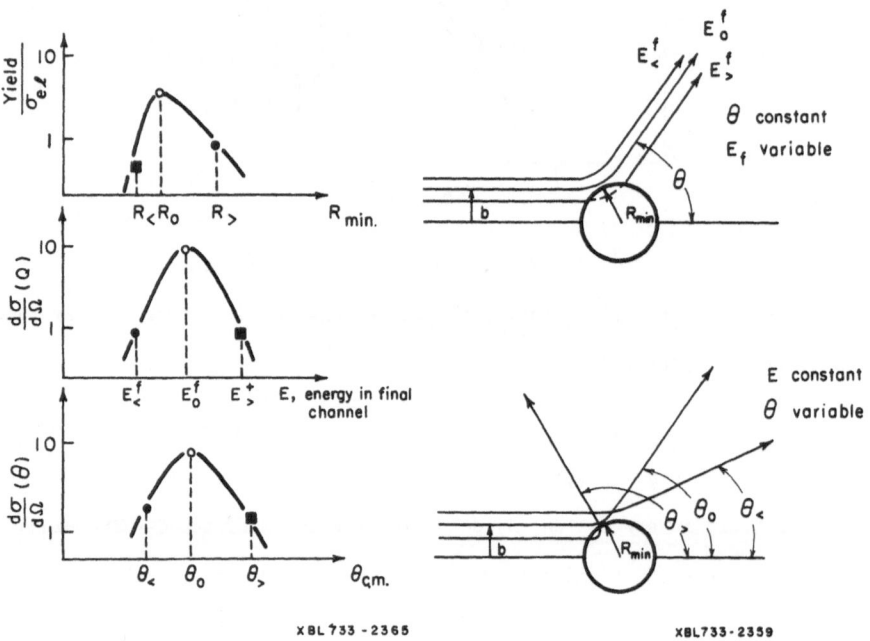

XBL 733-2365   XBL733-2359

Fig. 12.   Schematic presentation of variations of R_{min} with θ_f (E_f fixed) or with E_f (for fixed θ_f) and the corresponding yield curves

In a purely semi-classical framework the angular momentum transfer is fixed by the change of the parameters η, k, θ. However, in a reaction in which a determined amount of angular momentum is transferred during the quantum mechanical transfer process between definite states, the following prescription may be used to determine the final reaction yield. The grazing angular momentum in the initial channel is determined by:

$$L_{o_i} = k_i R_o \left(1 + \frac{2\eta_i}{k_i R_o} \right) , \tag{8}$$

θ_i is given by Eq. (1).

This angular momentum will give the largest contribution to the reaction; determine L_f by $\vec{L}_i - \vec{\ell} = \vec{L}_f$; then calculate θ_f using for example relation $L_f = \eta_f$ ctg $\theta_f/2$ for pure Coulomb fields. Having calculated θ_f the ratios $\sigma/\sigma_R(\theta)$ and the average minimum distance and $F(\Delta D)$ can be calculated. In cases of large mismatch for example transfer of two units of charge, we usually have $\eta_i \gg \eta_f$, $\ell = 0$ and $L_i \gg L_f$ and we obtain $\theta_f \ll \theta_i$. This actually implies that the absorption in the final channel will be smaller compared to that in the initial channel. The cross section will be small unless the grazing angle is much smaller than usually calculated (using R_o and Eq. (2)). The effects of this decrease in η in the final channel can already be seen in $(^{12}C,^{11}B)$ reaction on ^{208}Pb shown in fig. 13 (Ref. 15). The grazing angle θ_o stays constant as function of Q value. This is in contrast to DWBA calculations (and simple semi-classical considerations) with optical model parameters which allow sufficient averaging over the initial and final channel. The data actually indicate that the absorption in the incident channel dominates the process as discussed above. Similar conclusions were drawn from work done at Oak Ridge[16]; they actually found, that in the transfer of neutrons $^{208}Pb(^{12}C,^{13}C)^{207}Pb$ an average orbit including the final channel is followed - leading to a shift of θ_o as function of the Q-value.

Corresponding to Eq. (3) and fig. 12 there is always one side of the window which is determined by absorption with $R_{min} < R_o$, which implies $\theta > \theta_o$ or $E_f > E_o$; and another side which is determined by the overlap of the scattering waves (or classical orbits) with $R_{min} > R_o$, implying $\theta < \theta_o$ or $E_f < E_o$.

Generally, it can be concluded that the discussion of window effects as function of E_f and θ_f have to be made with clear reference to the value of the fixed variable (θ_f for Q-value window, E_f - for angular window), i.e., on which side of the window the relevant variable is kept fixed.

An application of the calculation of the Q-value dependence can be given by $(^{16}O,^{14}C)$ reactions on ^{140}Ce, ^{142}Nd and ^{144}Sm. The ground state Q-values change and the strength going to the ground states changes. In order to compare reactions

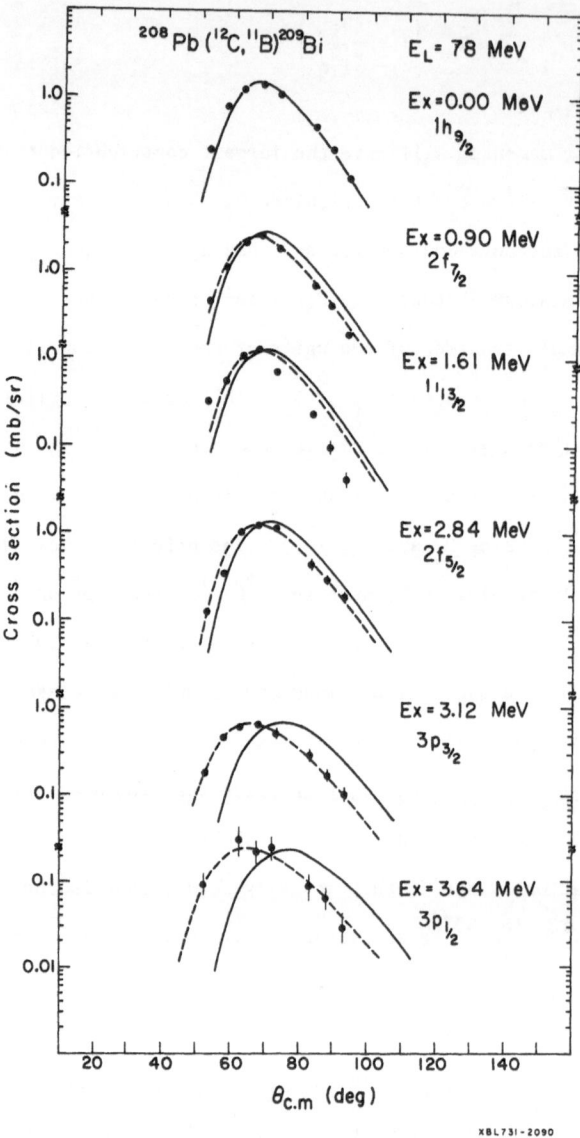

Fig. 13. Angular distributions of proton transfer (^{12}C, ^{11}B) on ^{208}Pb leading to different final states in ^{209}Bi. The full curves are DWBA calculations (dotted curves are shifted to fit the data) (Ref. 15)

with the same conditions the factors for absorption, and F(ΔD) have to be calcul-
ated. The result is $P_{tr}(R)$ or $P_{tr}(d_o)$ (with $R = d_o(A_1^{1/3} + A_2^{1/3})$) which can be compared for different target nuclei as the quantity which does not contain kinemat-
ical effects or nuclear size effects. Figure 14 shows the transfer probabilities
$P_{tr}(d_o)$ for the (^{16}O,^{14}C) ground state transitions for different target nuclei. It
is seen that the N = 82 nuclei Ce, Nd, Sm show an enhancement by factor 20-30
which is similar to that observed in (t,p) reactions on Sn-isotopes (neutron shells
in ^{108}Sn^{112}Sn correspond to proton shells in ^{140}Ce, ^{142}Nd, and ^{144}Sm).

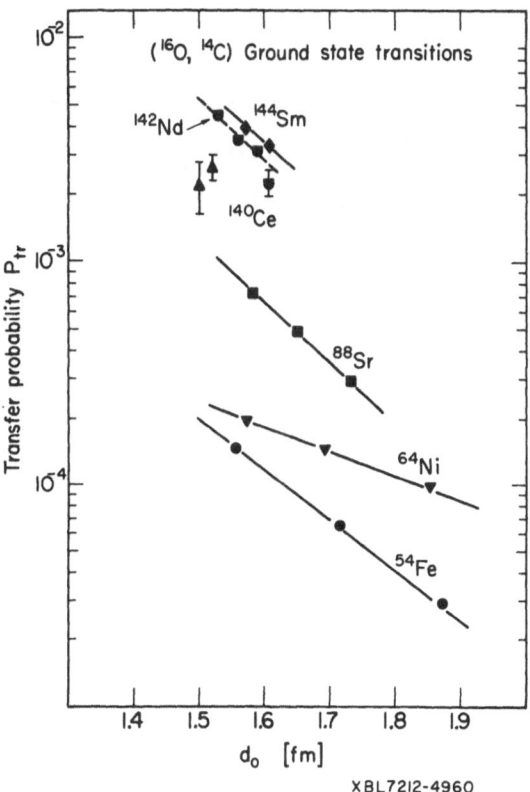

XBL7212-4960

Fig. 14. Transfer probabilities $P_{tr}(d_o)$ for (^{16}O,^{14}C) reactions on various target
nuclei deduced using semi-classical models. The reactions on Ce, Nd and
Sm are enhanced by a factor of ca 20-30 (Ref. 13)

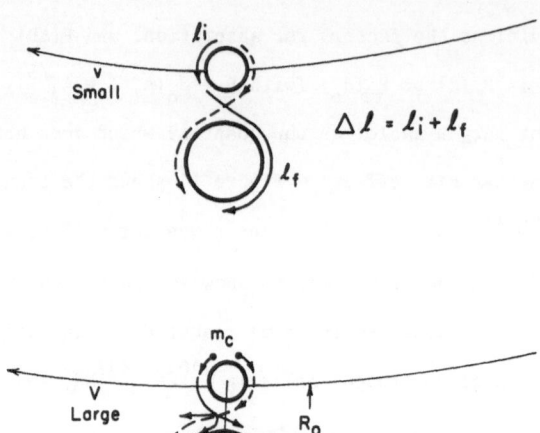

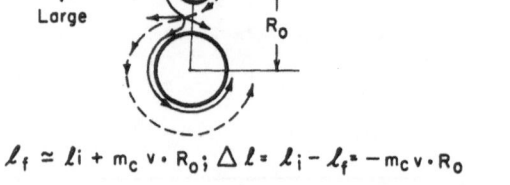

Fig. 15. Schematic illustration of particle transfer in a semi-classical scattering orbit for small relative velocities (top of figure) and large
relative velocities (lower part of figure)

B.   Finite Range Effects

Important changes in the reaction process compared to the previous discussion
are due to the finite mass of the transferred nucleons, also called recoil effects.
They are connected to the coordinates of relative motion $\vec{r}_i, \vec{r}_f$ in the initial and
final channels (reaction A+(bc) → (A+c)+b or A(a,b)B).

$$\vec{r}_i = \vec{r} + (m_c/M_a)\, \vec{r}_1 \approx \vec{r}$$

$$\vec{r}_f = \frac{M_A}{M_B}\, \vec{r} \; (m_c/M_B)\, \vec{r}_1 \approx \frac{M_A}{M_B}\, \vec{r}$$

which can be approximated by the distance between the two cores $\vec{r}$. This approximation has been used extensively, because it allows a simple calculation of the
transition amplitude. At energies near the Coulomb barrier when the wavelength of
relative motion is large compared to $\left(\dfrac{m_c}{M_a}\right)|r_1|$ the neglection of these terms
proportional to r_1 (or r_2) introduces only small errors because the change in phase

for the scattering waves is small. At higher energies, however, a complete treatment of all coordinates is necessary. There are two main effects which come into
play at higher energies[17,18]. A change in the contributing angular momentum
transfers and loss of the semi-classical conditions. At lower energies and large
Sommerfeld parameters η it was observed that the maximum angular momentum transfer
ℓ is favored by approximately a factor of 10 over the smaller values. This fact
can be understood semi-classically by taking as a condition for a large cross section that the velocity of the transferred nucleon is constant during the transfer
process $\frac{\lambda_1}{R_1} + \frac{\lambda_2}{R_2} = 0$; λ - projections of the internal angular momenta in the
initial (ℓ_1) and final channel (ℓ_2) on an axis perpendicular to the reaction plane

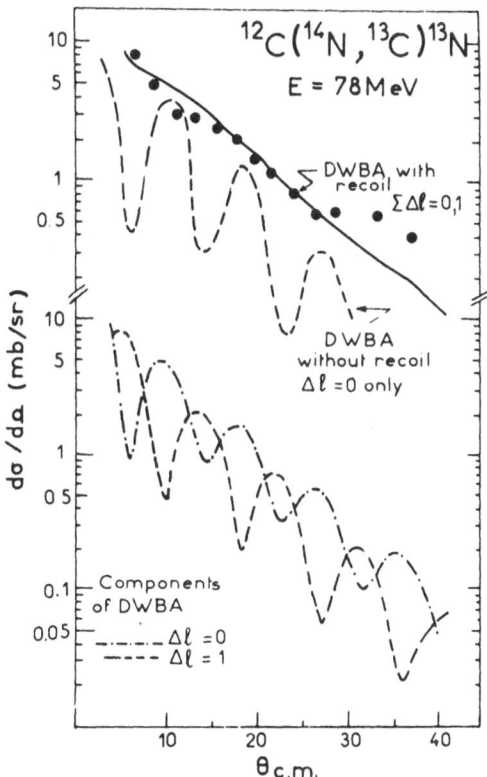

Fig. 16. Angular distributions of the reaction $^{12}C(^{14}N,^{13}C)^{13}N$ at 78 MeV and DWBA
calculations illustrating the damping of the diffraction structure due
to contributions of angular momenta with different parity ($\ell = 0$ and
$\ell = 1$) (Ref. 18)

$(\lambda_1/R_1$ is proportional to the momentum). Figure 15 gives a simplified illustration of the situation. This preference of the maximum ℓ transfer leads to j-dependence in single nucleon transfer[19]. Thus, for $(^{16}O, ^{15}N)$ reactions the proton starting with $j_< = \ell - 1/2$ preferentially populates $\ell + 1/2 = j$ states

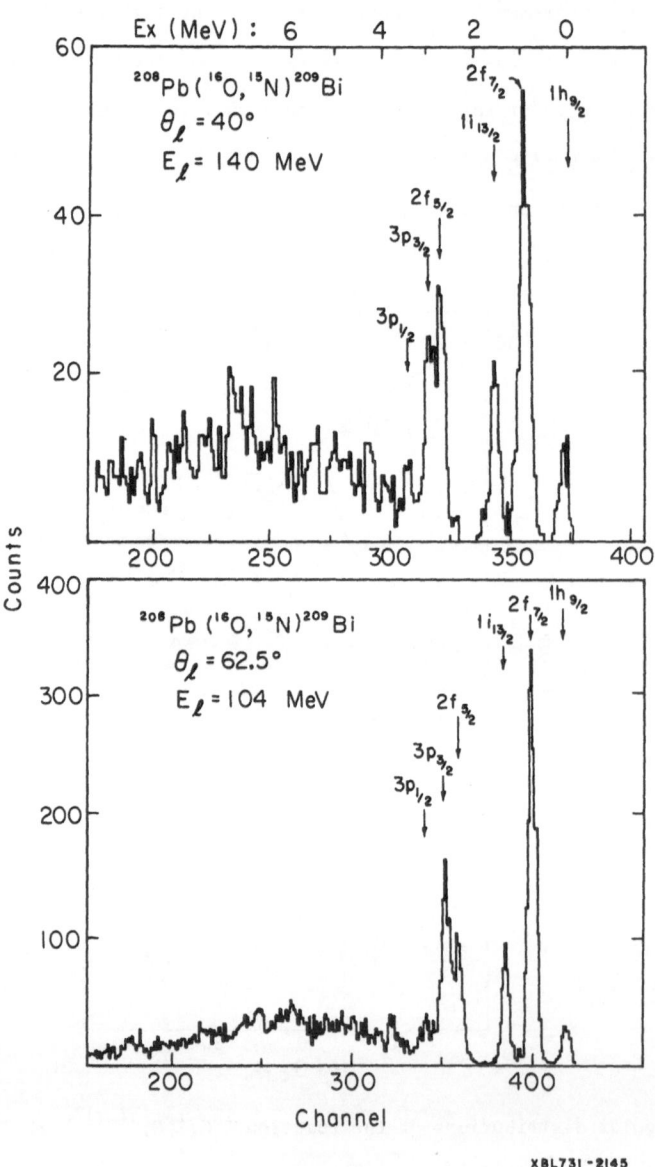

XBL731-2145

Fig. 17. Spectra of proton transfer $(^{16}O, ^{15}N)$ on ^{208}Pb at 104 MeV and 140 MeV (Ref. 15)

whereas (^{12}C,^{11}B) reactions preferentially populate $j_<$ states because the proton starts from a $p_{3/2}$ orbit (see fig. 1). Figure 16 gives an illustration of this j-dependence in single proton transfer reactions on ^{208}Pb for two energies. As the relative velocity increases the transferred particle carries an appreciable amount of the relative momentum (MeV/nucleon) and the picture is changed, the particle will be transferred preferentially in a different way as suggested in lower part of fig. 12. The j-dependence is thus mainly removed at higher energies (fig. 17). For the extraction of spectroscopic factors it becomes extremely important to calculate the recoil terms properly[15,18].

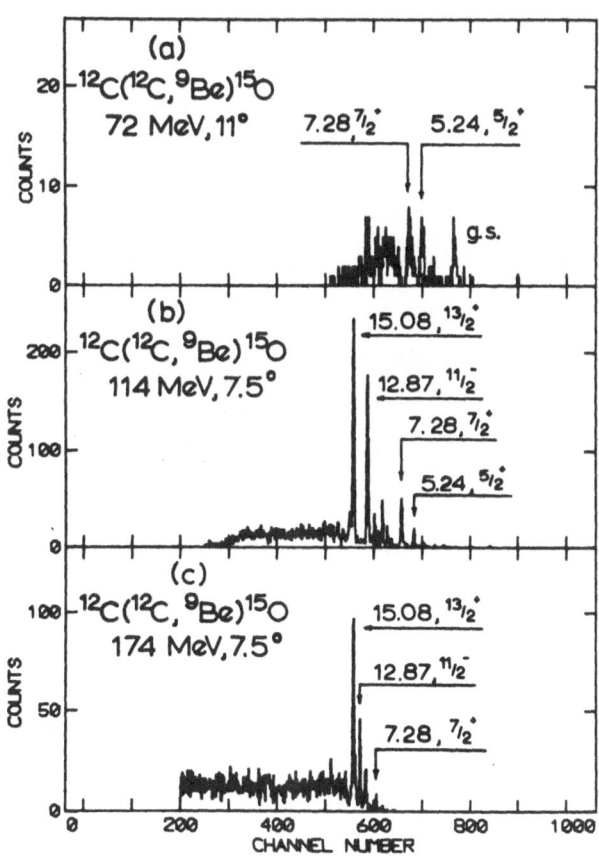

Fig. 18.   Selectivity in the three nucleon transfer reaction ^{12}C + ^{12}C → ^{9}Be + ^{15}N due to varying amounts of the angular momentum per nucleon at various incident energy (Ref. 20)

More precisely, the complete finite range description differs from the previously applied no-recoil approximations[17,18] in allowing the full space for the coupling of the intrinsic angular momenta in the initial and final channel $\vec{\ell} = \vec{j}_1 - \vec{j}_2$; $\vec{\ell} = \vec{\ell}_1 + \vec{\ell}_2$. Neglecting recoil terms, i.e., vectors proportional to $\frac{m_c}{M_b} \vec{r}_1$, reduces the non-local transfer operator to a local one, because the additional dependence on vector $\vec{r}_1$ is removed. The most conspicuous difference between the two operators is, that if the particle m_c is fixed on the inter-connection line between the two centers the local transfer operator has a "parity conserving" symmetry which yields the rule $\ell_1 + \ell_2 + \ell$ = even. This parity rule applies fairly well at energies below or at the Coulomb barrier, where the trans-ferred particle has to be on the interconnection line during the transfer process.

C. High Energy Reactions

The difference between full finite range calculations and no-recoil-approxima-tions becomes important at higher incident energies and is rather drastic as illustrated in fig. 16 by the reaction $^{12}C(^{14}N, ^{13}C)^{13}N$ (Ref. 18). The diffraction pattern which is observed in some cases in transfer reactions between light nuclei, where only one parity in the angular momentum transfer contributes, is damped due to the contributions of equal amounts of $\ell = 1$ and $\ell = 0$ in the present case. At very high energies above the barrier, i.e., large values of k_i, or large values of energy per nucleon the angular momentum per nucleon L_{oi}/M_a (this number depends on the size of the nuclei involved) becomes very large. It can be shown that in these cases final states are mainly populated with $\ell_2 \approx L_{oi}/M_a$. At small scatter-ing angles the main source of angular momentum transfer comes from the redistribu-tion of the masses. This selectivity in the population of final states is rather pronounced and was observed in one, two and three nucleon transfer reactions on light nuclei[20]. An example is shown in fig. 18 for the three nucleon transfer $^{12}C + ^{12}C \rightarrow ^{9}Be + ^{15}O$ at three different energies. At higher energies states involving large angular momentum transfer show up stronger and these states then have to be the aligned configurations of three individual nucleons with high spins. (The individual angular momenta are parallel $\ell_2' + \ell_2'' + \ell_2''' = \ell_2$; and $\ell_2' \approx \ell_2'' \approx \ell_2''' \approx L_{oi}/M_a$).

Other important features of high energy induced heavy ion reactions are
connected with the window effects discussed previously. Clearly the semi-classical
matching conditions involving Coulomb orbits and large η will change eventually to
the angular momentum matching conditions involving plane waves and small η. The
plane wave matching condition involves the condition $L_i \approx L_f$ for the relative
angular momenta, if the angular momentum transfer ℓ is small, or also $k_i \approx k_f$.
The wave number being $k = \sqrt{2E \mu h^2}$, a decrease of the reduced mass corresponding
to stripping reactions (increase - pickup) needs a positive Q-value (or negative
for pickup). This is in contrast to the situation with charged particle transfer
with large values of η where the optimum Q-value is negative. A broadening of
the Q-value window can be expected at energies high above the Coulomb barrier
(see fig. 19 as illustration).

Spectra of $(^{16}O,^{15}N)$, $(^{16}O,^{18}O)$ reactions[21] taken at 140 MeV ^{16}O on ^{208}Pb or
^{144}Sm show a continuous background (fig. 17) at higher excitation whose origin is

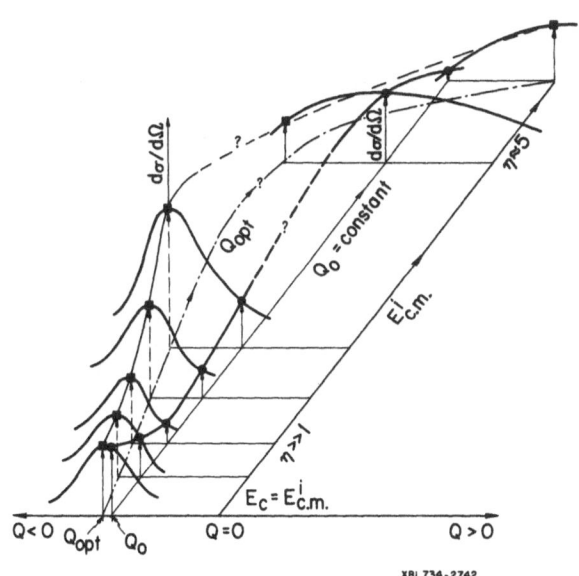

XBL 734-2742

Fig. 19. Yield of a transfer of charge from light to heavy nucleus (stripping
          reaction) as function of Q-value and incident energy at the angle of
          maximum cross section (grazing angle). At high energies the Coulomb
          dominated Q-value window has to change to corresponding plane wave
          conditions

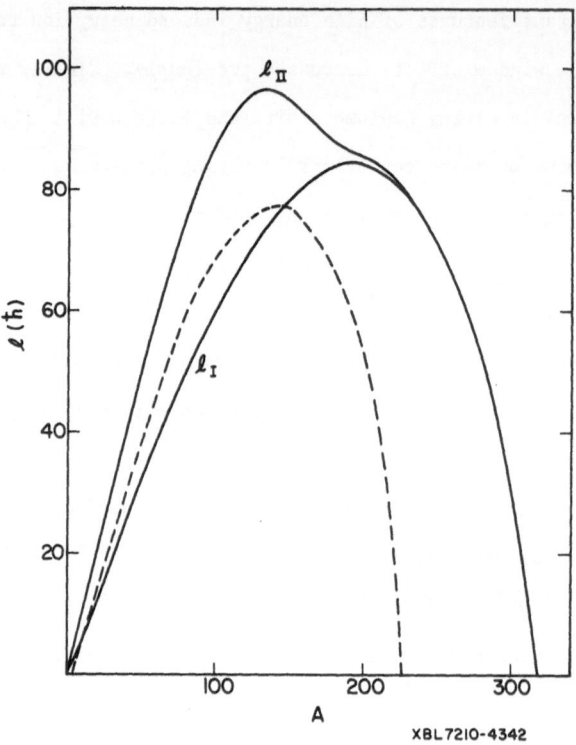

XBL7210-4342

Fig. 20. Limiting angular momenta for nuclei obtained from liquid drop calculations. O=I-designates the angular momentum at which the fission barrier
vanishes for an axially symmetric nucleus. (Ref. 23)

unexplained. There are indications that this background is particularly strong at
small angles[22]. This phenomenon could be connected with the fact that angular momenta in grazing collisions become rather large at energies well above the Coulomb
barrier. Liquid drop calculations[23] indicate that the maximum angular momentum at
which a nucleus becomes unstable to fission is approximately 80 $\hbar$ at mass 200. Fig.
20 illustrates the calculations for different values of the fission barrier. For
reactions where the surface angular momentum becomes comparable or larger than the
critical value of ℓ_{II} (fig. 20) complicated processes which involve many degrees of
freedom of many nucleons can occur. Single nucleon or two nucleon transfer reactions in these cases should show features which are not anymore compatible with
inert core models usually applied. Qualitatively new phenomena could be expected

in collisions and transfer processes, because the limiting angular momenta correspond to internuclear distances where nuclei overlap strongly.

ACKNOWLEDGMENTS

The author is very much indebted to his colleagues at Berkeley for discussions and the access to unpublished data. He thanks in particular B.G. Harvey for the kind hospitality at the 88-inch cyclotron and LBL.

REFERENCES

1. Harvey, B.G., et al.: Nucl. Instr. Methods 104, 21 (1972).

2. Haberkant, K.: Thesis MPJ Heidelberg, 1973 and Kaberkant, K., Grosse, E., et al.: to be published.

3. Broglia, R.A., Landowne, S., Winther, A.: Phys. Letters 40B, 293 (1972).

4. Videback, F., et al.: Phys. Rev. Letters 28, 1072 (1972) and Becchetti, F.: in Symposium on Heavy Ion Transfer Reactions, Argonne, 1973 (in press) and references therein.

5. von Oertzen, W.: Nucl. Phys. A148, 529 (1970); Bohlen H.G., von Oertzen, W.: Phys. Letters 37B, 451 (1971).

6. Gamp, A., et al.: Z. f. Physik (in press).

7. Artulck, A.G., et al.: Nucl. Phys. A168, 321 (1971) and references therein.

8. Glendenning, N.: in Symposium on Heavy Ion Transfer Reactions, Argonne, 1973 (in press).

9. Swiatecki, W.J.: in European Conference on Nuclear Physics, Aix en Prove (1972), J. de Physique C5-45 (1972).

10. Breit, G., et al.: Phys. Rev. 87, 74 (1952).

11. Broglia, R.A., Winther, A.: Phys. Rep. 4c, 155 (1972).

12. Christensen, P.R., Manko, V.I., Becchetti, F.D., Nickles, N.J.: NBI preprint, Nucl. Phys. (in press).

13. von Oertzen, W., Bohlen, H.G., Gebauer, B.: Nucl. Phys. (in press).

14. Buttle, P.J.A., Goldfarb, L.J.B.: Nucl. Phys. A115, 461 (1968).

15. Kovar, D., et al.: Phys. Rev. Letters (in press) and to be published.

16. Larson, J.S., Ford, J.L.C., Gaedke, R.M., et al.: Phys. Letters 42B, 205 (1972).

17. Brink, D.: Phys. Letters 40B, 37 (1972); also von Oertzen, W.: Single and Multinucleon Transfer Reactions in Nucl. Spectroscopy, ed., J. Cerny, Academic Press (in press).

18. de Vries, R.M., Kubo, K.I.: Phys. Rev. Letters 30, 325 (1973) and de Vries, R.: to be published.

19. Kovar, D., et al.: Phys. Rev. Letters 29, 1023 (1972).

20. Scott, D.K., et al.: Phys. Rev. Letters 28, 1659 (1972) and Anyas-Weiss, N., et al.: Oxford University, Nuclear Physics Laboratory, preprint 49/72.

21. Kovar, D.: in Symposium on Heavy Ion Transfer Reactions, Argonne, 1973, ANL (in press).

22. Galin, J., et al.: Nucl. Phys. A159, 461 (1970).

23. Cohen, S., Plasil, F., Swiatecki, W.J.: LBL-1502, preprint.

Lecture Notes in Physics

Selected Issues from

Lecture Notes in Mathematics

Selected Issues from

Springer Tracts in Modern Physics